Numerical and Evolutionary Optimization 2021

Numerical and Evolutionary Optimization 2021

Editors

Marcela Quiroz-Castellanos
Luis Gerardo de la Fraga
Adriana Lara
Leonardo Trujillo
Oliver Schütze

MDPI • Basel • Beijing • Wuhan • Barcelona • Belgrade • Manchester • Tokyo • Cluj • Tianjin

Editors

Marcela Quiroz-Castellanos
University of Veracruz
Mexico

Luis Gerardo de la Fraga
CINVESTAV-IPN
Mexico

Adriana Lara
Instituto Politécnico Nacional
ESFM-IPN
Mexico

Leonardo Trujillo
Instituto Tecnológico de Tijuana
Mexico

Oliver Schütze
CINVESTAV-IPN
Mexico

Editorial Office
MDPI
St. Alban-Anlage 66
4052 Basel, Switzerland

This is a reprint of articles from the Special Issue published online in the open access journal *Mathematical and Computational Applications* (ISSN 2297-8747) (available at: https://www.mdpi.com/journal/mca/special_issues/NEO21).

For citation purposes, cite each article independently as indicated on the article page online and as indicated below:

LastName, A.A.; LastName, B.B.; LastName, C.C. Article Title. *Journal Name* **Year**, *Volume Number*, Page Range.

ISBN 978-3-0365-7904-7 (Hbk)
ISBN 978-3-0365-7905-4 (PDF)

Contents

About the Editors

Marcela Quiroz-Castellanos

Marcela Quiroz-Castellanos is a full-time researcher with the Artificial Intelligence Research Institute at the Universidad Veracruzana in Xalapa City, Mexico. Her research interests include: Combinatorial Optimization, Metaheuristics, Experimental Algorithms, Characterization, and Data Mining. She received her Ph.D. in Computer Science from the Instituto Tecnológico de Tijuana, Mexico. She studied engineering in computer systems and received a master's degree in Computer Science at the Instituto Tecnológico de Ciudad Madero, Mexico. She is a member of the Mexican National Researchers System (SNI), and also a member of the directive committees of the Mexican Computing Academy (AMexComp) and the Mexican Robotics Federation (FMR).

Luis Gerardo de la Fraga

Luis Gerardo de la Fraga received his BS degree in Electrical Engineering from the Veracruz Institute of Technology (Veracruz, Mexico), in 1992, the MSc degree from the National Institute of Astrophysics, Optics, and Electronics (INAOE), Puebla, Mexico, in 1994, and a Ph.D. degree from the Autonomous University of Madrid, Spain, in 1998. He has developed his predoctoral work in the National Center of Biotechnology (CNB) in Madrid, Spain. Since 2000, he has been with the Computer Science Department at the Center of Research and Advanced Studies (Cinvestav) in Mexico City, Mexico. His research areas include Computer Vision, Application of Evolutionary Algorithms, Applied Mathematics, and Network Security. He is very enthusiastic about open software and GNU/Linux systems. Dr. de la Fraga has published more than 35 articles in international journals, 6 book chapters, 2 books, and more than 55 articles in international conferences. He has graduated 28 MSc and 4 PhD students. He is member of the ACM and IEEE societies since 2005.

Adriana Lara

Adriana Lara is a Full-Time Professor at the Physics and Mathematics School (ESFM) at the IPN in México City. Her research interests include Numerical Optimization and Mathematical Programming, Multi-criteria Decision Making, Bio-inspired heuristics, Memetic and Hybrid techniques, and Data Analysis. Her research work has received the IEEE Transactions on Evolutionary Computation Outstanding Paper Award (twice) in 2010 and 2012. She also received the 2010 Engineering Award granted by Mexico City's Science and Technology Institute (ICyTDF) and the Ph.D. Dissertation Award both by SMIA and ANIEI.

Leonardo Trujillo

Leonardo Trujillo is Professor at the Tecnológico Nacional de México/Instituto Tecnológico de Tijuana (ITT), working in the Department of Electrical and Electronic Engineering, and the Engineering Sciences Graduate Program. Dr. Trujillo received an Electronic Engineering degree and a Master's in Computer Science degree from ITT, as well as a doctorate in Computer Science from CICESE research center in Ensenada, Mexico. He is involved in interdisciplinary research in the fields of evolutionary computation, machine learning and pattern recognition. His research focuses on Genetic Programming (GP) and developing new learning and search strategies based on this paradigm. Dr. Trujillo has been the PI of several national and international research grants, receiving several distinctions from the Mexican science council (CONACYT). His work has been published in over 70 journal papers, 60 conference papers, 18 book chapters, and he has edited 6 books on EC and GP. He is on the Editorial Board of the journals GPEM (Springer) and MCA (MDPI), and associate editor of AI Communications, has been a Special Issue Guest Editor on 5 occasions, and regularly serves as a reviewer for highly respected journals in AI, EC and ML, is series co-chair of the NEO Workshop series, and has organized, been track chair or served as PC member of various prestigious conferences, including GECCO, GPTP, EuroGP, PPSN, CEC, CVPR and ECCV.

Oliver Schütze

Oliver Schütze is a Full Professor at the Cinvestav—IPN in Mexico City, Mexico. His main research interests are numerical and evolutionary optimization. He is the co-author of more than 150 publications, including 2 monographs, 5 school textbooks, and 10 edited books. Two of his papers have received the IEEE Transactions on Evolutionary Computation Outstanding Paper Award (in 2010 and 2012). He is the founder of the Numerical and Evolutionary Optimization (NEO) workshop series. He is Editor-in-Chief of the journal *Mathematical and Computational Applications*, and is a member of the Editorial Board of the journals *Engineering Optimization*, *Computational Optimization and Applications*, *IEEE Transactions on Evolutionary Computation*, *Research in Control and Optimization*, and *Applied Soft Computing*. He is a member of the Mexican Academy of Sciences and the National System of Researchers (SNI III).

Preface to "Numerical and Evolutionary Optimization 2021"

This volume was inspired by the 9th International Workshop on Numerical and Evolutionary Optimization (NEO 2021) held—due to the COVID-19 pandemic—as an online-only event from 8 to 10 September 2021. Solving scientific and engineering problems from the real world has always been a challenge, and the complexity of these tasks has only increased in recent years as more sources of data and information are continuously developed. That is why the development of powerful search and optimization techniques is of great importance. Two well-established fields focus on this duty; they are (i) traditional numerical optimization techniques and (ii) bio-inspired metaheuristic methods. Both general approaches have unique strengths and weaknesses, allowing researchers to solve some challenging problems but still failing in others. The goal of the NEO workshop series is to gather people from both fields to discuss, compare, and merge these complementary perspectives. Collaborative work allows researchers to maximize strengths and to minimize the weaknesses of both paradigms. NEO also intends to help researchers in these fields to understand and tackle real-world problems like pattern recognition, routing, energy, lines of production, prediction, modeling, among others.

This volume consists of 11 chapters that we will shortly summarize in the following. The order of the chapters is organized chronologically by the publication of the respective research papers in *Mathematical and Computational Applications* (MCA).

In Chapter 1, Dell'Amico and Magnani consider a pallet loading problem subject to stability requirements for which they propose a novel two phase metaheuristic approach. Computational experiments on real-life instances are used to assess the effectiveness of the algorithm.

In Chapter 2, Arcolezi et al. use machine learning techniques to predict the ambulances' response times (ART) of emergency medical services (EMS). Geo-indistinguishability was applied to sanitize each emergency location data. As shown in the results, the sanitization of the location data and the perturbation of its associated features (e.g., city, distance) had no considerable impact on predicting ARTs.

In Chapter 3, Pérez-Rodríguez proposes a new hybrid estimation of distribution algorithm (EDA) designed to tackle the quay crane scheduling problem (QCSP). The resulting hybrid algorithm uses a distance based ranking model together with the moth-flame algorithm. Numerical results indicate that this new approach yields better performance, or at least equal in effectiveness, than the so-called pure EDAs.

In Chapter 4, Pury proposes a new indicator to determine when emergency medical services (EMS) might reach a critical condition, when all service providers are busy and unable to respond to a new request. Such a scenario is crucial for EMS providers to take appropriate steps to avoid such a condition, with potentially life altering results. The usefulness of the proposed indicator is validated using simulations, with promising results that demonstrate its applicability in real-life scenarios.

In Chapter 5, Enríquez Zárate et al. study the problem of erosion in wind turbine blades, an important real-world issue regarding the efficiency and health of eolic energy generation. Determining the impact and presence of erosion, especially at the tip of the blades, is critical for proper maintenance and fault diagnosis. The work is based on the QBlade simulator to perform aerodynamic analysis and apply machine learning to predict and quantify the amount of erosion on different parts of the blade tip. The contribution is also unique because of the use of AutoML for the first time in this domain.

In Chapter 6, Cerrada et al. also study fault diagnosis in mechanical systems with AutoML, focusing on gearboxes that are widely used in industrial systems. Their work studies how different AutoML systems perform in generating machine learning pipelines to detect the severity level of different types of faults in these systems. It is shown that AutoML is competitive to the state-of-the-art, and the authors also analyze how AutoML performs feature selection in this domain, showing that different AutoML systems tend to converge towards very similar feature subsets.

In Chapter 7, Esqueda-Elizondo et al. present a methodology based on electroencephalographic (EEG) signals for attention measurement of a 13-year-old boy diagnosed with autism spectrum disorder (ASD). The authors claim that these findings allow to develop better learning scenarios according to the person's needs with ASD, and further, that it allows to obtain quantifiable information on their progress to reinforce the perception of the teacher or therapist.

In Chapter 8, Carmona-Arroyo et al. propose a grouping genetic algorithm (GGA) to deal with decomposition of the decision variables in order to efficiently tackle large-scale optimization problems. Although the cooperative co-evolution approach is widely used to deal with unconstrained optimization problems, there are few works related to constrained problems. The authors present results on 18 constrained functions with up to 1000 decision variables. These results indicate that a GGA is an appropriate tool to optimize the variable decomposition for large-scale constrained optimization problems, outperforming the decomposition obtained by a state-of-the-art genetic algorithm.

In Chapter 9, Contreras-Luján et al. consider several machine learning (ML) models in order to improve the diagnosis of deep venous thrombosis (DVT). In particular, the authors focus on their implementation on an edge device for the development of instruments that are smart, portable, reliable, and cost-effective. It is shown on data taken from literature that, compared to traditional methods, the best ML classifiers are effective at predicting DVT in an early and efficient manner.

In Chapter 10, Jain et al. consider ANFIS-Type methods in simulation of systems in marine environments. More precisely, the authors compare various artificial intelligence algorithms along with multivariate regression models to find the best fit model emulating effluent discharge and to determine the model with the least computational time. Tt is found that ANFIS-PSO performs better compared to the other considered models.

Finally, in Chapter 11, Pintér et al. address the problem to compute the conjectured sequence of largest small n-polygons. To this end, the authors develop high-precision numerical solution estimates of the maximal areas. Results are shown for n up to 1000, with demonstrably high precision.

We warmly thank all participants at NEO 2021 as well as all authors who submitted a work to this special issue. We hope that this book can be a contemporary reference regarding the field of numerical evolutionary optimization and its exciting applications.

Marcela Quiroz-Castellanos, Luis Gerardo de la Fraga, Adriana Lara, Leonardo Trujillo, and Oliver Schütze
Editors

Mathematical and Computational Applications

MDPI

Article

Solving a Real-Life Distributor's Pallet Loading Problem

Mauro Dell'Amico and Matteo Magnani *

Department of Sciences and Methods for Engineering, University of Modena and Reggio Emilia, Via Amendola 2, 42122 Reggio Emilia, Italy; mauro.dellamico@unimore.it

* Correspondence: ma.magnani@unimore.it

Abstract: We consider the distributor's pallet loading problem where a set of different boxes are packed on the smallest number of pallets by satisfying a given set of constraints. In particular, we refer to a real-life environment where each pallet is loaded with a set of layers made of boxes, and both a stability constraint and a compression constraint must be respected. The stability requirement imposes the following: (a) to load at level $k+1$ a layer with total area (i.e., the sum of the bottom faces' area of the boxes present in the layer) not exceeding α times the area of the layer of level k (where $\alpha \geq 1$), and (b) to limit with a given threshold the difference between the highest and the lowest box of a layer. The compression constraint defines the maximum weight that each layer k can sustain; hence, the total weight of the layers loaded over k must not exceed that value. Some stability and compression constraints are considered in other works, but to our knowledge, none are defined as faced in a real-life problem. We present a matheuristic approach which works in two phases. In the first, a number of layers are defined using classical 2D bin packing algorithms, applied to a smart selection of boxes. In the second phase, the layers are packed on the minimum number of pallets by means of a specialized MILP model solved with Gurobi. Computational experiments on real-life instances are used to assess the effectiveness of the algorithm.

Keywords: distributor's pallet loading problem; heuristics; bin packing; real-life instances

Citation: Dell'Amico, M.; Magnani, M. Solving a Real-Life Distributor's Pallet Loading Problem. *Math. Comput. Appl.* **2021**, *26*, 53. https://doi.org/10.3390/mca26030053

Academic Editor: Leonardo Trujillo

Received: 10 June 2021
Accepted: 15 July 2021
Published: 19 July 2021

1. Introduction

The *distributor's pallet loading problem* (DPLP) is a topic of wide interest for operational research and companies that deal with logistics, transport, and storage, in addition to production that leads to small and medium-sized packaging.

The problem is to find the optimal loading of parallelepiped-shaped boxes, not necessarily with the same sizes, on the fewest possible number of pallets, with predefined dimensions and weight limit. We consider the special case where the loading of each pallet is done by adding layers of boxes, one on top of the other. This case is very frequent in real-life applications.

Achieving the goal of minimizing the number of pallets built in an acceptable time means significantly reducing the costs due to the transport and storage of materials.

In practical problems, we have to deal with not only sizes but weights and the capacity of boxes to sustain other boxes. In this paper, we consider the following properties, which induce specific constraints:

- **stability**: that is, the property of a layer to sustain other layers, possibly with a larger area;
- **weight limit**: the sum of the weights of all boxes loaded on a pallet, which must not be greater than a certain limit given by the company;
- **compression limit**: capacity of a layer of boxes to support the weight of the boxes above it.

DPLP is strongly NP-hard, since it is a generalization of the *bin packing problem* (BPP) [1–6]. In the BPP we are given N items, each characterized by a weight, and an infinite number of bins of a given capacity. The problem is to load all the items in the minimum number of bins by respecting the capacity constraint. It is easy to see that the DPLP can model the BPP by disregarding all constraints but the one referring to the total weight of a pallet (bin capacity).

Since the 1960s, researchers and companies have deeply investigated the problem of cutting-stock, which is very similar to the BPP. In fact, in the cutting-stock problem, it is necessary to find a way to cut pieces of material, not necessarily with the same shapes and sizes, minimizing waste [7,8].

For a classification of some types of packing problems, refer to [9,10]; in [11], a review of the possible constraints most commonly used for packing and cargo problems is presented.

The DPLP is a generalization of the manufacturer's pallet loading problem (MPLP) [12,13], which deals with the same objective function but loads identical boxes on pallets.

For discussions of problems with constraints deriving from real applications, similar to the one in this paper, we refer to [14–21]. In particular, Ancora et al. [14] works with a hybrid genetic strategy; in [15–18], the authors use, respectively, heuristics, greedy approach, the genetic and differential evolution algorithm, and the branch and bound way to solve the DPLP; Gzara et al. [19] exploits a layer-based column generation; and Ancora et al. [14] works with a hybrid genetic strategy.

From a mathematical-modelling point of view, there is not much research that deals with the packing problem nor with real constraints (weight limit, compression, and stackability) as seen in this treatise. Nonetheless, the problem of 3D packing is usually dealt with by creating 2D layers, subsequently stacked on top of each other.

For a variant of this problem, the so-called container loading problem, which differs in the fact that, in general, constraints limiting the possibilities of layers overlapping are not explicit, we refer to [20–29]. (ILP strategy [23], GRASP method [20,24,25], heuristic way [27], heuristic-genetic algorithm [28], heuristics and MILP method [26], layer-based greedy strategy [21]).

We present in this paper a two-step algorithm to solve the pallet construction problem, basing our tests on real commercial orders from a logistics company using automated robots for creating and managing pallets. For similar work we refer to [30], which introduces visibility and contiguity constraints.

2. Materials and Methods

We are given a set $\mathscr{B}$ of 3D *boxes* partitioned into *types* associated with boxes of identical size, weight, and compression index. More specifically, let $\mathscr{I}$ denote the set of box types, and let n_i denote the number of identical boxes of type $i \in \mathscr{I}$ (with $\sum_{i \in \mathscr{I}} n_i = |\mathscr{B}|$). Each box of type i has width $w_i \in \mathbb{Z}^*_+$, depth $d_i \in \mathbb{Z}^*_+$, and height $h_i \in \mathbb{Z}^*_+$. Given a box $j \in \mathscr{B}$, we will denote with $i(j)$ the type of box j. We are also given an arbitrarily large set $\mathscr{P}$ of identical pallets. Each pallet has a two-dimensional loading surface of width $W \in \mathbb{Z}^*_+$ and depth $D \in \mathbb{Z}^*_+$, which can be used to load boxes up to a maximum height $H \in \mathbb{Z}^*_+$. We assume that $w_i \leq W$, $d_i \leq D$, $h_i \leq H$, for each $i \in \mathscr{I}$. Moreover, each box of type i has a *weight* p_i and a *compression index* c_i that will be used to define the maximum weight the box can support.

The problem requires assigning all the boxes to the pallets by restricting the loading to layered solutions. A feasible packing of boxes on a pallet can be decomposed into subsets of boxes each defining a layer, and the layers are loaded one over the other. More formally, a *layer* is a subset of boxes which can rotate 90 degrees on their support surface whose basis can be packed into a $W \times D$ rectangle.

Let us define as $\mathscr{L}$ the set of all the possible types of layers we could build with boxes $\mathscr{B}$. Observe that $\mathscr{L}$ has exponentially many elements, so we will adopt algorithms that only consider a subset of these layers. A layer $l \in \mathscr{L}$ is given by the set of boxes assigned to it, say $\mathscr{B}_l$, and by a specific 2D packing of these boxes (more precisely, of the basis of the boxes), whose total width must not exceed W, and whose total depth must not exceed

D. Let $H_l = \max_{j \in \mathscr{B}_l} h_{i(j)}$ denote the height of the layer, and let $A_l = \sum_{j \in B_l} h_{i(j)} d_{i(j)}$ denote the total area of the layer.

A layer built with boxes of the same height produces a planar surface (possibly with some holes), on which we can load another layer. However, the practical experience of the company has shown that it is not strictly necessary that the bottom layer has a perfectly planar upper surface to be used as support for another layer: it is enough that this surface is not too wavy. This can be translated into a simple requirement, imposing that the difference in height of its boxes is small enough. More precisely, given a layer l, we impose

$$|h_i - h_j| \leq \Delta h \quad \forall i, j \in \mathscr{B}_l, \tag{1}$$

in which parameter Δh defines the maximum height difference allowed between two boxes of a layer. We say that a layer for which (1) holds satisfies the **stackability constraint**. We will consider this as a unique exception to the last layer of a pallet (top layer): since no other layer will be loaded on the top one, the restriction on the difference of heights does not have to be considered.

Another constraint for feasible loading of one layer over another regards the ratio of the total areas of the boxes in the layer. It is obvious that if we load a layer with a large area on a layer with a very small area, there is an issue with the stability of the overlying layer. The **stability constraint**, defined by the inequality

$$A_l \geq \alpha A_m, \tag{2}$$

requires loading layer m immediately over layer l, where $\alpha \geq 1$ is a given parameter.

To each layer l, an overall *compression factor* is also associated:

$$C_l = \begin{cases} \min_{j \in \mathscr{B}_l} c_{i(j)} A_l & \text{if } l \text{ satisfies (1)} \\ 0 & \text{otherwise} \end{cases} \tag{3}$$

giving the maximum total weight the layer can support. We will call this requirement the **compression constraint** (see [14,18,19,21] for other studies that consider this constraint in a similar way). Note that giving a zero compression factor to layers not satisfying (1) implies that these layers can be only packed as the top layer of a pallet.

Resuming the above descriptions, the aim of the DPLP that we face is to load all boxes into the minimum number of pallets by ensuring that the following constraints are satisfied:

c1. Numerosity constraint: all boxes in $\mathscr{B}$ must be packed;
c2. Height constraint: the sum of the heights H_l of all layers loaded on a pallet must not exceed H;
c3. Stackability constraint: each layer, except the top one of each pallet, must be composed by boxes satisfying (1);
c4. Stability constraint: each pair of layers m, l, with m loaded immediately over l, must satisfy (2);
c5. Compression constraint: the total weight of all boxes in the layers loaded over a layer l cannot exceed the compression factor C_l.

2.1. Creating 2D Layers

For the creation of the layers' set $\mathcal{L}$, we again use a two-step method. In the first step, we partition the boxes into *families* of boxes with the procedure CREATEFAMILIES, which takes as input the number of families f in which we want to divide the set of box types and returns a partition $\mathscr{I}(1), \ldots, \mathscr{I}(f)$, where each family $\mathscr{I}(i)$ contains box types whose heights differentiate at most by $(1+\gamma)\Delta h$, and γ is a randomly selected small value. Therefore, we say that a family *almost* satisfies requirement **c3** (see (1)). In the second step, we apply to these families 2D packing methods from the literature to create the layers.

Our algorithm BUILDLAYERS of Algorithm 1 uses procedure CREATEFAMILIES (see Algorithm 2) and a set of heuristic algorithms $2D\text{-}H_1, \ldots, 2D\text{-}H_{a_{\max}}$.

Algorithm 1: Algorithm for creating the layers.

Algorithm BUILDLAYERS()
input: boxes $\mathscr{B}$ and their types $\mathscr{I}$, pallets' sizes
1. Sort the box types in $\mathscr{I}$ by non decreasing heights (i.e., $h_1 \leq h_2 \leq \ldots, \leq h_{|\mathscr{I}|}$)
2. $\mathcal{L} = \emptyset$;
3. **for** $f = f_{\min}$ **to** $f_{\max}$ **do**
4. $(\mathscr{I}(1), \ldots, \mathscr{I}(f))$ = CREATEFAMILIES(f);
5.   **for** $i = 1$ **to** f **do**
6.   **for** $a = 1$ **to** $a_{\max}$ **do**
7.   pack the boxes in $\mathscr{I}(i)$ with heuristic $2D$-H_a giving layers L
8. $\mathcal{L} = \mathcal{L} \cup L$;
9.   **endfor**
10.   **endfor**
11. **endfor**

return $\mathcal{L}$

Algorithm 2: Procedure for creating the box types families.

Procedure CREATEFAMILIES(number of families f)
1. Let $n = |\mathscr{I}|$
2. Choose randomly $\gamma \in [0,3;0,5]$
3. **for** $j = 1, \ldots, f$ **do** let $\nu(j) = $ RANDOM $((j-1)\lceil n/f \rceil, \min(j\lceil n/f \rceil, |\mathscr{I}|))$
4. $\mathscr{I}(1) = \{1, \ldots, \bar{\imath}\}$ with $\bar{\imath} = \arg\max\{h_i \leq h_{\nu(1)} + \frac{1}{2}(1+\gamma)\Delta h\}$
5. **for** $j = 2$ **to** $f - 1$ **do**
6. $\mathscr{I}(j) = \{\underline{\imath}, \ldots, \bar{\imath}\}$ with $\underline{\imath} = \arg\min\{h_i \geq h_{\nu(j)} - \frac{1}{2}(1+\gamma)\Delta h\}$
    $\bar{\imath} = \arg\max\{h_i \leq h_{\nu(j)} + \frac{1}{2}(1+\gamma)\Delta h\}$
7. **endfor**
8. $\mathscr{I}(f) = \{\underline{\imath}, \ldots, n\}$ with $\underline{\imath} = \arg\min\{h_i \geq h_{\nu(|\mathscr{I}|)} - \frac{1}{2}(1+\gamma)\Delta h\}$
9. **for** $j = 1$ **to** $f - 1$ **do**
10.   **if** $\mathscr{I}(j) \cap \mathscr{I}(j+1) \neq \emptyset$ **then**
11. $\mathscr{I}(j+1) = \mathscr{I}(j+1) \setminus \mathscr{I}(j)$
12.   **endif**
13. **endfor**
14. **foreach** $i \in \mathscr{I}$ **do**
15.   **if** $i \notin \cup_{j=1}^{f} \mathscr{I}(j)$ **then**
16.   assign i to the set with nearest pivot height
17.   **endif**
18. **endfor**

return$(\mathscr{I}(1), \ldots, \mathscr{I}(f))$

The loop 'for' at line 3 of algorithm BUILDLAYERS calls CREATEFAMILIES a few times with different values of the partition's number, f, to create different families. At each family, the 2D packing heuristics $2D$-$H_1, \ldots, 2D$-$H_{a_{\max}}$ are applied in turn, which produces layers that are added to the layer set $\mathcal{L}$. Note that due to the 'almost' satisfaction of requirement **c3**, set $\mathcal{L}$ will contain both layers satisfying (1) or not.

Procedure CREATEFAMILIES starts by dividing the interval of the box types into f subinterval of identical length, except the last one, which could have, in some cases, fewer elements than the other subinterval, and selects a pivot box type index $\nu(i)$ from each subinterval i (with $i \in \{1, \ldots, f\}$). Each family $\mathscr{I}(i)$ is defined as the set of box types with absolute height difference from $h_{\nu(i)}$ not exceeding $\frac{1}{2}(1+\gamma)\Delta h$. Given this first selection of box types, the possible intersections are then removed, and the box types not inserted into any subset, if any, are assigned to the set with closest pivot height.

The need to insert the random parameter γ and to choose the random pivot derives from the fact that we want to create slightly different families at each iteration. By doing so, we provide the possibility of creating different layers at each iteration, keeping the families unchanged for most of the elements (because γ is small).

For the same reason, in our experiments, we used some 2D packing methods from [31], named the maximal rectangle and skyline algorithms.

In particular, the maximal rectangle algorithm works by storing a list of free rectangles, which represent the free area of the layer. Every time a rectangle is placed in a layer (if possible, otherwise a new one is opened), the list of free rectangles of that layer is updated, adding 2 rectangles that form an L-shaped region as shown in Figure 1.

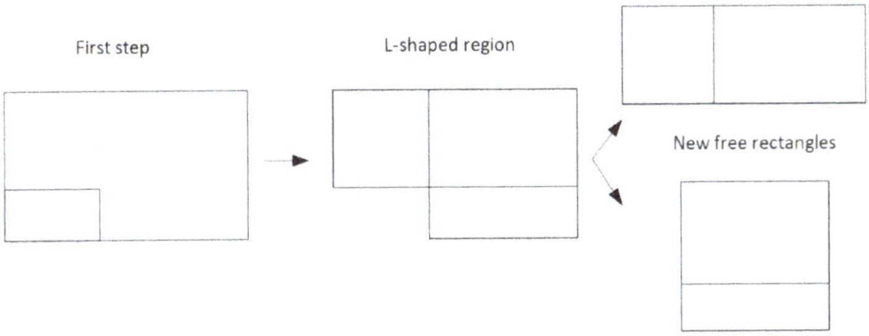

Figure 1. Free Rectangles.

Any overlaps between free rectangles or degenerate free rectangles are eliminated each time the list of free rectangles is updated.

The skyline structure is the same as that of the *envelope* in [32]: a list containing the high edges of the already packed rectangles is updated every time a new rectangle is placed in the current layer (if possible, otherwise a new one is opened). Due to the simplicity of the structure at the base of the skyline, it is easy to memorize the areas that would otherwise no longer be used during packing (any holes in the packing); let us call these areas waste map. Before looking for new locations for the new rectangles, we try to place them on the waste map.

For both structures, we used different approaches for choosing where to place the rectangles:

- MaxRectBL: maximal rectangle with bottom-left strategy (place each rectangle in the position where the y-coordinate of the top side of the rectangle is the smallest, and if there are several such valid positions, pick the one that has the smallest x-coordinate value);
- MaxRectBLR: maximal rectangle with bottom-left strategy and rotation allowed;
- MaxRectBssfR: maximal rectangles best short side fit strategy chooses to pack the current rectangle into the free rectangle, which minimizes the differences between the dimensions of the rectangle and the free one;
- SkylineBlWm: skyline with bottom-left and waste map strategy;
- SkylineBlWmR: skyline with bottom-left and waste map strategy with rotation allowed;
- SkylineMwfWm: skyline with min waste fit with low profile heuristic, minimizing the area wasted below the rectangle; at the same time, it tries to keep the height minimal;
- SkylineMwfWmR: skyline with min waste fit with low profile heuristic and rotation allowed.

2.2. A Mathematical Model for Loading Layers

In this section, we present a mathematical model for the optimal loading of layers in the minimum number of pallets.

We represent the layers' types obtained in the first phase by a $|\mathscr{I}| \times |\mathcal{L}|$ matrix A where a_{il} denotes the number of boxes of type i packed in layer l.

We will use two sets of binary variables. Variable y_p will take value 1 if the pallet $p \in \mathscr{P}$ is used, and zero otherwise. Variable x_{lpk} will have value 1 if layer l is stacked on pallet p at level k, and 0 otherwise.

Parameter $\kappa = \lfloor H/min_{i \in \mathscr{I}} h_i \rfloor$ denotes the maximum number of layers that can be loaded on any pallet.

Finally, we calculate the weight of each layer as the sum of the weights of the items present on each one:

$$P_l = \sum_{i \in \mathscr{I}} a_{il} p_i$$

Then, the mathematical model for stacking on pallets is as follows:

$$\min \sum_{p \in \mathscr{P}} y_p \tag{4}$$

$$\sum_{k=1}^{\kappa} \sum_{p \in \mathscr{P}} \sum_{l \in \mathcal{L}} a_{il} x_{lpk} = n_i \qquad i \in \mathscr{I} \tag{5}$$

$$\sum_{k=1}^{\kappa} \sum_{l \in \mathcal{L}} H_l x_{lpk} \leq H y_p \qquad p \in \mathscr{P} \tag{6}$$

$$\sum_{k=1}^{\kappa} \sum_{l \in \mathcal{L}} P_l x_{lpk} \leq P y_p \qquad p \in \mathscr{P} \tag{7}$$

$$x_{lpk} + x_{mp(k+1)} \leq 1 \qquad l, m \in \mathcal{L}, : A_l \geq \alpha A_m,\ p \in \mathscr{P},\ k \in \{1, \ldots, \kappa - 1\} \tag{8}$$

$$\sum_{h=k+1}^{\kappa} \sum_{\ell \in \mathcal{L}} x_{\ell ph} P_\ell \leq C_l + M(1 - x_{lpk}) \qquad l \in \mathcal{L}, p \in \mathscr{P}, k \in \{1, \ldots, \kappa - 1\} \tag{9}$$

$$\sum_{l \in \mathcal{L}} x_{lpk} \leq 1 \qquad p \in \mathscr{P}, k \in \{1, \ldots, \kappa\} \tag{10}$$

$$\sum_{l \in \mathcal{L}} (x_{lpk} - x_{lp(k-1)}) \leq 0 \qquad k \in \{2, \ldots, \kappa\}, p \in \mathscr{P} \tag{11}$$

$$y_p \leq y_{p-1} \qquad p \in \{2, \ldots, |\mathscr{P}|\} \tag{12}$$

$$y_p \in \{0, 1\} \qquad p \in \mathscr{P} \tag{13}$$

$$x_{lpk} \in \{0, 1\} \qquad l \in \mathcal{L}, p \in \mathscr{P}, k \in \{1, \ldots, \kappa\} \tag{14}$$

The objective function (4) minimizes the number of pallets used. Constraint (5) requires that each box is packed once on the pallet, thus implementing requirement **c1**. Constraints (6) implement requirement **c2** by imposing that the sum of the heights of the layers on each pallet does not exceed the available height H. Constraints (7) require that the sum of the weights of the boxes on each pallet do not exceed P, where P is the maximum weight that can be loaded on each pallet. The stability requirement **c4** is satisfied by constraints (8), while the compression requirement **c5** is guaranteed by (9) (M being, as usual, a very large positive number). Note that requirement **c3** is satisfied by definition (3) for all layers with $C_l > 0$, while when $C_l = 0$, constraints (9) impose that the layer be packed as the top one of a pallet.

To conclude, the model constraints (10), (11), and (12) respectively impose the following: (a) to load at most one layer per level, (b) to load a level k only if level $k - 1$ has been loaded, and (c) to use a pallet p only if pallet $p - 1$ has been used. The definition of the domains of the variables follows.

3. Results

All experiments have been conducted on a PC with Intel Core i7-10510U CPU 2.30 GHz, 16 GB RAM, and Windows 10 Operating System. The algorithms have been implemented in Python 3.8 and run using PyCharm 2021.1.2.

We solved the MILP model with Gurobi 9.1.1. We considered a set of five instances from real orders within the corresponding company manual solutions. For each instance, we set the time limit to 120 min for the Gurobi solver, running the algorithm five times, using all variants of the skyline and maximal rectangle algorithms we presented in Section 2.1.

For all instances, the pallet dimensions were set to 1650, 1200, and 800 for height, width, and length, respectively. The maximum weight P loadable on every pallet was set to 4000, and the Δh was set to 20 for every layer in $\mathcal{L}$. Finally, α in (2) was set to 1.1.

In the following Tables 1–7, we show some experimental results, all based on real commercial orders of a logistics company.

Table 1. Instance details.

Instance	N° Items	N° Items Type	Tot. Weight	Min. Compr.	Max. Height	Min. Height
A	332	53	2967.58	87.5	305	150
B	136	22	1564.56	87.5	305	150
C	349	70	3272.756	87.5	305	150
D	669	68	6901.96	87.5	305	150
E	83	14	464.83	75	265	150

Table 2. First run for all instances.

Instance	$\lvert\mathcal{L}\rvert$	Best Bound	Best Solution	Comp. Solution	Time (min)
A	139	5	6	7	120
B	65	3	3	4	3
C	180	7	8	8	120
D	220	11	12	13	120
E	34	1	1	2	2

Table 3. Second run for all instances.

Instance	$\lvert\mathcal{L}\rvert$	Best Bound	Best Solution	Comp. Solution	Time (min)
A	141	5	6	7	120
B	68	3	3	4	4
C	174	7	9	8	120
D	228	11	12	13	120
E	34	1	1	2	2

Table 4. Third run for all instances.

Instance	$\lvert\mathcal{L}\rvert$	Best Bound	Best Solution	Comp. Solution	Time (min)
A	136	5	6	7	120
B	67	3	3	4	3
C	178	7	8	8	120
D	222	11	12	13	120
E	32	1	1	2	2

Table 5. Fourth run for all instances.

Instance	$\lvert\mathcal{L}\rvert$	Best Bound	Best Solution	Comp. Solution	Time (min)
A	132	5	7	7	120
B	66	3	3	4	3
C	174	7	8	8	120
D	229	11	12	13	120
E	33	1	1	2	2

Table 6. Fifth run for all instances.

Instance	$\lvert\mathcal{L}\rvert$	Best Bound	Best Solution	Comp. Solution	Time (min)
A	137	5	6	7	120
B	64	3	3	4	3
C	174	7	8	8	120
D	231	11	12	13	120
E	32	1	1	2	2

Table 7. Average of computational results.

Instance	$\|\mathcal{L}\|$	Best Bound	Best Solution	Comp. Solution	Time (min)
A	137	5	6	7	120
B	66	3	3	4	3
C	176	7	8	8	120
D	226	11	12	13	120
E	33	1	1	2	2

We ran the algorithm BUILDLAYERS F times, where $F = f_{\max} - f_{\min} + 1$, as the creation of the layers based on families is partially randomized. In particular, we set $f_{\min} = 2$ and $f_{\max} = |\mathcal{H}|$, where $\mathcal{H}$ is the set of different heights of boxes which are present in the commercial order. The use of multiple layerization methods, multiple division into families, and partial randomization allows for the creation of different layers that can expand the pool of layers (see Figure 2 for our results). It is possible that by adding some layers to $\mathcal{L}$, even if they are slightly different from each other, the general solution to the problem is greatly improved. The algorithm always improves the manual solutions, on average. In very few cases (2 out of 25), the proposed solution equals the manual one, but never exceeds it (see Figure 3).

The time columns refer to the time required by Gurobi to solve the problem. Small commercial orders (Instances B and E) are processed in a short time, as few layers are created, making the minimum convergence of the palletizing model fast. On the other hand, orders that have many boxes and many box types (Instances A, C, and D) require more computing time, often reaching the time limit.

These computing times are compatible with the practical management of large orders when loading can be planned early. From the computational point of view, the most onerous process is represented by the resolution of the palletization model. In fact, even for the largest orders, the filling of the layers' pool ends in a maximum of 5 min.

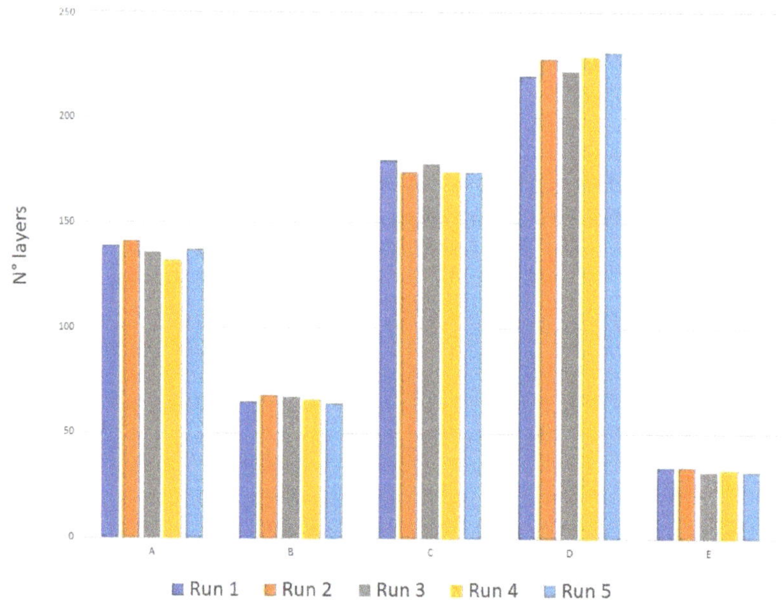

Figure 2. Layer for every run, for every instance.

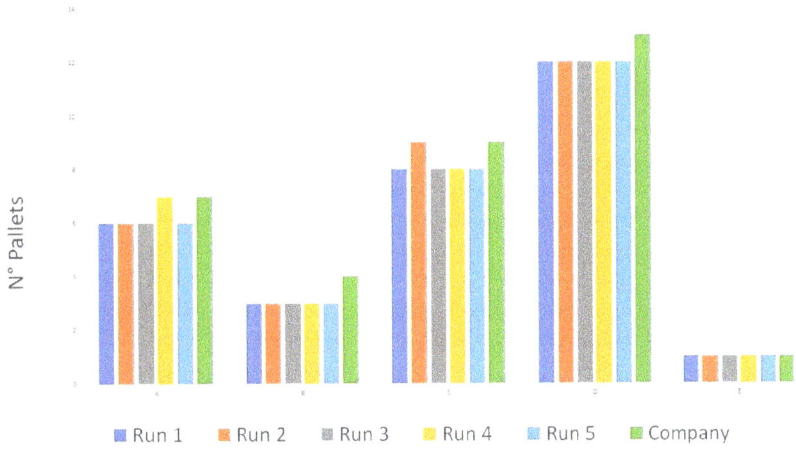

Figure 3. Pallets for every run, for every instance, with company results.

For the sake of privacy required by the company, we are not able to publish the complete database we used for the experiments.

We want to highlight that this algorithm has some advantages: it solves a real problem with constraints deriving from real experiences; it is possible to obtain substantial modifications: for example, changing the algorithm that solves the 2D layer creation problem or varying the γ hyperparameter of the families' creation. It can be sped up by dividing into fewer families or by using fewer 2D packing algorithms.

We are confident that in the future, with experiments on more orders and with changes to the hyperparameters, this algorithm can achieve even more satisfactory results than those obtained in this paper.

Author Contributions: Conceptualization, M.M. and M.D.; methodology, M.M. and M.D.; software, M.M.; validation, M.M. and M.D.; formal analysis, M.M. and M.D.; investigation, M.M.; resources, M.M.; data curation, M.M.; writing—original draft preparation, M.M.; writing—review and editing, M.D.; visualization, M.M.; supervision, M.D.; project administration, M.M. All authors have read and agreed to the published version of the manuscript.

Funding: This research received no external funding.

Conflicts of Interest: The authors declare no conflict of interest.

References

1. Hu, H.; Zhang, X.; Yan, X.; Wang, L.; Xu, Y. Solving a new 3D bin packing problem with deep reinforcement learning method. *arXiv* **2017**, arXiv:1708.05930.
2. Maarouf, W.; Barbar, A.; Owayjan, M. A new heuristic algorithm for the 3D bin packing problem. In *Innovations and Advanced Techniques in Systems, Computing Sciences and Software Engineering*; Springer: Dordrecht, The Netherlands, 2008; pp. 342–345.
3. Martello, S.; Pisinger, D.; Vigo, D.; Boef, E.; Korst, J. Algorithm 864: General and robot-packable variants of the three-dimensional bin packing problem. *ACM Trans. Math. Softw.* **2007**, *33*, 7-es. [CrossRef]
4. Paquay, C.; Schyns, M.; Limbourg, S. A mixed integer programming formulation for the three-dimensional bin packing problem deriving from an air cargo application. *Int. Trans. Oper. Res.* **2016**, *23*, 187–213. [CrossRef]
5. Garey, M.R.; Johnson, D.S. *Computers and Intractability: A Guide to the Theory of NP-Completeness*; W. H. Freeman & Co.: New York, NY, USA, 1979.
6. Korte, B.; Vygen, J. "Bin-Packing". In *Combinatorial Optimization: Theory and Algorithms. Algorithms and Combinatorics*; Springer: Berlin/Heidelberg, Germany, 2006; Volume 21, pp. 426–441.
7. Gilmore, R.; Gomory, R. A Linear Programming Approach to the Cutting Stock Problem. *Oper. Res.* **1961**, *9*, 849–859 [CrossRef]
8. Gilmore, P.; Gomory, R. Multi-stage cutting stock problems of two or more dimensions. *Oper. Res.* **1965**, *13*. [CrossRef]
9. Dowsland, K.A.; Dowsland, W.B. Packing problems. *Eur. J. Oper. Res.* **1992**, *56*, 2–14. [CrossRef]
10. Egeblad, J. Heuristics for Multidimensional Packing Problems. Ph.D. Thesis, Department of Computer Science, University of Copenhagen, Copenhagen, Denmark, 2008.

11. Bortfeldt, A.; Wascher, G. Constraints in container loading a state-of-the-art review. *Eur. J. Oper. Res.* **2013**, *229*, 1–20. [CrossRef]
12. Morabito, R.; Morales, S. A simple and effective recursive procedure for the manufacturer's pallet loading problem. *J. Oper. Res. Soc.* **1998**, *49*, 819–828. [CrossRef]
13. Silva, E.; Oliveira, J.F.; Wascher, G. The pallet loading problem: A review of solution methods and computational experiments. *Int. Trans. Oper. Res.* **2016**, *23*, 147–172. [CrossRef]
14. Ancora, G.; Palli, G.; Melchiorri, C. A hybrid genetic algorithm for pallet loading in real-world applications. *IFAC-PapersOnLine* **2020**, *53*, 10006–10010. [CrossRef]
15. Bischoff, E.E.; Ratcliff, M.S.W. Loading multiple pallets. *J. Oper. Res. Soc.* **1995**, *46*, 1322–1336. [CrossRef]
16. Bischoff, E.E.; Janetz, F.; Ratcliff, M.S.W. Loading pallets with non-identical items. *Eur. J. Oper. Res.* **1995**, *84*, 681–692. [CrossRef]
17. Piyachayawat, T.; Mungwattana, A. A hybrid algorithm application for the multi-size pallet loading problem case study: Lamp and lighting factory. In Proceedings of the 4th International Conference on Industrial Engineering and Applications (ICIEA), Nagoya, Japan, 21–23 April 2017; pp. 100–105.
18. Scheithauer, G.; Terno, J. A heuristic approach for solving the multi-pallet packing problem. In *Decision Making under Conditions of Uncertainty (Cutting–Packing Problems)*; Mukhacheva, E.A., Ed.; Ufa State Aviation Technical University, Ufa, Ruassia, 1997; pp. 140–154.
19. Gzara, F.; Elhedhli, S.; Yildiz, B.C. The pallet loading problem: Three-dimensional bin packing with practical constraints. *Eur. J. Oper. Res.* **2020**, *287*, 1062–1074. [CrossRef]
20. Moura, A.; Oliveira, J.F. A GRASP approach to the container-loading problem. *IEEE Intell. Syst.* **2005**, *20*, 50–57. [CrossRef]
21. Saraiva, R.D.; Nepomuceno, N.; Pinheiro, P-R. A layer-building algorithm for the three-dimensional multiple bin packing problem: A case study in an automotive company. *IFAC-PapersOnLine* **2015**, *48*, 490–495. [CrossRef]
22. Singh, M.; Almasarwah, N.; Suer, G. A two-phase algorithm to solve a 3-dimensional pallet loading problem. *Procedia Manuf.* **2019**, *39*, 1474–1481. [CrossRef]
23. Alonso, M.T.; Alvarez-Valdes, R.; Iori, M.; Parreño, F. Mathematical models for multi container loading problems with practical constraints. *Comput. Ind. Eng.* **2019**, *127*, 722–733. [CrossRef]
24. Alonso, M.T.; R. Alvarez-Valdes, R.; Parreño, F.; Tamarit, J.M. Algorithms for pallet building and truck loading in an interdepot transportation problem. *Math. Probl. Eng.* **2016**, *2016*, 3264214. [CrossRef]
25. Alvarez Martinez, D.; Alvarez-Valdes, R.; Parreno, F. A GRASP algorithm for the container loading problem. *Pesqui. Oper.* **2015**, *35*, 1–24. [CrossRef]
26. Ranck Júnior, R.; Yanasse, H.H.; Morabito, R.; Junqueira, L. A hybrid approach for a multi-compartment container loading problem. *Expert Syst. Appl.* **2019**, *137*, 471–492. [CrossRef]
27. Jens, E.; Garavelli, C.; Lisi, S.; Pisinger, D. Heuristics for container loading of furniture. *Eur. J. Oper. Res.* **2010**, *3*, 881–892.
28. Olsson, J. Solving a Highly Constrained Multi-Level Container Loading Problem from Practice. Bachelor's Thesis, Linköping University, Linköping, Sweden, 2017.
29. Zhao, X.; Bennell, J.A.; Bektaş, T.; Dowsland, K. A comparative review of 3D container loading algorithms. *Int. Trans. Oper. Res.* **2016**, *23*, 287–320. [CrossRef]
30. Iori, M.; Locatelli, M.; Moreira, M.C.; Silveira, T. Reactive GRASP-based algorithm for pallet building problem with visibility and contiguity constraints. In Proceedings of the 11th International Conference on Computational Logistics, Enschede, The Netherlands, 28–30 September 2020.
31. Jylanki, J. A Thousand Ways to Pack the Bin—A Practical Approach to Two-Dimensional Rectangle Bin Packing. 2010. Available online: http://clb.demon.fi/files/RectangleBinPack.pdf (accessed on 15 July 2021).
32. Wei, L.; Zhang, D.; Chen, Q. A least wasted first heuristic algorithm for the rectangular packing problem. *Comput. Oper. Res.* **2009**, *36*, 1608–1614. [CrossRef]

Mathematical and Computational Applications

MDPI

Article

Preserving Geo-Indistinguishability of the Emergency Scene to Predict Ambulance Response Time

Héber H. Arcolezi *, Selene Cerna, Christophe Guyeux * and Jean-François Couchot

FEMTO-ST Institute, UMR 6174 CNRS, Université Bourgogne Franche-Comté (UBFC), 90000 Belfort, France; selene_leya.cerna_nahuis@univ-fcomte.fr (S.C.); jean-francois.couchot@univ-fcomte.fr (J.-F.C.)
* Correspondence: heber.hwang_arcolezi@univ-fcomte.fr (H.H.A.); christophe.guyeux@univ-fcomte.fr (C.G.)

Abstract: Emergency medical services (EMS) provide crucial emergency assistance and ambulatory services. One key measurement of EMS's quality of service is their ambulances' response time (ART), which generally refers to the period between EMS notification and the moment an ambulance arrives on the scene. Due to many victims requiring care within adequate time (e.g., cardiac arrest), improving ARTs is vital. This paper proposes to predict ARTs using machine-learning (ML) techniques, which could be used as a decision-support system by EMS to allow a dynamic selection of ambulance dispatch centers. However, one well-known predictor of ART is the location of the emergency (e.g., if it is urban or rural areas), which is *sensitive data* because it can reveal who received care and for which reason. Thus, we considered the 'input perturbation' setting in the privacy-preserving ML literature, which allows EMS to sanitize each location data independently and, hence, ML models are trained only with sanitized data. In this paper, geo-indistinguishability was applied to sanitize each emergency location data, which is a state-of-the-art formal notion based on differential privacy. To validate our proposals, we used retrospective data of an EMS in France, namely Departmental Fire and Rescue Service of Doubs, and publicly available data (e.g., weather and traffic data). As shown in the results, the sanitization of location data and the perturbation of its associated features (e.g., city, distance) had no considerable impact on predicting ARTs. With these findings, EMSs may prefer using and/or sharing sanitized datasets to avoid possible data leakages, membership inference attacks, or data reconstructions, for example.

Keywords: emergency medical services; emergency medicine; decision-support system; pre-hospital emergency care; ambulance response time; machine learning; geo-indistinguishability; differential privacy; privacy-preserving machine learning; input perturbation

Citation: Arcolezi, H.H.; Cerna, S.; Guyeux, C.; Couchot, J.-F. Preserving Geo-Indistinguishability of the Emergency Scene to Predict Ambulance Response Time. *Math. Comput. Appl.* **2021**, *26*, 56. https://doi.org/10.3390/mca26030056

Academic Editor: Leonardo Trujillo

Received: 2 July 2021
Accepted: 2 August 2021
Published: 4 August 2021

1. Introduction

Ambulance response time (ART) is a key component for evaluating pre-hospital emergency medical services (EMS) operations. ART refers to the period between the EMS notification and the moment an ambulance arrives at the emergency scene [1,2], and it is normally divided into two periods: the pre-travel delay, from the notification to the ambulance dispatch, and the travel time, from the ambulance dispatch to arrival on-scene. In many urgent situations (e.g., cardiovascular emergencies, trauma, or respiratory distress), the victims need first-aid treatment within adequate time to increase survival rate [1–6] and, hence, improving ART is vital.

In many parts of the world, such as France, fire departments are responsible for many critical situations, including fires, hazards, severe storms, floods, as well as non-urgent and urgent EMS calls (e.g., traffic accidents, drowning). In this paper, we analyzed EMS operations of the Departmental Fire and Rescue Service of Doubs (SDIS 25), which has 71 centers currently deployed across the Doubs region in France to attend to its population. As noticed in [7,8], the SDIS 25 and fire departments in general, have been facing a continuous increase in the number of interventions over the years, which may have adverse consequences on ARTs. For instance, the pre-travel delay directly affects ARTs if there is a lack of human and material resources when a call is received. This means, if there is a lack of firefighters, ambulances, or both, ART may be higher than allowed and, hence, a breakdown in the SDIS 25 service occurs [9]. This inability to assist within the time limits impacts negatively both EMS and victims because the safety of a certain area or population will be at risk. Thus, there is a need for an intelligent ART prediction system, which can assist SDIS 25 (and EMS, in general) in the dispatching of ambulances.

Indeed, predicting ART is useful for many reasons. First, it can help in choosing the best center to provide the ambulance. At present, for SDIS 25, each city in the department is associated with an ordered list of centers with the needed engine to respond, so that the first centers are the most likely to provide a rapid and adequate response. This structure is mainly defined by the administrative policies of the organization, which considers, for example, the operational load (number of interventions) that the city represents, and according to this, the necessary armament that its nearest center should have, as well as the shortest distances and times between the centers and the cities. However, this structure varies very little over time, for example, when there is a creation or territorial modification of a city. Although it takes into account the actual travel distance (considering street structures, highways, etc.), it does not take into account the real-time state of road traffic, weather conditions, etc. Predicting ART would therefore make it possible to move from static center scheduling to dynamic scheduling. It would also make it possible to estimate the pre-travel delay partially and to see in advance whether, at a given moment, a center is at risk of running out of ambulances. In other words, it enables the anticipation of breakdowns and the redeployment of resources. Lastly, in the long term, it can be an element of a simulator to determine the evolution of response time and breakdowns during the creation or relocation of a center, the modification of resources by the center, etc.

One important factor of ART is the *location* of the intervention [2,3,10–12], e.g., in dense urban areas, the distance may be short, but the travel time may be longer due to traffic congestion. On the other hand, travel distance and travel time may be longer for rural areas. In other words, the location information is of great importance for the prediction of travel time and, naturally, ART [10,12]. However, the location of an emergency is also regarded as *sensitive data* because it can reveal who received care and for which reason. For example, by knowing that one intervention took place in front of the house of a debilitated person, attackers with auxiliary information may accurately infer that this person received care and (mis)use this information for their own good. Indeed, location privacy is an emerging and active research topic in the literature [13–15] as publicly exposing users' location raises major privacy issues. A common way to achieve location privacy is by applying a *location obfuscation* mechanism. In [15], the authors proposed geo-indistinguishability (GI), which is based on the state-of-the-art differential privacy (DP) [16] model, to protect the location privacy of users. GI has received considerable attention due to its effectiveness and simplicity of implementation (e.g., Location Guard [17]).

In this paper, we propose to sanitize, independently, each emergency location data with GI before training any ML techniques to predict ARTs to protect the ML model against, e.g., membership inference attacks and data reconstruction attacks [18,19]. In our context, besides the own location, with the exact coordinates of both SDIS 25 centers and the emergency scenes, one can retrieve important features such as the distance and estimated travel time. However, if the location is sanitized via GI, many other explanatory variables (e.g., distance, travel time, city) would be 'perturbed' too. In the privacy-preserving ML literature, training ML models with sanitized data is common practice [7,20–25], which is also known as *input perturbation* [26]. In contrast to objective [27] and gradient [28] perturbation settings, input perturbation is the easiest method to apply, and it is independent of any ML and post-processing techniques. We also remark that input perturbation is in accordance with real-world applications where EMS would only use and/or share sanitized data with third parties to *train* and develop ML-based decision-support systems.

To summarize, this paper proposes the following contributions:

- Recognize the most influential variables when building accurate ML-based models to predict ART. This would allow other EMS to collect these variables and recreate our methodology or develop their own taking into account their policies.
- Evaluate the effectiveness of several values of ϵ (i.e., the privacy budget) to sanitize emergency location data with GI and train ML-based models to predict ART. To the author's knowledge, this is the first work to assess the impact of geo-indistinguishability on sanitizing the location of emergency scenes when training the ML model for such an important task. Although predicting ART is a means to allow EMS to save more lives, we notice that it is also possible to do so while preserving the victims' location privacy.

Outline: The remainder of this paper is organized as follows. In Section 2, we describe the material and methods used in this work, i.e., the geo-indistinguishability privacy model, the data presentation (context, collection, and analysis), the sanitization of emergency scenes with GI, the ML models, and the experimental setup. In Section 3, we present the results of our experiments and our discussion. Lastly, in Section 4, we present the concluding remarks and future directions.

2. Materials and Methods

In this section, we revise the geo-indistinguishability privacy model (Section 2.1), we provide a description of the processing of interventions by SDIS 25 (Section 2.2), the data collection process (Section 2.3), the analysis of SDIS 25 ARTs (Section 2.4), the GI-based sanitization of emergency location data (Section 2.5), the ML models used for predicting ARTs (Section 2.6), and the experimental setup (Section 2.7).

2.1. Geo-Indistinguishability

Differential privacy [16] has been accepted as the *de facto* standard for data privacy. DP was developed in the area of statistical databases, but it is now applied to several fields. Furthermore, DP has also been extended to a local model (*a.k.a.* LDP [26]) in which users sanitize their data before sending it to the server. Although DP is well-suited to the case of trusted curators, with LDP, users do not need to trust the curator.

Geo-indistinguishability [15] is based on a generalization of DP developed in [29] and has been proposed for preserving location privacy without the need for a trusted curator (e.g., a malicious location-based service – LBSs). A mechanism satisfies ϵ-GI if for any two locations x_1 and x_2 within a radius r, the output y of them is (ϵ, r)-geo-indistinguishable if we have:

$$\frac{\Pr(y|x_1)}{\Pr(y|x_2)} \leq e^{\epsilon r}, \forall r > 0, \forall y, \forall x_1, x_2 : d(x_1, x_2) \leq r.$$

Intuitively, this means that for any point x_2 within a radius r from x_1, GI forces the corresponding distributions to be at most $l = \epsilon r$ distant. In other words, the level of distinguishability l increases with r, e.g., an attacker can distinguish that the user is in Paris rather than London but can hardly (controlled by ϵ) determine the user's exact location. Although both GI and DP use the notation of ϵ to refer to the privacy budget, they cannot be compared directly because ϵ in GI contains the unit of measurement (e.g., meters).

On the continuous plane (as we consider in this paper), an intuitive polar Laplace mechanism has been proposed in [15] to achieve GI, which is briefly described in the following. Rather than reporting the user's true location $x \in \mathbb{R}^2$, we report a point $y \in \mathbb{R}^2$ generated randomly according to $D_\epsilon(y) = \frac{\epsilon^2}{2\pi} e^{-\epsilon d_2(x,y)}$. Algorithm 1 shows the pseudocode of the polar Laplace mechanism in the continuous plane. More specifically, the noise is drawn by first transforming the true location x to polar coordinates. Then, the angle θ is drawn randomly between $[0, 2\pi)$ (line 3), and the distance r is drawn from $C_\epsilon^{-1}(p)$ (line 5), which is calculated using the negative branch W_{-1} of the Lambert W function. Finally, the generated distance and angle are added to the original location.

Algorithm 1 Polar Laplace mechanism in continuous plane [15]

1: **Input:** $\epsilon > 0$, real location $x \in \mathbb{R}^2$.
2: **Output:** sanitized location $y \in \mathbb{R}^2$.
3: Draw θ uniformly in $[0, 2\pi)$
4: Draw p uniformly in $[0, 1)$
5: Set $r = C_\epsilon^{-1}(p) = -\frac{1}{\epsilon}\left(W_{-1}\left(\frac{p-1}{e}\right) + 1\right)$
6: **Return:** $y = x + \langle r\cos(\theta), r\sin(\theta)\rangle$

2.2. Process Flow Description

The Departmental Fire and Rescue Service of Doubs currently has 71 centers deployed throughout the region of Doubs, France, serving a population of around 540,000 people. The focus of this paper is on interventions with *victims* that were further transported to hospitals. In these interventions, there was a need for an *emergency and victim assistance vehicle* (*a.k.a.* Véhicule de Secours et d'Assistance aux Victimes - VSAV). VSAVs are equipped with adequate material and personnel for first-aid treatment in urgent situations. In this paper, we interchangeably use the term 'ambulance' when referring to VSAV.

The process of an intervention is briefly described in the following. First, an emergency call is received and treated by an operator. Next, the adequate crew/engine is notified (i.e., the starting date—*SDate*). Once the sufficient armament is gathered, the ambulance goes to the emergency scene. Upon arriving on-scene, the crew uses a mechanical system to report their arrival (i.e., the arrival date—*ADate*). We focus on the ART period, which is calculated as: $ART = ADate - SDate$.

The operation process to decide the adequate SDIS 25 center to attend the intervention depends on the exact *location* of the intervention. As stated previously, there is a city, a district, and a zone that jointly define a list of priority centers, which are responsible for the call. The reason for such a list is because a single center may not have sufficient resources at time *SDate* to attend an intervention. In this case, if the first center of the list does not have sufficient resources, another center(s) would be in charge of the call. Additionally, many situations may generate several victims (e.g., traffic accidents, floods). In these cases, a single intervention can require more than one ambulance, which can come from different centers depending on the availability of resources. This means different ARTs for the same intervention and, therefore, we focus on each ambulance in our analysis and predictions.

In addition, although in some countries the *reason* of the emergency may require a recommended ART [30], for SDIS 25, ART depends on the *Zone* as detailed in [9]. There are three zones: Z1 refers to urban areas, Z2 refers to semi-urban areas, and Z3 refers to rural ones. Therefore, SDIS 25 ambulances should arrive on-scene with $ART \leq 10$ minutes (min) on Z1 and with $ART \leq 25$ min on Z2 and Z3, i.e., including the pre-travel delay (gathering armament) and travel time. If these time limits are not reached, a breakdown in SDIS 25 services is generated [9]. The victim state may also be impacted negatively with high ARTs [1,5]. Lastly, SDIS 25 may also help other EMS outside the Doubs region, and in this case, there is no pre-defined ART limit by SDIS 25.

2.3. Data Collection

We used retrospective data of EMS operations recorded by SDIS 25. All interventions with *victim* that were attended by SDIS 25 centers with a VSAV and further transported to hospitals were eligible for inclusion. These data covered the period of January 2006 to June 2020. The main attributes of these data are described in the following:

- *ID* is a unique identifier for each intervention;
- *SDate* is the "Starting Date" of the intervention, which represents the time SDIS 25 took charge of the intervention after processing the call;
- *ADate* is the "Arrival Date" of an ambulance on the emergency scene;
- *Center* is the SDIS 25 center from which the ambulance left;
- *Location* is the precise location (latitude, longitude) of the intervention;

- *Zone* is either urban (Z1), semi-urban (Z2), or rural (Z3);
- *City* is the municipality where the intervention took place. A city may have zero or more *Districts*.

Each ambulance represents one sample, i.e., a single intervention may have received one or more ambulances. The ART variable was calculated as $ART = ADate - SDate$. We excluded outlying observations with ART of less than 1 min and with ART of more than 45 min, which represented less than 1.4% of the original number of samples.

Using *SDate*, we have added temporal information such as: year, month, day, weekday, hour, and categorical indicators to denote holidays, end/start of the month, and end/start of the year. Moreover, with the exact coordinates from both *Center* and emergency's *Location*, we calculated the great-circle distance (https://en.wikipedia.org/wiki/Great-circle_distance) to add as a feature, which is the shortest distance between two points on the surface of a sphere. We used the great-circle distance since it is faster to be calculated than the Geodesic distance and more accurate than the Euclidean distance. Moreover, we have added the number of interventions in the past hour and the number of active interventions in the current hour. As also remarked in the literature [3,10], the number of interventions on previous hours might impact ART. In addition, external data that may affect ART were gathered from the following sources:

- Bison-Futé [31] provides prediction of traffic level for the Doubs region as indicators ranging from 1 (regular flow) to 4 (extremely difficult flow) per day. We added these indicators according to *SDate*;
- Météo-France [32] supplies historical weather information such as precipitation, temperature, wind speed, and gust speed. We added weather data per hour according to *SDate*;
- OSRM API [33] gives the driving distance on the fastest route and its travel time duration. This way, with the coordinates from both *Center* and emergency's *Location*, we added these two features, i.e., estimated travel time in minutes and driving distance in kilometers (km), for each ambulance.

2.4. Data Analysis

After removing outlying observations, the dataset at our disposal has 186,130 dispatched ambulances from SDIS 25 centers that attended 182,700 EMS interventions. The frequency on the number of dispatched ambulances per zone is 39.62% (Z1), 33.38% (Z2), 26.71% (Z3), and 0.29% (outside the Doubs region), respectively. Figure 1 illustrates the distribution of our variable of interest, namely ART, via three histograms with bins of 1 min for each zone within the Doubs region. One can notice that the ART distributions follow a typical right-skewed distribution also observed in other works/countries [3,34,35]. The mean and standard deviation (std) values for zones Z1, Z2, and Z3 are 8.79 ± 5.66 min, 11.43 ± 6.15 min, and 15.38 ± 6.41 min, respectively. SDIS 25 had about 79.52% of the time $ART \leq 10$ min on zone Z1, and had about 95.76% and 92.50% of the time $ART \leq 25$ min on zones Z2 and Z3, respectively.

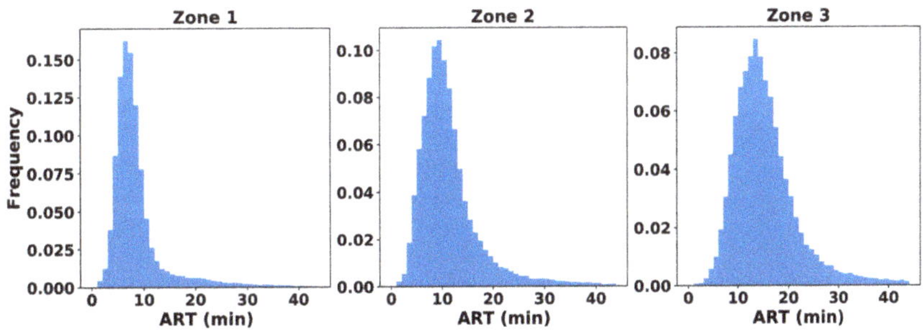

Figure 1. Distribution of the ART variable for zones Z1, Z2, and Z3, respectively.

Figure 2 illustrates the total number of dispatched ambulances per hour (left-hand plot) and the cumulative ART in hours per day of the week and hour in the day (right-hand plot). One can notice that the total number of dispatched ambulances is notably related to the hour in the day, i.e., there were more interventions in working periods rather than between 0 h to 6 h. This behavior is also noticed in the works [12,35]. Moreover, as one can notice with the right-hand plot of Figure 2, from 8 h in the morning on, the cumulative ART starts to increase and remains high up to 19 h when it starts to decrease. Although this high cumulative ART can be linked with the high hourly demand, ambulances dispatched during working periods are also more likely to traffic congestion and, naturally, to undergo through longer travel time. Secondly, due to the number of interventions in a given hour, SDIS 25 centers may have taken more time to dispatch ambulances if their resources were in use in other incidents. A slightly different profile can be seen on weekends, with noticeable higher cumulative ARTs in the late night (0–6 h) and during some hours of the day too.

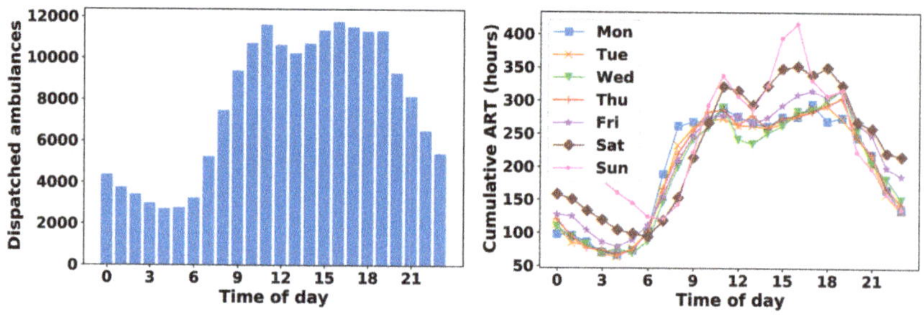

Figure 2. Histogram of the total number of dispatched ambulances per hour in the day (**left-hand plot**) and cumulative ART in hours per day of the week and hour in the day (**right-hand plot**).

Summary statistics per year and per zone are shown in Table 1. The metrics in this table includes the total number of dispatched ambulances (Nb. Amb.), and descriptive statistics such as mean and standard deviation (std) values for the ART variable. We recall that for 2020, these statistics are up to June 2020 only. As also noticed in [7,8], the number of interventions increases throughout the years. The year 2010 presented high values in comparison with all other years, e.g., for Z1, the average ART was above the 10 min recommendation.

2.5. *Preserving Emergency Location Privacy with Geo-Indistinguishability*

To preserve the privacy of each emergency scene, we apply the polar Laplace mechanism in Algorithm 1 to the *Location* attribute of each intervention. This means, even if our dataset is per ambulance dispatch (i.e., 186,130 ambulances), we used the same sanitized value per intervention (i.e., 182,700 unique interventions). Although in [15] the authors propose two further steps to Algorithm 1, i.e., discretization and truncation, both steps can be neglected in our context. This is, first, because SDIS 25 may also help other EMS outside the Doubs region as we discussed in Section 2.2, and second, we assume that any location in the continuous plane can be an emergency scene. Although reporting an approximate location in the middle of a river may not have much sense in LBSs, in an emergency dataset with approximate locations, this may indicate an urgency for someone who drowned in the river, for example.

We used five different levels for the privacy budget $\epsilon = l/r$, where l is the privacy level we want within a radius r. Table 2 exhibits the five different levels of privacy. For the sake of illustration, Figure 3 exhibits three maps of the Doubs region with the points of original location (left-hand plot), $\epsilon = 0.005493$-GI location (middle plot), and $\epsilon = 0.002747$-GI location (right-hand plot). As one can notice, with an intermediate privacy level ($l = \ln(3), r = 400$), locations are more spread throughout the map while with a lower privacy level ($l = \ln(3), r = 200$), locations approximate the real clusters.

Table 1. Mean and std values for the ART variable and the total number of dispatched ambulances (Nb. Amb.) per year in zones Z1, Z2, and Z3, respectively. For 2020, we only consider cases of the first semester.

Year	Z1			Z2			Z3		
	Nb. Amb.	Mean	Std	Nb. Amb.	Mean	Std	Nb. Amb.	Mean	Std
2006	197	9.23	4.41	367	11.25	5.50	354	14.27	5.40
2007	236	7.39	3.05	671	10.79	5.04	595	14.35	5.52
2008	799	8.69	6.04	1055	11.19	5.32	911	14.53	6.02
2009	1363	8.76	6.05	2087	11.08	5.67	1872	14.94	6.46
2010	2643	10.08	7.23	2797	12.48	6.85	2483	16.01	7.22
2011	5971	8.26	5.61	4276	11.24	6.13	3295	14.50	6.25
2012	6078	8.66	5.89	4661	11.18	6.39	3602	14.86	6.24
2013	6780	8.82	5.72	5048	11.03	6.11	3972	15.07	6.30
2014	6847	8.37	5.23	5481	10.80	5.86	4240	14.91	6.34
2015	7226	8.46	5.50	5596	10.86	5.78	4643	15.02	6.12
2016	7510	8.50	5.35	6179	11.19	5.92	4861	15.32	6.35
2017	8650	8.76	5.32	7251	11.49	6.01	5523	15.51	6.36
2018	9051	8.90	5.46	7641	11.64	6.11	5956	15.59	6.23
2019	7030	9.42	6.02	6238	12.29	6.66	5016	16.60	6.88
2020	3397	9.73	5.87	2843	12.59	6.56	2449	16.46	6.44

With the new *Location* values of each intervention, we also reassigned the city, the district, and the zone when applicable. In addition, we recalculated the following features associated with it: the great-circle distance, the estimated driving distance, and estimated travel time. The latter two features were recalculated with OSRM API, which only considers roads, i.e., if the obfuscated location is in the middle of a farm, the closest route estimates the driving distance and travel time until the closest road. We also highlight that if the new coordinates of the emergency scene indicate a location closer to another SDIS 25 center, even in real life, it would not imply that this center took charge of the intervention. Therefore, the *center* attribute was not 'perturbed'.

Table 2. Values of $\epsilon = l/r$ for sanitizing emergency location data with GI.

$\epsilon = l/r$	l	r (m)
0.005493	$\ln(3)$	200
0.002747	$\ln(3)$	400
0.001155	$\ln(2)$	600
0.000866	$\ln(2)$	800
0.000693	$\ln(2)$	1000

To show the impact of the noise added to the *Location* attribute, Table 3 exhibits the percentage of time that categorical attributes (zone, city, and district) were 'perturbed' (i.e., reassigned); the mean and std values of the great-circle distance attribute and its correlation with the ART variable (Corr. ART). In Table 3, we report the mean(std) values since we repeated our experiments with 10 different seeds (i.e., DP algorithms are randomized). Although we did not include the estimated driving distance and estimated travel time from OSRM API in this analysis, in preliminary tests, we noticed that these two features follow a similar pattern as the great-circle distance attribute.

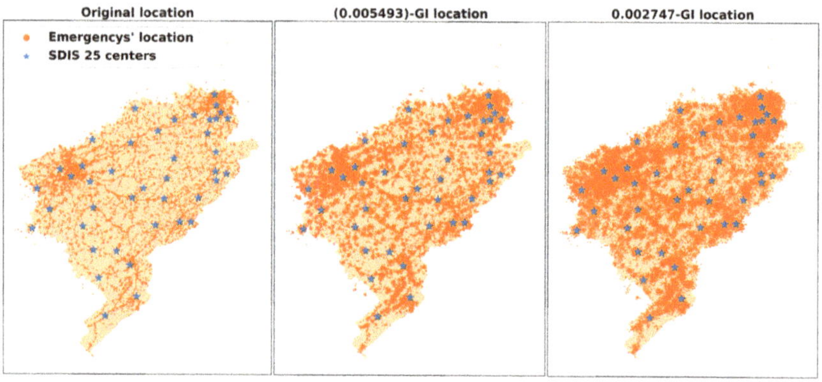

Figure 3. Emergency locations and SDIS 25 centers throughout the Doubs region: original data (left-hand plot), $\epsilon = 0.005493$-GI data (middle plot), and $\epsilon = 0.002747$-GI data (right-hand plot).

From Table 3, one can notice that many features are perturbed due to sanitization of emergency's location with GI. With high levels of ϵ (i.e., less private), the city and the zone suffer low 'perturbation'. On the other hand, district is reassigned many times as it is geographically smaller than the others. When $\epsilon = 0.000866$, the city is already reassigned more than 50% of the time and the district about 75% of the time. Moreover, one can notice that the mean and std values of the great-circle distance increase as the ϵ parameter decreases (i.e., more private). Because $\epsilon = l/r$, making l smaller and/or r higher, the stricter ϵ becomes, and therefore more noise is added to the original locations. Moreover, the correlation between the great-circle distance with the ART variable decreases proportionally as ϵ becomes smaller.

Table 3. Percentage of perturbation for categorical attributes (city, zone, and district) according to ϵ and statistical properties (mean and std values and correlation with ART) of the original and GI-based datasets for the great-circle distance attribute. Mean(std) values are reported since we repeated our experiments with 10 different seeds.

Data	Zone	City	District	Great-circle Dist. (km)		
		'Perturbation' (%)		Mean	std	Corr. ART
Original	-	-	-	3.44	3.72	0.369
$\epsilon = 0.005493$	5.20 (0.05)	7.68 (0.06)	25.8 (0.05)	3.48 (1×10^{-3})	3.72 (7×10^{-4})	0.367 (2×10^{-4})
$\epsilon = 0.002747$	11.3 (0.05)	17.6 (0.10)	41.5 (0.12)	3.57 (1×10^{-3})	3.72 (1×10^{-3})	0.362 (2×10^{-4})
$\epsilon = 0.001155$	28.1 (0.06)	42.3 (0.10)	66.2 (0.09)	4.03 (3×10^{-3})	3.74 (3×10^{-3})	0.335 (5×10^{-4})
$\epsilon = 0.000866$	35.5 (0.10)	52.4 (0.11)	74.0 (0.11)	4.38 (3×10^{-3})	3.81 (4×10^{-3})	0.313 (1×10^{-3})
$\epsilon = 0.000693$	41.4 (0.12)	60.3 (0.09)	79.4 (0.05)	4.77 (6×10^{-3})	3.92 (5×10^{-3})	0.288 (1×10^{-3})

2.6. Machine-Learning Models

Four state-of-the-art ML techniques have been used in our experiments, to predict the scalar ART outcome in a regression framework. More precisely, we compared the performance of two state-of-the-art ML techniques based on decision trees, which are known for their high performance (and speed) with tabular data; a traditional and well-known neural network, and a classical statistical method that can perform both variable selection and regularization. These methods are briefly described in the following:

- Extreme Gradient Boosting (XGBoost) [36] is a decision-tree-based ensemble ML algorithm that produces a forecast model based on an ensemble of weak forecast models (decision trees). XGBoost uses a novel regularization approach over standard gradient boosting machines, which significantly decreases model's complexity. The system is optimized by a quick parallel tree construction and adapted to be fault-tolerant under distributed environments.
- Light Gradient Boosted Machine (LGBM) [37] is a novel gradient boosting framework, which implemented a leaf-wise strategy. This strategy significantly reduces computational speed and resource consumption in comparison to other decision-tree-based algorithms.
- Multilayer Perceptron (MLP) is an artificial neural network of the feedforward type [38]. These algorithms are based on the interconnection of several units (neurons) to transmit signals, which are normally structured into three or more layers, input, hidden(s), and output. We used the Keras library [39] to implement our deep learning models.
- Least Absolute Shrinkage and Selection Operator (LASSO), a method of contracting the coefficients of the regression, whose ability to select a subset of variables is due to the nature of the constraint on the coefficients. Originally proposed by Tibshirani [40] for models using the standard least squares estimator, it has been extended to many statistical models such as generalized linear models, etc. We used the LASSO implementation from the Scikit-learn library [41].

2.7. Experiments

All algorithms were implemented in Python 3.8.8. To run our codes, we used a machine with Intel (R) Core (TM) i7-10750 CPU @ 2.60 GHz, 16 GB RAM, and a GPU with 1920 cores and 6 GB of RAM using Windows 10. Because in Table 3 there are low variations (i.e., small std values) on all features that depend on the sanitized location, we ran our experimental validation only once. In our experiments, each sample corresponds to one ambulance dispatch, in which we included temporal features (e.g., hour, day), weather data (e.g., pressure, temperature), traffic data, the emergency's location (latitude and longitude in radians), and computable features (e.g., distance, travel time). The scalar target variable is the ART in minutes, which is the time measured from the EMS notification to the ambulance's arrival on-scene. All numerical features (e.g., temperature) were standardized using the *StandardScaler* function from the Scikit-Learn library. Categorical features (e.g., center, zone, hour) were encoded using mean encoding, i.e., the mean value of the ART variable with respect to each feature. The target variable, namely ART, was kept in its original format (minutes) since no remarkable improvement was achieved with scaling.

Our experimentation considers the scenario in which EMS would perform both the sanitization of the dataset and the development of ML models. In this case, the objective is to have all ML models to be trained with ϵ-GI data to prevent, for example, membership inference attacks and data reconstruction attacks [18,19]. This also means that ML models will be trained with sanitized data and the testing set will use original data, as it would be if EMS deployed a decision-support system in real life. On the one hand, this would prevent having in real-life a sanitized location that would compromise the EMS response time. On the other hand, each time the model is re-fitted (or retrained), the new known data should also be sanitized with ϵ-GI. A different scenario could consider that both training and testing sets are sanitized, which corresponds to the case where EMS published the data openly or transmitted it to an untrusted party. This latter scenario was out of the scope of this paper and, thus, is left as future work.

With these elements in mind, we divided our dataset into training (years 2006–2019) and testing (six months of 2020) sets to evaluate our models. Thus, five models per ML technique (i.e., XGBoost, LGBM, MLP, and LASSO) were built to predict ART on each month of 2020 using the sanitized (training) datasets with different levels of ϵ-GI location data (*cf.* Table 2). All models were trained continuously, i.e., at the end of each month, the new known data were added to the training set after sanitization with ϵ-GI. Lastly, **all**

models were tested with original data. In addition, for comparison, we also trained and evaluated one additional model per ML technique with original data. In this paper, the models were evaluated using the following regression metrics:

- Root mean squared error (RMSE) measures the square root average of the squares of the errors and is calculated as: $RMSE = \frac{1}{n}\sqrt{\sum_{i=1}^{n}(y_i - \hat{y}_i)^2}$;
- Mean absolute error (MAE) measures the averaged absolute difference between real and predicted values and is calculated as: $MAE = \frac{1}{n}\sum_{i=1}^{n}|y_i - \hat{y}_i|$;
- Mean absolute percentage error (MAPE) measures how far the model's predictions are off from their corresponding outputs on average and is calculated as: $MAPE = \frac{1}{n}\sum_{i=1}^{n}\left|\frac{y_i-\hat{y}_i}{y_i}\right| \cdot 100\%$;
- Coefficient of determination (R^2) measures the proportion of the variance in the dependent variable that is predictable from the independent variable(s). An $R^2 = 1$ would indicate a model that fully captures the variation in ARTs;

in which y_i is the real output, $\hat{y}_i$ is the predicted output, and n is the total number of samples, for $i \in [1, n]$. Results for each metric were calculated using data from the 6 months evaluation period. The RMSE metric was also used during the hyperparameters tuning process via Bayesian optimization (BO). To this end, we used the HYPEROPT library [42] with 100 iterations for each model. Table A1 in Appendix A displays the range of each hyperparameter used in the BO, as well as the final configuration used to train and test the models.

3. Results and Discussion

In this section, we present the results of our experimental validation (Section 3.1) and a general discussion (Section 3.2) including related work and limitations.

3.1. Privacy-Preserving ART Prediction

Figure 4 illustrates the impact of the level of GI for each ML model to predict ART according to each metric. As one can notice in this figure, for XGBoost, LGBM, and LASSO, there were minor differences between training models with original location data or sanitized ones. On the other hand, models trained with MLP performed poorly with GI-based data. In addition, by analyzing models trained with original data, while the smaller RMSE for LASSO is about 5.65, for more complex ML-based models, RMSE is less than 5.6, achieving 5.54 with XGBoost and LGBM. In comparison with the results of existing literature, lower R^2 scores and similar RMSE and MAE results were achieved in [11] to predict ART while using original location data only. With more details, Table A2 in Appendix A numerically exhibits the results from Figure 4.

Indeed, among the four tested models, LGBM and XGBoost achieve similar metric results while favoring the LGBM model. Thus, Figure 5 illustrates the BO iterative process for LGBM models trained with original and sanitized data according to the RMSE metric (left-hand plot); and ART prediction results for 50 dispatched ambulances in 2020 out of 8709 ones (right-hand plot) with an LGBM model trained with original data (Pred: original) and with two LGBM models trained sanitized data, i.e., with $\epsilon = 0.005493$ (low privacy level) and with $\epsilon = 0.000693$ (high privacy level).

As one can notice in the left-hand plot of Figure 5, once data are sanitized with different levels of ϵ-GI, the hyperparameters optimization via BO is also perturbed. This way, local minimums were achieved in different steps of the BO (i.e., the last marker per curve indicates the local minimum). For instance, even though $\epsilon = 0.002747$ is stricter than $\epsilon = 0.005493$, results were still better for the former since, in the last steps of BO, three better local minimums were found. Moreover, prospective predictions were achieved with either original or sanitized data. For instance, in the right-hand plot of Figure 5, even for the high peak-value of ART around 40 min, LGBM's prediction achieved some reasonable estimation. Although several features were perturbed due to the sanitization of the emergency scene (e.g., city, zone, etc.), the models could still achieve similar predictions as the model trained with original location data.

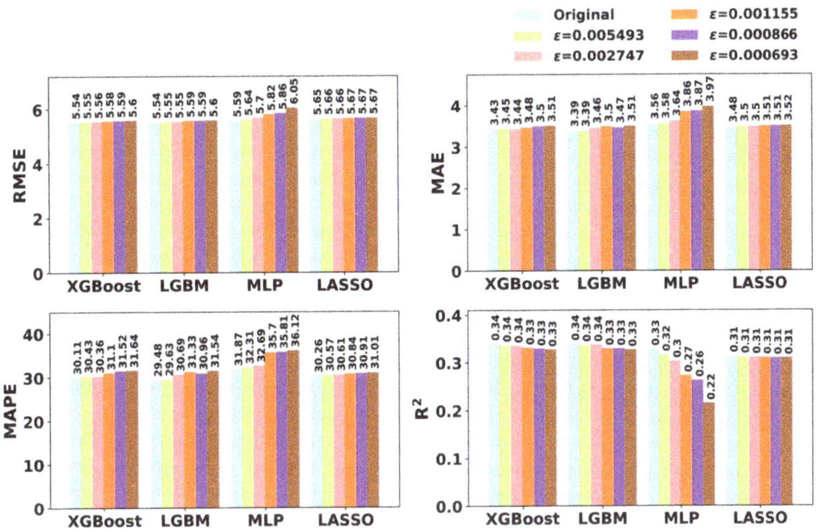

Figure 4. Impact of the level of ϵ-geo-indistinguishability for each ML model to predict ART according to each metric.

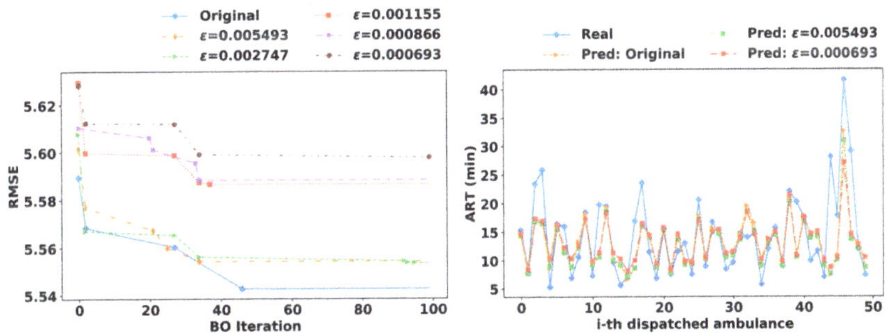

Figure 5. The left-hand plot illustrates the hyperparameters tuning process via Bayesian optimization with 100 iterations for LGBM models trained with original data and sanitized ones. The right-hand plot illustrates the prediction of ARTs with LGBM models trained with original data and with sanitized ones.

Furthermore, in terms of training time, for both original and sanitized datasets, the LASSO method was the fastest to fit our data. On the other hand, MLP models took the longest time to execute than all other methods. Between both decision-tree methods, LGBM models were faster than XGBoost ones. Lastly, the importance of the features, taking into account LASSO coefficients and decision trees' importance scores were: averaged ART per categorical features (e.g., center, city, hour); OSRM API-based features (i.e., estimated driving distance and estimated travel time); the great-circle distance between the center and the emergency scene; the number of interventions in the previous hour, and the number of interventions still active. Immediately thereafter, it appeared the weather data, which were added as "real-time" features, i.e., using the date of the intervention to retrieve the features. Penultimate, the traffic data, which are indicators provided by [31] at the beginning of each year and, which might have shown more influence if they had been retrieved in real time.

Finally, it appeared some temporal variables such as weekend indicators, start/end of the month, and the day of the year.

3.2. Discussion

The medical literature has mainly focused attention on the analysis of ART [3,34,43] and its association with trauma [2,30] and cardiac arrest [1,4,6], for example. To reduce ART, some works propose reallocation of ambulances [5,44], operation demand forecasting [5,7,8,22,45], travel time prediction [12], simulation models [35,46], and EMS response time predictions [11,12]. The work in [11] propose a real-time system for predicting ARTs for the San Francisco fire department, which closely relates to this paper. The authors processed about 4.5 million EMS calls using original location data to predict ART using four ML models, namely linear regression, linear regression with elastic net regularization, decision-tree regression, and random forest. However, no privacy-preserving experiment was performed because the main objective of their paper was proposing a scalable, ML-based, and real-time system for predicting ART. Besides, we also included weather data that the authors in [11] did not consider in their system, which could help to recognize high ARTs due to bad weather conditions, for example.

Currently, many private and public organizations collect and analyze data about their associates, customers, and patients. Because most of these data are personal and confidential (e.g., location), there is a need for privacy-preserving techniques for processing and using these data. Location privacy is an emergency research topic [13,14] due to the ubiquity of LBSs. Within our context, using and/or sharing the exact location of an emergency raises many privacy issues. For instance, the Seattle Fire Department [47] displays live EMS response information with the precise location and reason for the incident. Although the intention of some fire departments [11,47] is laudable, there are many ways for (mis)using this information, which can jeopardize users' privacy. Even if the intervention's *reason* could be an indicator of the call urgency, we did not consider this sensitive attribute in our data analysis nor privacy-preserving prediction models. This is because, for SDIS 25, the ARTs limits are defined by the zone [9]. Additionally, we also did not include the victims' personal data (e.g., gender, age) in our predictions or analysis since, during the calls, the operator may not acquire such information, e.g., when a third party activates the SDIS 25 for unidentified victims. This way, we focused our attention on the *location privacy* of each intervention.

To address location privacy, the authors in [15] proposed the concept of GI, which is based on a generalization [29] of the state-of-the-art DP [16] model. As highlighted in [15], attackers in LSBs may have side information about the user's reported location, e.g., knowing that the user is probably visiting the Eiffel Tower instead of swimming in the Seine river. However, this does not apply in our context because someone may have drowned, and EMS had to intervene. Similarly, even for the dataset with intermediate (and high) privacy in which locations are spread out in the Doubs region (*cf.* map with 0.005493-GI location in Figure 3), someone may have been lost in the forest and EMS would have to interfere. For these reasons, using (or sharing datasets with) approximate emergency locations (e.g., sanitized with GI) is a prospective direction since many locations are possible emergency scenes. Indeed, we are not interested in hiding the emergency's location completely since some approximate information is required to retrieve other features (e.g., city, zone, estimated distance) to use for predicting ART.

Moreover, learning and extracting meaningful patterns from data, e.g., through ML, play a key role in advancing and understanding several behaviors. However, on the one hand, storing and/or sharing original personal data with trusted curators may still lead to data breaches [48] and/or misuse of data, which compromises users' privacy. On the other hand, training ML models with original data can also leak private information. For instance, in [18] the authors evaluate how some models can memorize sensitive information from the training data, and in [19], the authors investigate how ML models are susceptible to membership inference attacks. To address these problems, some works [7,20–25,49] propose to train ML models with sanitized data, which is also known as input perturbation [26] in the privacy-preserving ML literature.

Input perturbation-based ML and GI are linked directly with local DP [26] in which each sample is sanitized independently, either by the user during the data collection process or by the trusted curator, which aims to preserve privacy of each data sample. This way, data are protected from data leakage and are more difficult to reconstruct, for example. In [23,49], the authors investigate how input perturbation through applying controlled Gaussian noise on data samples can guarantee (ϵ, δ)-DP on the final ML model. This means, since ML models are trained with perturbed data, there is a perturbation on the gradient and on the final parameters of the model too.

In this paper, rather than Gaussian noise, the emergency scenes were sanitized with Algorithm 1, i.e., adding two-dimensional Laplacian noise centered at the exact user location $x \in \mathbb{R}^2$. In addition, this sanitization also perturbs other associated and calculated features such as: city, district, zone (e.g., urban or not), great-circle distance, estimated driving distance, and estimated travel time (*cf.* Table 3). As well as the optimization of hyperparameters, i.e., once data are differentially private, one can apply any function on it and, therefore, we also noticed perturbation on the BO procedure. Yet, as shown in the results, prospective ART predictions were achieved with either original or sanitized data. Furthermore, even with a high level of sanitization ($\epsilon = 0.000693$) there was a good privacy-utility trade-off. According to [50], if the mean absolute percentage error (i.e., MAPE) is greater than 20% and less than 50%, the forecast is reasonable, which is the results we have in this paper with MAPE around 30%.

Lastly, some limitations of this work are described in the following. We analyzed ARTs using the data and operation procedures of only one EMS in France, namely SDIS 25. Although it may represent a sufficient number of samples, other public and private organizations are also responsible for EMS calls, e.g., the SAMU (Urgent Medical Aid Service in English) analyzed in [46]. Moreover, there is the possibility of human error when using the mechanical system to report (i.e., record) the arrival on-scene time "*ADate*". For instance, the crew may have forgotten to record status on arrival and may have registered later, or conversely, where the crew may have accidentally recorded before arriving at the location. Additionally, it is noteworthy to mention that the arrival on-scene does not mean arriving at the victim's side, e.g., in some cases the real location of a victim is at the n-th stage of a building as investigated in [43].

4. Conclusions

In the event of an acute medical event such as a respiratory crisis or cardio-respiratory arrest, the time an ambulance takes to arrive on-scene has a direct impact on the quality of service provided. Ambulance response time is a fundamental indicator of the effectiveness of EMS systems [1,2,4–6,30]. For this reason, an intelligent decision-support system is necessary to help minimize overall EMS response times. The present work first analyzes historical records of ARTs to find correlations between their extracted features and explain the trends through the 15 years of collected data. Then, we sought to predict the response time that each center equipped with ambulances had to an event, but not only that, because we also consider that the ML models could be subject to attacks, which would compromise the victims' privacy. Therefore, the joint work aimed to evaluate the effectiveness of predicting ARTs with ML models trained over sanitized location data with different levels of ϵ-geo-indistinguishability. As shown in the results, the sanitization of location data and the perturbation of its associated features (e.g., city, distance) had no considerable impact on predicting ART. With these findings, EMS may prefer using and/or sharing sanitized datasets to avoid possible data leakages, membership inference attacks, or data reconstructions, for example.

For future work, we aim to extend the analysis and predictions to different operation times such as the pre-travel delay (i.e., gathering personnel and ambulances) and travel times (e.g., from the center to the emergency scene, from the emergency scene to hospitals), while respecting users' privacy. In addition, new variables will be added such as the number of dispatched ambulances registered in a previous or current time, and the number of ambulances and firefighters available in each center at a given time, given that while there are few resources available, ART may be longer. Indeed, the aim is to build an intelligent system capable of predicting ARTs while respecting victims' privacy. This way, this system would allow reinforcing SDIS 25 centers with the necessary firefighters to attend incidents faster; to create a new center according to the concurrence and high average ARTs for a given area; as well as to convert a static resource deployment plan into a dynamic one, which would be based on the selection of the center with shorter response times taking into account the community the emergency took place, traffic and weather conditions, and so on. Lastly, we would like to evaluate, in practice, the trade-off between such an ART prediction decision-support system with the victims' privacy, on using ϵ-GI location data.

Author Contributions: Conceptualization, H.H.A. and S.C.; methodology, H.H.A. and S.C.; software, H.H.A. and S.C.; validation, H.H.A. and S.C.; formal analysis, H.H.A. and S.C.; investigation, H.H.A., S.C., C.G., and J.-F.C.; resources, C.G. and J.-F.C.; data curation, H.H.A. and S.C.; writing—original draft preparation, H.H.A., S.C., C.G., and J.-F.C.; writing—review and editing, H.H.A., S.C., C.G., and J.-F.C.; visualization, S.C. and C.G.; supervision, C.G. and J.-F.C.; project administration, C.G. and J.-F.C.; funding acquisition, C.G. and J.-F.C. All authors have read and agreed to the published version of the manuscript.

Funding: This research received no external funding.

Acknowledgments: This work was supported by the Region of Bourgogne Franche-Comté CADRAN Project, by the EIPHI-BFC Graduate School (contract "ANR-17-EURE-0002"), and by SDIS du Doubs, with the support of the French Ministry of Higher Education and Research (managed by the National Association of Research and Technology (ANRT) for the CIFRE thesis (N 2019/0372). The authors would also like to thank SDIS 25 Commander Guillaume Royer-Fey and Captain Céline Chevallier for their great collaborations and continuous feedback. All computations have been performed on the "Mésocentre de Calcul de Franche-Comté".

Conflicts of Interest: The authors declare no conflict of interest.

Abbreviations

The following abbreviations are used in this manuscript:

ART	Ambulance response time
BO	Bayesian optimization
DP	Differential privacy
EMS	Emergency medical services
GI	Geo-Indistinguishability
LASSO	Least Absolute Shrinkage and Selection Operator
LBSs	Location-based services
LDP	Local differential privacy
LGBM	Light Gradient Boosted Machine
MLP	Multilayer Perceptron
MAE	Mean absolute error
MAPE	Mean absolute percentage error
RMSE	Root mean squared error
SDIS 25	Departmental Fire and Rescue Service of Doubs
XGBoost	Extreme Gradient Boosting
Z1	Zone urban
Z2	Zone semi-urban
Z3	Zone rural

Appendix A. Complementary Results

Table A1. Search space for hyperparameters by ML model and the best configuration obtained for predicting ARTs per dataset.

Model	Search Space	Best Configuration per Dataset					
		Original	$\epsilon = 0.005493$	$\epsilon = 0.002747$	$\epsilon = 0.001155$	$\epsilon = 0.000866$	$\epsilon = 0.000693$
XGBoost	max_depth: [1, 10]	9	9	6	6	9	9
	n_estimators: [50, 500]	465	465	130	235	465	465
	learning_rate: [0.001, 0.5]	0.0265	0.0265	0.0858	0.0486	0.0265	0.0265
	min_child_weight: [1, 10]	5	5	7	7	5	5
	max_delta_step: [1, 11]	4	4	3	4	4	4
	gamma: [0.5, 5]	3	3	0	2	3	3
	subsample: [0.5, 1]	0.8	0.8	1	1	0.8	0.8
	colsample_bytree: [0.5, 1]	0.5	0.5	0.5	0.5	0.5	0.5
	alpha: [0, 5]	2	2	1	2	2	2
LGBM	max_depth: [1, 10]	7	8	10	8	8	6
	n_estimators: [50, 500]	355	326	477	250	80	441
	learning_rate: [1×10^{-4}, 0.5]	0.0188	0.0098	0.0164	0.0285	0.0586	0.0300
	subsample: [0.5, 1]	0.54066	0.5228	0.6138	0.6699	0.6732	0.5812
	colsample_bytree: [0.5, 1]	0.5160	0.5575	0.5204	0.6870	0.5507	0.5451
	num_leaves: [31, 400]	400	192	245	398	132	95
	reg_alpha: [0, 5]	4	0	5	0	1	4
MLP	Dense layers: [1, 7]	7	3	4	6	6	6
	Number of neurons: [2^8, 2^{13}]	2^{10}	2^{12}	2^{12}	2^9	2^{12}	2^9
	Batch size: [32, 168]	140	80	48	82	70	44
	Learning rate: [1×10^{-5}, 0.01]	0.00265	0.00124	0.0099	0.0099	0.0094	0.0077
	Optimizer: Adam	Adam	Adam	Adam	Adam	Adam	Adam
	Epochs: 100	100	100	100	100	100	100
	Early stopping: 10	10	10	10	10	10	10
LASSO	alpha: [0.01, 2]	0.0205	0.0307	0.0105	0.0100	0.0112	0.0107

Table A2. Metrics results for each ML model trained with original data and sanitized ones.

Data	Metric	XGBoost	LGBM	MLP	LASSO
Original	RMSE	5.5398	5.5427	5.5916	5.6511
	MAE	3.4286	3.3880	3.5623	3.4760
	MAPE	30.114	29.476	31.867	30.260
	R^2	0.3412	0.3405	0.3289	0.3145
$\epsilon = 0.005493$	RMSE	5.5547	5.5544	5.6401	5.6596
	MAE	3.4515	3.3915	3.5773	3.4960
	MAPE	30.432	29.628	32.307	30.571
	R^2	0.3377	0.3378	0.3172	0.3124
$\epsilon = 0.002747$	RMSE	5.5617	5.5536	5.6959	5.6636
	MAE	3.4430	3.4628	3.6357	3.4991
	MAPE	30.364	30.688	32.687	30.606
	R^2	0.3360	0.3379	0.3036	0.3115
$\epsilon = 0.001155$	RMSE	5.5788	5.5867	5.8184	5.6671
	MAE	3.4803	3.4991	3.8550	3.5094
	MAPE	31.097	31.327	35.704	30.835
	R^2	0.3319	0.3300	0.2733	0.3106
$\epsilon = 0.000866$	RMSE	5.5892	5.5885	5.8575	5.6716
	MAE	3.5033	3.4702	3.8736	3.5134
	MAPE	31.515	30.964	35.810	30.907
	R^2	0.3295	0.3296	0.2635	0.3095
$\epsilon = 0.000693$	RMSE	5.5962	5.5978	6.0463	5.6717
	MAE	3.5119	3.5087	3.9704	3.5171
	MAPE	31.638	31.543	36.122	31.007
	R^2	0.3278	0.3274	0.2153	0.3095

References

1. Bürger, A.; Wnent, J.; Bohn, A.; Jantzen, T.; Brenner, S.; Lefering, R.; Seewald, S.; Gräsner, J.T.; Fischer, M. The Effect of Ambulance Response Time on Survival Following Out-of-Hospital Cardiac Arrest. *Deutsches Aerzteblatt Online* **2018**. [CrossRef]
2. Byrne, J.P.; Mann, N.C.; Dai, M.; Mason, S.A.; Karanicolas, P.; Rizoli, S.; Nathens, A.B. Association Between Emergency Medical Service Response Time and Motor Vehicle Crash Mortality in the United States. *JAMA Surg.* **2019**, *154*, 286. [CrossRef] [PubMed]
3. Do, Y.K.; Foo, K.; Ng, Y.Y.; Ong, M.E.H. A Quantile Regression Analysis of Ambulance Response Time. *Prehosp. Emerg. Care* **2012**, *17*, 170–176. [CrossRef] [PubMed]
4. Holmén, J.; Herlitz, J.; Ricksten, S.E.; Strömsöe, A.; Hagberg, E.; Axelsson, C.; Rawshani, A. Shortening Ambulance Response Time Increases Survival in Out-of-Hospital Cardiac Arrest. *J. Am. Heart Assoc.* **2020**, *9*. [CrossRef] [PubMed]
5. Chen, A.Y.; Lu, T.Y.; Ma, M.H.M.; Sun, W.Z. Demand Forecast Using Data Analytics for the Preallocation of Ambulances. *IEEE J. Biomed. Health Inform.* **2016**, *20*, 1178–1187. [CrossRef] [PubMed]
6. Lee, D.W.; Moon, H.J.; Heo, N.H. Association between ambulance response time and neurologic outcome in patients with cardiac arrest. *Am. J. Emerg. Med.* **2019**, *37*, 1999–2003. [CrossRef] [PubMed]
7. Arcolezi, H.H.; Couchot, J.F.; Cerna, S.; Guyeux, C.; Royer, G.; Bouna, B.A.; Xiao, X. Forecasting the number of firefighter interventions per region with local-differential-privacy-based data. *Comput. Secur.* **2020**, *96*, 101888. [CrossRef]
8. Cerna, S.; Guyeux, C.; Arcolezi, H.H.; Couturier, R.; Royer, G. A Comparison of LSTM and XGBoost for Predicting Firemen Interventions. In *Trends and Innovations in Information Systems and Technologies*; Springer International Publishing: Cham, Switzerland, 2020; pp. 424–434. [CrossRef]
9. Cerna, S.; Guyeux, C.; Royer, G.; Chevallier, C.; Plumerel, G. Predicting Fire Brigades Operational Breakdowns: A Real Case Study. *Mathematics* **2020**, *8*, 1383. doi:10.3390/math8081383. [CrossRef]
10. Nehme, Z.; Andrew, E.; Smith, K. Factors Influencing the Timeliness of Emergency Medical Service Response to Time Critical Emergencies. *Prehosp. Emerg. Care* **2016**, *20*, 783–791. [CrossRef]
11. Lian, X.; Melancon, S.; Presta, J.R.; Reevesman, A.; Spiering, B.; Woodbridge, D. Scalable Real-Time Prediction and Analysis of San Francisco Fire Department Response Times. In Proceedings of the 2019 IEEE SmartWorld, Ubiquitous Intelligence & Computing, Advanced & Trusted Computing, Scalable Computing & Communications, Cloud & Big Data Computing, Internet of People and Smart City Innovation (SmartWorld/SCALCOM/UIC/ATC/CBDCom/IOP/SCI), Leicester, UK, 19–23 August 2019. [CrossRef]
12. Aladdini, K. EMS Response Time Models: A Case Study and Analysis for the Region of Waterloo. Master's Thesis, University of Waterloo, Waterloo, ON , Canada, 2010.
13. Shokri, R.; Theodorakopoulos, G.; Boudec, J.Y.L.; Hubaux, J.P. Quantifying Location Privacy. In Proceedings of the 2011 IEEE Symposium on Security and Privacy, Oakland, CA, USA, 22–25 May 2011. [CrossRef]
14. Chatzikokolakis, K.; ElSalamouny, E.; Palamidessi, C.; Pazii, A. Methods for Location Privacy: A comparative overview. *Found. Trends Priv. Secur.* **2017**, *1*, 199–257. [CrossRef]
15. Andrés, M.E.; Bordenabe, N.E.; Chatzikokolakis, K.; Palamidessi, C. Geo-indistinguishability. In Proceedings of the 2013 ACM SIGSAC Conference on Computer & Communications Security, Berlin, Germany, 4–8 November 2013. [CrossRef]
16. Dwork, C.; Roth, A. The algorithmic foundations of differential privacy. *Found. Trends Theor. Comput. Sci.* **2014**, *9*, 211–407. [CrossRef]
17. Location Guard. Available online: https://github.com/chatziko/location-guard (accessed on 2 August 2021).
18. Song, C.; Ristenpart, T.; Shmatikov, V. Machine Learning Models that Remember Too Much. In Proceedings of the 2017 ACM SIGSAC Conference on Computer and Communications Security, Dallas, TX, USA, 30 October–3 November 2017. [CrossRef]
19. Shokri, R.; Stronati, M.; Song, C.; Shmatikov, V. Membership Inference Attacks Against Machine Learning Models. In Proceedings of the 2017 IEEE Symposium on Security and Privacy, San Jose, CA, USA, 22–24 May 2017. [CrossRef]
20. Chamikara, M.A.P.; Bertok, P.; Khalil, I.; Liu, D.; Camtepe, S. Privacy Preserving Face Recognition Utilizing Differential Privacy. *Comput. Secur.* **2020**, *97*, 101951. [CrossRef]
21. Fan, L. Image Pixelization with Differential Privacy. In Proceedings of the 32nd IFIP Annual Conference on Data and Applications Security and Privacy, Bergamo, Italy, 16–18 July 2018; pp. 148–162. [CrossRef]
22. Couchot, J.F.; Guyeux, C.; Royer, G. Anonymously forecasting the number and nature of firefighting operations. In Proceedings of the 23rd International Database Applications & Engineering Symposium, Athens, Greece, 10–12 June 2019. [CrossRef]
23. Fukuchi, K.; Tran, Q.K.; Sakuma, J. Differentially Private Empirical Risk Minimization with Input Perturbation. In Proceedings of the International Conference on Discovery Science, Kyoto, Japan, 15–17 October 2017; pp. 82–90. [CrossRef]
24. Agrawal, D.; Aggarwal, C.C. On the design and quantification of privacy preserving data mining algorithms. In Proceedings of the Twentieth ACM SIGMOD-SIGACT-SIGART Symposium on Principles of Database Systems, Santa Barbara, CA, USA, 21–23 May 2001. [CrossRef]
25. Agrawal, R.; Srikant, R. Privacy-preserving data mining. In Proceedings of the 2000 ACM SIGMOD International Conference on Management of Data, Dallas, TX, USA, 15–18 May 2000. [CrossRef]
26. Kasiviswanathan, S.P.; Lee, H.K.; Nissim, K.; Raskhodnikova, S.; Smith, A. What Can We Learn Privately? In Proceedings of the 2008 49th Annual IEEE Symposium on Foundations of Computer Science, Philadelphia, PA, USA, 26–28 October 2008. [CrossRef]
27. Chaudhuri, K.; Monteleoni, C.; Sarwate, A.D. Differentially private empirical risk minimization. *J. Mach. Learn. Res.* **2011**, *12*, 1069–1109. [PubMed]

28. Abadi, M.; Chu, A.; Goodfellow, I.; McMahan, H.B.; Mironov, I.; Talwar, K.; Zhang, L. *Deep Learning with Differential Privacy*; Association for Computing Machinery: New York, NY, USA, 2016; pp. 308–318. [CrossRef]
29. Chatzikokolakis, K.; Andrés, M.E.; Bordenabe, N.E.; Palamidessi, C. Broadening the Scope of Differential Privacy Using Metrics. In *Privacy Enhancing Technologies*; Springer: Berlin/Heidelberg, Germany, 2013; pp. 82–102. [CrossRef]
30. Pons, P.T.; Markovchick, V.J. Eight minutes or less: Does the ambulance response time guideline impact trauma patient outcome? *J. Emerg. Med.* **2002**, *23*, 43–48. [CrossRef]
31. Bison-Futé. Les Prévisions de Trafic. Available online: https://www.bison-fute.gouv.fr (accessed on 2 February 2021).
32. Météo-France. Données Publiques. Available online: https://donneespubliques.meteofrance.fr/?fond=produit&id_produit=90&id_rubrique=32 (accessed on 2 February 2021).
33. Luxen, D.; Vetter, C. Real-time routing with OpenStreetMap data. In Proceedings of the 19th ACM SIGSPATIAL International Conference on Advances in Geographic Information Systems, Chicago, IL, USA, 1–4 November 2011; pp. 513–516. [CrossRef]
34. Austin, P.C. Quantile Regression: A Statistical Tool for Out-of-Hospital Research. *Acad. Emerg. Med.* **2003**, *10*, 789–797. [CrossRef]
35. Peleg, K.; Pliskin, J.S. A geographic information system simulation model of EMS: Reducing ambulance response time. *Am. J. Emerg. Med.* **2004**, *22*, 164–170. [CrossRef]
36. Chen, T.; Guestrin, C. XGBoost: A Scalable Tree Boosting System. In Proceedings of the 22nd ACM SIGKDD International Conference on Knowledge Discovery and Data Mining, San Francisco, CA, USA, 13–17 August 2016. [CrossRef]
37. Ke, G.; Meng, Q.; Finley, T.; Wang, T.; Chen, W.; Ma, W.; Ye, Q.; Liu, T.Y. LightGBM: A Highly Efficient Gradient Boosting Decision Tree. In *Advances in Neural Information Processing Systems 30*; Guyon, I., Luxburg, U.V., Bengio, S., Wallach, H., Fergus, R., Vishwanathan, S., Garnett, R., Eds.; Curran Associates, Inc.: Red Hook, NY, USA, 2017; pp. 3146–3154.
38. Goodfellow, I.; Bengio, Y.; Courville, A.; Bengio, Y. *Deep Learning*; MIT Press: Cambridge, MA, USA, 2016; Volume 1.
39. Keras. Available online: https://keras.io (accessed on 2 August 2021).
40. Tibshirani, R. Regression Shrinkage and Selection via the Lasso. *J. R. Stat. Soc. B Methodol.* **1996**, *58*, 267–288. [CrossRef]
41. Pedregosa, F.; Varoquaux, G.; Gramfort, A.; Michel, V.; Thirion, B.; Grisel, O.; Blondel, M.; Pedregosa, P.; Weiss, R.; Dubourg, V.; et al. Scikit-learn: Machine Learning in Python. *J. Mach. Learn. Res.* **2011**, *12*, 2825–2830.
42. Bergstra, J.; Yamins, D.; Cox, D.D. Making a Science of Model Search: Hyperparameter Optimization in Hundreds of Dimensions for Vision Architectures. In Proceedings of the 30th International Conference on International Conference on Machine Learning, Atlanta, GA, USA, 16–21 June 2013; pp. I-115–I-123.
43. Silverman, R.A.; Galea, S.; Blaney, S.; Freese, J.; Prezant, D.J.; Park, R.; Pahk, R.; Caron, D.; Yoon, S.; Epstein, J.; et al. The "Vertical Response Time": Barriers to Ambulance Response in an Urban Area. *Acad. Emerg. Med.* **2007**, *14*, 772–778. [CrossRef]
44. Carvalho, A.; Captivo, M.; Marques, I. Integrating the ambulance dispatching and relocation problems to maximize system's preparedness. *Eur. J. Oper. Res.* **2020**, *283*, 1064–1080. [CrossRef]
45. Lin, A.X.; Ho, A.F.W.; Cheong, K.H.; Li, Z.; Cai, W.; Chee, M.L.; Ng, Y.Y.; Xiao, X.; Ong, M.E.H. Leveraging Machine Learning Techniques and Engineering of Multi-Nature Features for National Daily Regional Ambulance Demand Prediction. *Int. J. Environ. Res. Public Health* **2020**, *17*, 4179. [CrossRef] [PubMed]
46. Aboueljinane, L.; Jemai, Z.; Sahin, E. Reducing ambulance response time using simulation: The case of Val-de-Marne department Emergency Medical service. In Proceedings of the 2012 Winter Simulation Conference (WSC), Berlin, Germany, 9–12 December 2012. [CrossRef]
47. Seattle Fire Department: Real-Time 911 Dispatch. Available online: http://www2.seattle.gov/fire/realtime911/ (accessed on 18 February 2021).
48. McCandless, D.; Evans, T.; Quick, M.; Hollowood, E.; Miles, C.; Hampson, D.; Geere, D. World's Biggest Data Breaches & Hacks. 2021. Available online: https://www.informationisbeautiful.net/visualizations/worlds-biggest-data-breaches-hacks/ (accessed on 18 February 2021).
49. Kang, Y.; Liu, Y.; Niu, B.; Tong, X.; Zhang, L.; Wang, W. Input Perturbation: A New Paradigm between Central and Local Differential Privacy. *arXiv* **2020**, arXiv:2002.08570.
50. Lewis, C. *Industrial and Business Forecasting Methods: A Practical Guide to Exponential Smoothing and Curve Fitting*; Butterworth Scientific: London, UK, 1982.

Mathematical and Computational Applications

MDPI

Article

A Hybrid Estimation of Distribution Algorithm for the Quay Crane Scheduling Problem

Ricardo Pérez-Rodríguez

Circuito Universitario, Faculty of Engineering, CONACYT—UAQ, Autonomous University of Queretaro, Cerro de las Campanas s/n, Santiago de Queretaro 76010, Mexico; dr.ricardo.perez.rodriguez@gmail.com

Abstract: The aim of the quay crane scheduling problem (QCSP) is to identify the best sequence of discharging and loading operations for a set of quay cranes. This problem is solved with a new hybrid estimation of distribution algorithm (EDA). The approach is proposed to tackle the drawbacks of the EDAs, i.e., the lack of diversity of solutions and poor ability of exploitation. The hybridization approach, used in this investigation, uses a distance based ranking model and the moth-flame algorithm. The distance based ranking model is in charge of modelling the solution space distribution, through an exponential function, by measuring the distance between solutions; meanwhile, the heuristic moth-flame determines who would be the offspring, with a spiral function that identifies the new locations for the new solutions. Based on the results, the proposed scheme, called QCEDA, works to enhance the performance of those other EDAs that use complex probability models. The dispersion results of the QCEDA scheme are less than the other algorithms used in the comparison section. This means that the solutions found by the QCEDA are more concentrated around the best value than other algorithms, i.e., the average of the solutions of the QCEDA converges better than other approaches to the best found value. Finally, as a conclusion, the hybrid EDAs have a better performance, or equal in effectiveness, than the so called pure EDAs.

Keywords: estimation of distribution algorithm; Mallows model; moth-flame algorithm; job shop scheduling problem; quay crane scheduling problem

Citation: Pérez-Rodríguez, R. A Hybrid Estimation of Distribution Algorithm for the Quay Crane Scheduling Problem. *Math. Comput. Appl.* **2021**, *26*, 64. https://doi.org/10.3390/mca26030064

Academic Editors: Marcela Quiroz, Luis Gerardo de la Fraga, Adriana Lara, Leonardo Trujillo and Oliver Schütze

Received: 27 July 2021
Accepted: 7 September 2021
Published: 10 September 2021

1. Introduction

1.1. Recent Research on Seaport Operations

Approximately 90% of products that are globally commercialized are transported by sea, and in one decade, the average capacity of cargo ships has doubled, with ports being the ones that allow the execution of exchanges. Seaports facilitate trading and make logistics costs more competitive. They allow the transportation of products without several checkpoints, making the process and the supply chain more efficient.

The geographic location of each seaport, combined with the quantity of active seaports, provides significant advantages in this industry for any economy.

Seaports are geographical areas and economic units of the specific place where the terminals may be found, the terminals are operating units of a seaport, which are actively able to provide modal interchange and seaport services.

The connectivity among seaports must be constantly improved to facilitate the movement of products by sea or land, and, within this connectivity, intermodal schemes must be defined to establish, for example, the transportation of products by rail or carrier.

The International Seaport System has been constantly modernized with a vision to improve logistics and multimodal connectivity, where the projects regarding seaport, road and rail infrastructures are more integrated in order to respond to the increasing demand of national and international trade. Currently, the capacity of seaports has been increased by millions of tons per year.

As was previously mentioned, seaports have been, are and shall be, a great entry gate to the different continents of the world. Supported by the great advantage of having access to the oceans, countries have an extensive growth potential regarding foreign trade; it is only a matter of continuously improving governmental management and support.

One of the greatest challenges of trading by sea is the existence of the international market, countries have reached the maximum uses of their seaports, the demand for infrastructure and comprehensive logistic service offered at lower costs is still increasing.

Additionally, trained staff are required to put into operation the resources acquired in an efficient and safe manner. Another significant challenge is environmental protection, for example, in the Caribbean Sea, a Mesoamerican Reef System was designed, which is a sensitive area for navigating and preventing pollution from ships.

There is still a gap to consider: cabotage routes must be reinforced and transport routes must be developed; however, maritime transport is encouraged every year, the development of the industry is increasing and governments are considering it as a significant matter.

Therefore, the role of international transportation, maritime shipping, and marine container terminals in the economic development of numerous countries has been widely studied using different approaches. Seaport problems continue to be attracted to developing new optimization techniques. Recently, important contributions have been published in order to improve the performance of seaport operations. A representative example is the study of [1]. This study details a tailored global optimization algorithm for deploying, scheduling, and sequencing heterogeneous vessels in a container liner shipping route. The key advantage is that the weekly dependent shipping demand is considered to incorporate fluctuations in the planning horizon. In addition, the authors consider the distinct fuel efficiencies, cost structures of the ships, and capacities. These conditions determine the number of containers transported, the bunker fuel consumption, and the operating cost of a shipping route. A mixed integer-programming model minimizes the total cost by considering, from a set of candidate ships, the optimal ships. In addition, the proposed model integrates the sequences, schedules, and sailing speeds of the shipping route.

Ref. [2] is a seaport study that considers emergent ship arrivals for building berth schedules. The authors propose an optimization procedure to schedule ships by reducing disturbances in the service and maximizing the number of emergent ships served. A case study illustrates the aforementioned optimization procedure. The proposed procedure provides key concerns to decision makers dealing with the berth scheduling issues of emergent ship arrivals.

Ref. [3] attends to real world situations, such as water depth and tide conditions, in seaport operations. An enriched particle swarm optimization procedure is depicted to solve the continuous berth allocation, quay crane assignment and quay crane scheduling problems. The authors incorporate, in the solution, safety operations between quay cranes. Their results are compared with the results of the exact solution and the basic particle swarm optimization procedure.

Ref. [4] concerns the increase in seaport operations in the last thirty years, and argues that berth scheduling can improve the throughput of marine container terminals. The authors detail an innovative evolutionary algorithm to minimize handling costs, waiting costs, and the late departure costs of the vessels that are to be served at a marine container terminal. The scheme of the authors relies on an augmented self adaptive parameter control strategy, which changes algorithmic parameters throughout the search process.

Another important example is the research of [5]. The author develops a novel memetic algorithm to help the marine container terminal operators to build proper schedules, and to tackle congestion issues caused by the increasing number of large size vessels. Although this study does not account for uncertainty in vessel arrivals, the proposed algorithm serves as an efficient planning tool for marine container terminal operators and assists with efficient the berth scheduling.

1.2. The Quay Crane Scheduling Problem

Based on the preliminary review, it is clear that the performance of seaport operations, for the international supply chain, continues to be highly important. Solving problems, such as the quay crane scheduling problem (QCSP), a combinatorial problem, contributes to improving the efficiency of any seaport. The QCSP is selected in this research due to its complex nature.

Quay cranes are very important resources at container terminals. They are used to load containers onto, and discharge containers from vessels at, the quayside of terminals. Quay cranes along the same berth operate on a common set of rails.

Jobs represent container unloading/loading work, and arise at specific locations along the quayside. The collection of jobs may represent work for a single ship, or multiple ships.

Normally, an optimization formulation for the QCSP considers reducing the makespan, i.e., the total completion time. Any proposed solution should consider scheduling operations restrictions, such as

- A minimum distance being left between quay cranes to avoid boom collision.
- Each quay crane should work when it will not be busy.
- Like a traditional shop environment, precedence between operations may exist.

A mathematical formulation considers

- p_i the processing time of task i.
- r_k the release time of quay crane k.
- t_{ij} the travel time of a quay crane from the bay position of task i to the bay position of task j.
- M a sufficient large constant.
- Ω a set of all tasks.
- Φ a set of ordered pairs of tasks between which there is a precedence relationship.
- $x_{ijk} = 1$ if task j immediately follows task i on quay crane k; 0, otherwise. Tasks 0 and T will be considered the initial and final states of each quay crane, respectively. Thus, when task j is the first task of quay crane k, $x_{0jk} = 1$. In addition, when task j is the last of quay crane k, $x_{jTk} = 1$.
- Q_k the completion time of quay crane k.
- C_i the completion time of task i.
- C_{max} makespan.

Therefore, the QCSP can be formulated as follows

$$\text{Minimize } \frac{\sum_i C_i}{\Omega} \tag{1}$$

subject to

$$C_i \leq C_{max} \quad \forall k = 1, \ldots, K \tag{2}$$

$$\sum_{j \in \Omega} x_{0jk} = 1 \quad \forall k = 1, \ldots, K \tag{3}$$

$$\sum_{i \in \Omega} x_{jTk} = 1 \quad \forall k = 1, \ldots, K \tag{4}$$

$$\sum_k \sum_{i \in \Omega} x_{ijk} = 1 \quad \forall j \in \Omega \tag{5}$$

$$C_i + t_{ij} + p_j - C_j \leq M\left(1 - x_{ijk}\right) \quad \forall i, j \in \Omega, \quad \forall k = 1, \ldots, K \tag{6}$$

$$C_i + p_j \leq C_j \quad \forall i, j \in \Phi \tag{7}$$

$$C_j + t_{jTk} + Q_k \leq M\left(1 - x_{jTk}\right) \quad \forall i \in \Omega, \forall k = 1, \ldots, K \tag{8}$$

$$r_k + C_j + t_{0jk} + p_j \leq M\left(1 - x_{0jk}\right) \quad \forall i \in \Omega, \forall k = 1, \ldots, K \tag{9}$$

$$x_{ijk} = 0 \text{ or } 1 \quad \forall i, j \in \Omega, \forall k = 1, \ldots, K \tag{10}$$

$$Q_k, C_i \geq 0 \quad \forall i \in \Omega, \forall k = 1, \ldots, K \tag{11}$$

The objective function (1) minimizes the average waiting time. Constraint (2) determines the makespan, which corresponds to the completion time of all tasks. Constraints (3) and (4) define the first and the last tasks for each quay crane, respectively. Constraint (5) ensures that every task must be completed by exactly one crane. Constraints (6) and (7) determine the completion time for each task. Constraint (8) defines the completion time of each quay crane. Constraint (9) ensures that the first task of a crane is not started before the crane is ready. Finally, (10) and (11) define the domains of the decision variables.

The objective of our study is to identify the best schedule to minimize the quays' overall average waiting time. The methodology used in this research details how a solution can be built by two decisions, i.e., operation scheduling and crane assignment.

1.3. Solving Combinatorial Problems through Evolutionary Algorithms

In a wide set of studies, evolutionary algorithms have been proposed to offer useful solutions for high scale optimization problems. The solutions are required in different engineering fields. Particularly, evolutionary algorithms are competitive in the combinatorial optimization field. Some relevant studies of evolutionary algorithms can be appreciated in the research of [6], for job shop operations and process integration. The authors use and apply the concept of interaction between different species to tackle the problem. This interaction is not considered in any traditional genetic algorithm. The study of [7] is representative of the evolutionary algorithms family for tackling job shop problems. In this applied case, an ant colony optimisation based approach is proposed to address flexible job shop scheduling with routing flexibility and setup times problems. Recent publications are found in [8]. This research group employs a genetic algorithm with the Pareto front technique. The approach is applied on a mail-order pharmacy system. Ref. [9] utilizes a bee colony method with the Pareto front technique for solving the single machine group-scheduling problem with sequence dependent setup times. Ref. [10] uses an enriched discrete particle swarm optimization method to solve a flexible job shop scheduling problem. Ref. [11] detail a bat based approach to tackle a dual flexible job shop scheduling problem. Moreover, recently, the use of hybrid evolutionary algorithms is remarkable. The aim is to offer more efficient methods, or better solutions, to those previously found. The work of [12] falls into this category for solving the vehicle routing problem. The author employs a genetic algorithm and local search. The research of [13] also utilizes a genetic technique with a variable neighborhood descent method to address flexible job shop scheduling problems. The study of [14] integrates a genetic algorithm with a neural network to improve container-loading sequences in seaports. The practical implementation of their algorithm is confirmed by using a simulation in the research. The work of [15] uses a genetic technique with a minimum cost flow network model to solve a dispatching problem for vehicles in a transshipment hub scenario. Their experiments show that the proposal is more effective than a neighborhood search algorithm. The research of [16] details an efficient parameter free greedy randomized adaptive search method, plus a variable neighborhood descent technique, to tackle a school bus routing problem with bus stop selection. The study of [17] proposes a genetic technique with a tabu search to address more than 200 flexible job shop scheduling open problems. Following the previous review, we can appreciate that there exist different types of evolutionary algorithms for solving different combinatorial problems.

1.4. Estimation of Distribution Algorithms

This article focuses on the use of estimation of distribution algorithms (EDAs) as a section of the classification of evolutionary algorithms. There exists a vast literature about EDAs. Studies, such as [18], designed for solving permutation flowshop scheduling problems. The research of [19] contributes guidelines for developing effective EDAs for single machine scheduling problems. Furthermore, the work of [20] proposes an EDA for solving lot-streaming flowshop problems with setup times. These investigations are robust in the solution of combinatorial problems using EDAs.

When using EDAs, as with other evolutionary algorithms, the main task is to produce a population of solutions through some generations. However, with EDAs, a probability model, i.e., a distribution, is built based on a set of solutions to the problem, to produce new solutions. Traditionally, the performance of EDAs lies in how efficient the probability distribution is. The probability distribution is built using the members of the population to obtain better solutions. It has been widely documented, in papers such as [21,22], that the probability model should consider the structure of a problem with more precision. The aim of EDAs is to consider high order interactions between the variables of the problem. Thus, complex probability models are employed to improve the efficiency of EDAs. Nonetheless, one of the drawback of EDAs is the lack of the diversity of the solutions and the poor ability of exploitation [22]. There is more than one way to counter the drawbacks of EDAs. Nevertheless, most studies suggest hybridization. In the literature review section, some of the studies are listed.

In this paper, we can visualize the importance of the effective estimation of the solution space distribution to build a competitive EDA. In addition, if we analyze the results shown in the corresponding results and comparison section, we are going to reach the conclusion that the hybridization of EDAs with other methods should practically always be considered in the design and development of such algorithms. According to the results of this research, hybridization is not only a reason for publication, it also avoids the necessity of requiring building complex probability models. If this occurs, the algorithm could be difficult to understand.

The main reason for performing this research was to determine what is most useful, i.e., the development of complex probability models or the hybridization of the EDA with other methods to obtain the same or better solutions. Most published articles, related to EDAs and to solve optimization problems, have a lack of diversity in the process of the generation of solutions. The aforementioned drawback can be tackled through the hybridization of an EDA and other methods. However, there exists a wide field of development to determine which methods are more suitable to obtain the better performance of an EDA. This research aims to use bioinspired techniques to help to reduce the deficiencies of an EDA. In addition, it is preferable that the methods used, in the hybridization process, should contain some well defined math expressions. This helps to understand how the solutions are generated. Therefore, the aim is to use new methods and to establish the search process of explicitly new solutions.

The hybridization approach used for this investigation considers a distance based ranking model coupled with a moth-flame algorithm, a novel nature inspired heuristic paradigm [23]. Therefore, the QCSP is solved by the proposed approach in order to show how hybridization helps to enhance the performance of the EDA. Distance based ranking models appear sparingly in the literature to tackle the drawback of the EDAs. There are few papers available about it. For example, Ref. [24] introduces an EDA based on a distance based ranking model for the flowshop scheduling problem. The distance based ranking model is used as a probability model in the permutations field.

1.5. The Proposed Hybrid Approach, a Brief Explanation

In this research, the members of the population are considered as rankings. Table 1 shows a ranking of five items or elements as an example.

Table 1. A ranking example.

Rank	1st	2nd	3rd	4th	5th
Element	Π	B	α	∞	μ

In this sense, the members of the population are permutations of elements. A ranking defines a solution for the problem to solve. Table 2 depicts some rankings as flowshop-scheduling solutions, where each item represents a job, and then a processing sequence is executed according to each ranking.

Table 2. Rankings for the flowshop as an example.

Processing Sequence	1st	2nd	3rd	4th	5th
Rank 1	J_5	J_3	J_1	J_4	J_2
Rank 2	J_5	J_1	J_4	J_3	J_2
Rank 3	J_3	J_1	J_2	J_4	J_5

Then, a distance metric is computed between rankings, by the algebra of permutations. After that, a distribution probability is built, i.e., an exponential distribution. The aforementioned exponential distribution occupies, as an input parameter, the distance between each of the permutations and a reference permutation. A known distance based ranking model is the model of [25]. In general terms, it concerns assigning a probability to each permutation that declines exponentially based on its distance from a reference permutation. It is then said that such a permutation is more (or less) likely to be chosen as a good solution, according to the [25]'s model. Figure 1 illustrates such a distribution, with five rankings.

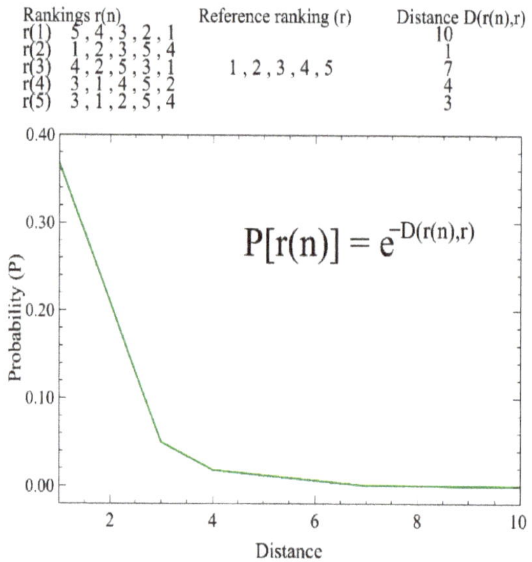

Figure 1. Exponential distribution using a distance between rankings.

However, [25]'s model is not able to generate offspring by itself. The reason is because there can be many permutations with the same distance to the reference permutation. This causes confusion regarding which permutation is the offspring. It is solved by the factorization (decomposition) of the distance previously obtained. The factorization is computed by a procedure called GMD (the Generalized Mallows Model), proposed by [26,27]. Such authors propose how to factorize in $n-1$ elements, the distance between two rankings where n is the size of the rankings. With that factorization, it is possible to obtain the offspring. Table 3 details such factorization.

Table 3. Factorization of the distance between rankings.

Rankings	Reference Ranking	Distance	Factorization Based on Fligner & Verducci [26]
4, 2, 5, 3, 1	1, 2, 3, 4, 5	7	[3, 1, 2, 1]
5, 4, 1, 2, 3		7	[4, 3, 0, 0]

Then, such factorization (decomposition) serves as input parameter in the moth-flame heuristic to generate new offspring. Based on the results obtained in this study, it is more efficient to produce new and better offspring through such hybridization. The novelty of this study is to show how this hybridization is better than the direct use of a distance based ranking model.

Roughly, the factorization (decomposition) obtained in the previous step is used to determine where a moth is located in the solution space. That is, the decomposition of the distance between rankings (permutations) is seen as a moth's location coordinate in the feasible space of solutions. Figure 2 shows such factorized rankings, and the moths′ location coordinates.

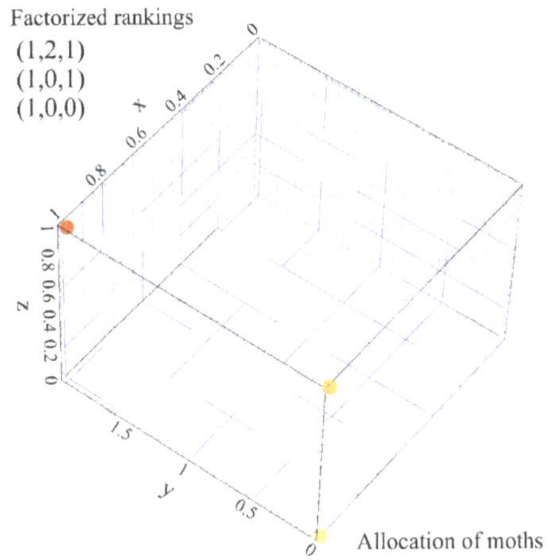

Figure 2. Relation between factorized rankings and the moths' location coordinates.

A moth is a search agent that moves in a feasible space, whereas a flame is located at the best position obtained so far [23]. A spiral-flying path, using a defined mathematical model, of moths around flames is the way to execute the movement of the moths to new locations. Thus, the GMD is in charge of modelling the solution space distribution; meanwhile the heuristic moth-flame is in charge of determining who would be the offspring. That is how this proposed algorithm works to enhance the results obtained from those algorithms that use complex probability models.

The rest of the paper is organized as follows. Section 2 reviews the literature of EDAs and different approaches to tackle the problem. Section 3 presents the inspiration of this work and proposes the quay crane estimation of distribution algorithm, with the Mallows model and a moth-flame structure, called QCEDA. The experimental setup and results are provided in Section 4. Section 5 concludes the work and suggests several directions for future studies.

2. Literature Review

It is of particular interest to present current research of the use of EDAs to solve combinatorial problems. Thus, the first part of the literature review is focused in that direction.

2.1. Complex Problems, Complex Probability Models

An important direction in the development of EDAs focuses on building complex probability models. Such models consider high order interactions between the variables of a problem. The challenge, in this group of EDAs, is how well such interactions are estimated and how to produce better results. We can start with the case where there exists no interaction between variables. Algorithms such as the univariate marginal distribution algorithm (UMDA), detailed in [28], fall into this category. If there exists interaction between a pair of variables, an algorithm with a good performance is the mutual information maximization for input clustering (MIMIC), presented in [29]. If we try to model high order interactions between variables, probability models become more complex to understand and compute. Moreover, a large number of solutions might be required to estimate, in the best way, such interactions. Some important algorithms in this category are the combining optimizers with mutual information trees (COMIT), published by [30], and the Bayesian optimization algorithm (BOA) proposed by [31]; to mention these as the most frequent and cited.

The development of EDAs, which considers high order interactions between variables, does not finish here. Conversely, different combinations of complex probability models have been used for solving combinatorial problems, such as in [32]. The authors compute more than one complex model to address a flexible job shop scheduling problem.

Ref. [33] offer a clear revision of the use of the complex models used in EDAs to solve combinatorial problems. In such a revision, we can appreciate that the complexity of the probability model there exists when we estimate higher order interactions between variables.

2.2. The Hybridization Approach

As discussed in the introduction section, hybridization is an efficient way to counter the weakness of EDAs. Practically, hybrid EDAs include some improvement components. The support method used to improve the performance of EDAs differs between authors, but the most common are heuristics or other evolutionary algorithms. Some studies that use hybrid evolutionary algorithms to solve optimization problems are found in [34]. The author presents a hybrid genetic algorithm coupled with a gravitational search algorithm, called the GA-GSA algorithm, to optimize the metrics of a real industrial case, employing uncertain data. The application of the proposed algorithm can save manpower, budget, and time, and improve the metrics of the company. In addition, Ref. [35] presents a hybrid technique. It uses a particle based swarm method and a genetic algorithm, named PSO-GA, to tackle the constrained optimization problems. The particle based swarm method focuses on enhancing the solution vector, while the genetic technique is employed to modify the offspring. Furthermore, Ref. [36] proposes a hybrid gravitational search algorithm coupled with a genetic algorithm, labeled as the GSA-GA algorithm. The aforementioned algorithm is developed to solve nonlinear optimization problems, and it includes mixed variables. Similarly, each solution is generated through a gravitational search approach, and then each offspring is updated by the genetic technique.

Other studies try to integrate the concept of hybridization between evolutionary algorithms and EDAs. The most representative studies are found in [37]. The authors present a hybrid algorithm through genetics and the estimation of distribution algorithms. A key point is taking advantages from both procedures. The authors apply the hybrid EDA for solving synthetic optimizations problems and two real world cases. Ref. [38] uses an EDA with a 2-opt local search, in a hybridization phase, for addressing a quadratic assignment problem. Ref. [39] considers a permutation flowshop scheduling problem using a particle swarm optimization method with an EDA. Ref. [22] confronts a flexible job shop scheduling problem through the hybridization of an EDA and a local search approach to improve the exploitation process of the EDA. Ref. [40] proposes an EDA to tackle a stochastic resource constrained project-scheduling problem. The hybridization phase uses a permutation based local search. Ref. [41] presents a fuzzy logic based hybrid EDA for addressing distributed permutation flowshop scheduling problems with machine breakdown. The hybridization phase uses genetic variation operators to create offspring. The articles mentioned above are current and representative of this group of hybrid EDAs.

2.3. Distance Based Ranking Models

Another approach for the development of EDAs is to search and use already defined probability models for the permutations field. These models should be able to adapt themselves to the structure of the representation of the members of the population. The main goal is to model the solution space from another perspective. Examples include the EDAs proposed by [24], for a flowshop scheduling problem; Ref. [42], detailing the school bus routing problem with bus stop selection; Ref. [43], depicting a flexible job shop scheduling problem with process plan flexibility; and [44], illustrating the vehicle routing problem with time windows. For the aforementioned papers, an interaction estimation between variables is not required. A distance based ranking model is preferable; specifically, the GMD.

2.4. Quay Crane Scheduling, Recent Literature

Ref. [45] offers an exact and computationally fast solution technique to solve the QCSP. The technique combines a partitioning heuristic with a branch and price algorithm. Finally, a traveling salesman formulation is utilized to minimize crane repositioning movements.

Ref. [46] describes a mathematical model for the QCSP. The authors tackle the non-crossing constraints by addressing the structure of workload assignments. The proposed mathematical formulation is based on the logic based Benders decomposition.

Ref. [47] finds a compact mathematical formulation of the unidirectional cluster based quay crane scheduling problem that can be easily solved by a standard optimization solver.

Ref. [48] presents a revised optimization model for the scheduling of quay cranes and proposes a heuristic solution procedure. The authors propose a branch and bound algorithm, which searches a subset of above average quality schedules; meanwhile, the heuristic considers the impact of crane interference.

Ref. [49] details a genetic algorithm that it is built through a novel workload balancing heuristic, and the loading conditions of different quay cranes during the reassignment of task to quay crane are considered. The idea is modelled as a fuzzy logic controller to guide the mutation rate and mutation mechanism of the genetic algorithm. As a result, the proposed algorithm does not require any predefined mutation rate. Meanwhile, the genetic algorithm can more adequately reassign tasks to quay cranes according to the quay cranes' loading conditions throughout the evolution.

Ref. [50] improves the efficiency of a genetic algorithm by (1) using an initial solution based on a heuristics rule, (2) using a new approach for defining chromosomes (i.e., solution representation) to reduce the number of decision variables, and (3) using new procedures for calculating tighter lower and upper bounds for the decision variables.

Ref. [51] tackles the issue of the interference among cranes by a modified genetic algorithm to deal with the QCSP.

Ref. [52] develops a two quay-crane schedule with noninterference constraints for the port container terminal of Narvik. First, a mathematical formulation of the problem is provided, and, then, a genetic algorithm approach is developed to obtain near optimal solutions.

It is clear, from this review, that the QCSP has been studied intensively in a recent stream of research. In addition, based on the current research, it is important to handle the correct treatment of crane interference constraints. In addition, there is still a gap to improve the performance of EDAs, and the hybridization approach continues to be a useful way to enhance EDAs. In addition, from the exposed review, it is of interest to know how much better hybrid EDAs can be, than other pure EDAs and recent algorithms.

The contributions of this article are

- To propose a hybrid EDA, not only a reason for publication, this also avoids the necessity of requiring building complex probability models.
- To determine what is most useful, i.e., the development of complex probability models or the hybridization of an EDA with other methods to obtain the same or better solutions.
- As a reason for doing this research, there exists a wide field of development to determine which methods are more suitable to obtain the better performance of an EDA.
- The research motivation is to show how bioinspired techniques help to reduce the deficiencies of an EDA.

- Based on the results obtained in this study, it is more efficient to produce new and better offspring through such hybridization.

3. The QCEDA Approach

This section may be divided by subheadings. It should provide a concise and precise description of the experimental results, their interpretation, as well as the experimental conclusions that can be drawn.

3.1. Initial Population

A representation of solutions to the QCSP is detailed through the processing sequence of containers on the available quay cranes and the establishment of containers for the quay cranes.

For the processing sequence of containers, a first vector is proposed, i.e., the containers sequence vector. It has a length that equals the number of containers moving through the quay cranes. Thus, any vector of this type is simply a sequence of n containers.

An example is depicted in Table 4, with five containers indexed from 1 to 5.

Table 4. A task sequence vector example.

Ranking	1st	2nd	3rd	4th	5th
Task(container)	3	1	4	2	5

Where container number three should be executed (moved) at the beginning, after that, container number one, container number four, and so on. Each element in the containers sequence vector shown above might be in any place on the representation, or as any permutation based solution, i.e., as any ranking.

As in other EDAs, the aforementioned representation is suitable for using in the set of combinatorial problems used in the comparison section. With this representation, for the solution vectors, the GMD can be directly applied in the distance-based ranking model phase. The initial population contains 1000 members. Each member has a length, n, where n is the number of elements in the permutation, i.e., containers. All this is in each generation. The number of members is a fixed parameter.

For the assignment of containers, a second vector is built, i.e., the quay crane assignment vector. It also has a length that equals the number of containers. Every element represents the selected quay crane for every container. An example is depicted in Table 5.

Table 5. A quay crane assignment vector.

Ranking	1st	2nd	3rd	4th	5th
Quay crane	2	2	1	1	3

The first container, based on the containers sequence vector shown in Table 4 and labeled with (3), is moved by the crane labeled with the number two, according to the crane assignment solution vector. The second container, labeled with (1), is moved by the same crane, the third container, labeled with (4), is moved by the crane labeled with the number one, and so on.

3.2. Fitness

The average waiting time that is required to move all the tasks (containers) is the fitness for the QCSP in this research. The fitness is computed for each member of the population. Once the completion time for each task is obtained, all the times are accumulated and divided by the number of tasks to obtain the average waiting time, i.e., the fitness.

To ensure compliance with the quay crane constraints for each crane and avoid any collision between cranes, the Delmia-Quest® simulation language is preferred in this research. Delmia-Quest® avoids any collisions between cranes in each simulation run. A controller establishes the order of movements, in each simulation run, to satisfy the aforementioned constraints. Using the Delmia-Quest® simulation language, the main constraints related to the cranes are satisfied. Furthermore, considering Delmia-Quest®, the fitness is obtained directly from the simulation model and the QCEDA is in charge of modelling the solution space distribution.

For each solution, the corresponding values are obtained from the simulation model, built on Delmia Quest®. The main details of the simulation model are outlined below

- A crane system is used to move containers through the ship.
- The crane system is multidirectional.
- Different cranes can service any container according to a predefined sequence.
- Containers can receive service from different cranes based on the predefined sequence.
- The movement of containers is only possible by cranes.

With all these features, the simulation model is able to integrate operation times and workflows. Finally, the fitness is used by the QCEDA to obtain the best solution at the end of execution.

3.3. Probability Model for the Quay Crane Assignment Vectors

A matrix, q, is built to describe the probability model for the quay crane assignment vectors. That is, each q_i value in the matrix mentioned details the number of times the quay crane, i, is elected with a container. Again, the vectors depicted below are utilized to create a probability matrix, q, for the first position in the sequences, i.e., q_i. In this example, six vectors (as a population), with a length that equals five containers, are shown in Table 6.

Table 6. Quay crane assignment vectors.

Ranking	**1st**	**2nd**	**3rd**	**4th**	**5th**
V_1	2	2	1	1	3
V_2	3	2	1	1	1
V_3	1	2	2	2	1
V_4	3	1	3	3	2
V_5	2	2	1	2	2
V_6	2	3	2	3	3
q_1 matrix					
Quay crane	$p(X_1 = x)$				
1	1/6				
2	3/6				
3	2/6				

New quay crane assignment vectors are constructed as follows: in each position on the new individual, the quay crane i is elected by the cumulative probability of q_i. Each q_i cumulative value considers the opportunity of the quay crane, i, be elected for the place, j, in the offspring.

Firstly, a random value should be generated in every place on the offspring. Then, every random value is interpolated in the cumulative probability of q_i, to identify which quay crane should be selected. Figure 3 depicts an illustration of this process.

Therefore, the offspring, i.e., the new quay crane assignment vectors, are sampled according the cumulative probability of q_i.

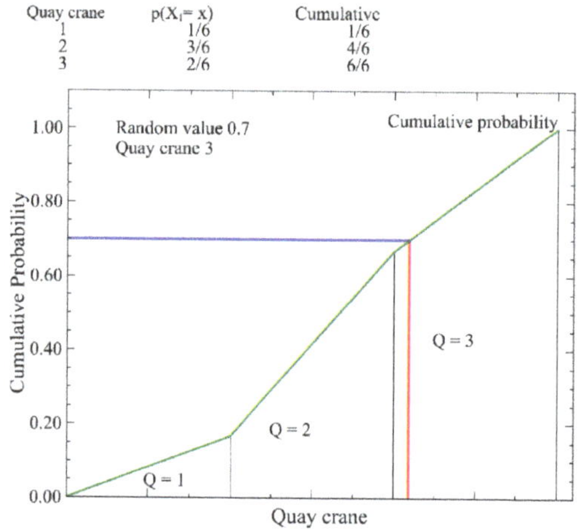

Figure 3. Sampling example.

3.4. *Probability Model for the Task Sequence Vectors*

3.4.1. The Distance Based Ranking Model Phase

- Central ranking computing

With all members of the population, i.e., the task sequence vectors, the central (reference) permutation is computed. The central permutation is a parameter inside of the GMD process. In this research, the procedure based on [53] is used to compute the reference permutation. Firstly, the mean for each position, considering the members of the population, is computed. Secondly, the smallest mean, from the all the positions, is identified and the smallest element of the reference permutation is assigned in the smallest mean position, identified previously. This continues when the second smallest mean, from all the positions, is identified and the second smallest element of the reference permutation is assigned in the second smallest mean position identified previously, and so on until the procedure finishes. This procedure is widely used to define a consensus of the all rankings. Figure 4 presents a simple example with seven vectors of length five, the corresponding means and the result, i.e., the central permutation.

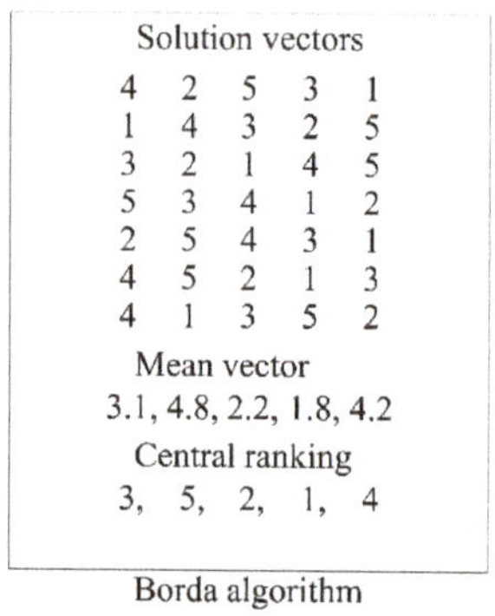

Figure 4. Central ranking computing example.

With the consensus ranking (central permutation), it is possible to compute the distance between each member of the population and the consensus ranking. The procedure defined by [26,27] compute the distance. Firstly, the composition (product) between the inverse of each permutation and the consensus ranking should be computed. Figure 5 details an example of the composition mentioned between two vectors with length four.

Vector 1 - π
2 4 1 3
Central ranking - σ
4 3 1 2
Inverse of the Vector 1 - $\pi^{(-1)}$
3 1 4 2
Composition - $\sigma\pi^{(-1)}$
2 4 3 1

Figure 5. Composition between permutations.

For further details, the reader is referred to the algebra of permutations literature. With the composition previously detailed, it is possible to decompose in $n-1$ items, i.e., to factorize such composition (see Figure 6).

Composition
(2, 4, 3, 1)
Factorization process ...
1st element
= 1, because there is one value smaller than number 2 on its right
2nd element
= 2, because there are two values smaller than number 4 on its right
3rd element
= 1, because there is one value smaller than number 3 on its right
Output: Factorized vector
(1, 2, 1)

Figure 6. Factorization of the distance between permutations.

3.4.2. The Moth-Flame Phase

- Initialization of moths in the feasible space

The factorization obtained in the previous step now represent coordinates where a moth is located in an initial search phase. Then, there are M members, i.e., M solution vectors, in the initial population and M moths for the moth-flame phase. Figure 7 shows an allocation of a moth. Based on this idea, the members of the population each contain n elements, and the moth-flame phase uses $n-1$ elements for each moth. Therefore, the search space is stablished in dimension R^{n-1}.

- Fitness of the moths

In this study, the fitness, obtained for each individual, is the same fitness for every moth in the moth-flame phase.

- Moths sorting

The population of moths is sorted in descending order, according to the fitness.

- Flames amount computing

The equation published in the work of [23] is used to determine the number of flames in each generation. The equation is below

$$\text{number of flames} = \text{round}\left(N - l \cdot \frac{N-1}{T}\right) \tag{12}$$

where l is the current generation, N is the maximum number of flames, and T expresses the maximum number of generations. The number of flames gradually decreases when Equation (12) is implemented. This permits the right exploration and exploitation of solutions [23]. Figure 8 shows how the number of flames gradually decreases when the number of generations increases.

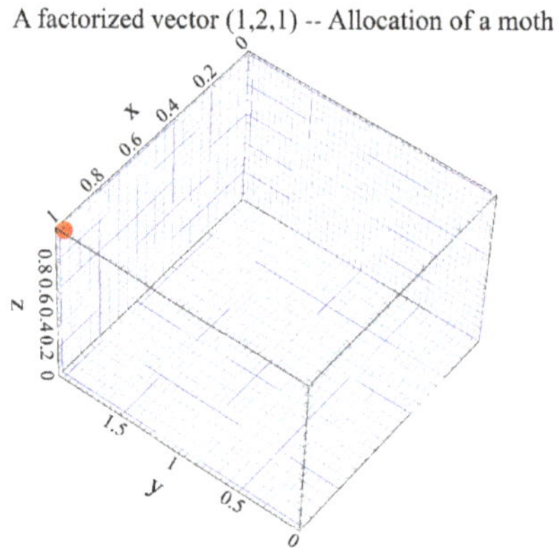

Figure 7. A moth allocation in the space.

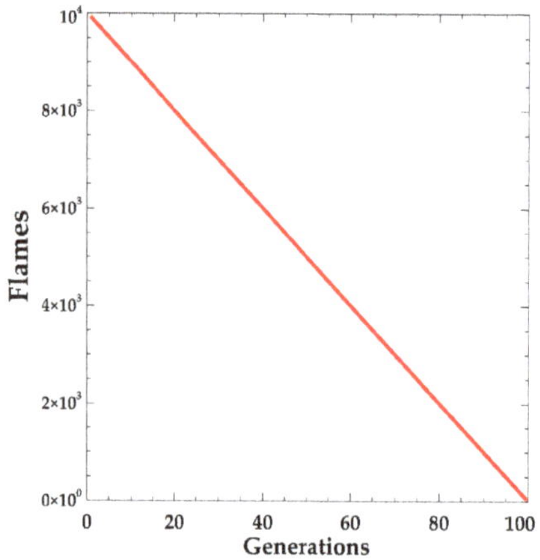

Figure 8. Flames amount computing.

- Flames setting

  The coordinates of the moths, already sorted, are considered to determine the locations of the flames. Table 7 shows an example in R^3, with three flames and nine moths.

- Flames' fitness setting

  The fitness for the flames is initialized with the fitness of the corresponding moth.

- Moth-flame assignment

  Before generating offspring, i.e., the movements of the moths in the search space, each moth should move only with respect to a specific flame. That is, each moth should be assigned to a specific flame in each generation. Table 8 depicts an example in R^3, with three flames and nine moths.

Table 7. Flames allocation example.

Flames	Flame Coordinates	Sorted Moths	Moth Coordinates
1	(0, 0, 1)	1	(0, 0, 1) *
2	(3, 1, 0)	2	(3, 1, 0) *
3	(2, 1, 2)	3	(2, 1, 2) *
		4	(0, 1, 0)
		5	(1, 1, 1)
		6	(3, 0, 0)
		7	(2, 2, 0)
		8	(2, 3, 0)
		9	(1, 0, 0)

* the best moths.

Table 8. Moth-flame assignment example.

Moths	Flame
1	1
2	1
3	1
4	2
5	2
6	2
7	3
8	3
9	3

- Moth movement

Each moth achieves a new location in the search space by the logarithmic function defined in [23].

$$\text{Spiral function} = D_i \cdot e^{bt} \cdot \cos(2\pi t) + F_j \tag{13}$$

where D_i indicates the distance between the i-th moth and the j-th flame, i.e., $D_i = |M_i - F_j|$, b is a value that depicts the shape of the function, t is a random value between [–1, 1]. Figure 9 depicts an example, in R^2, for a moth after computation of the spiral movement.

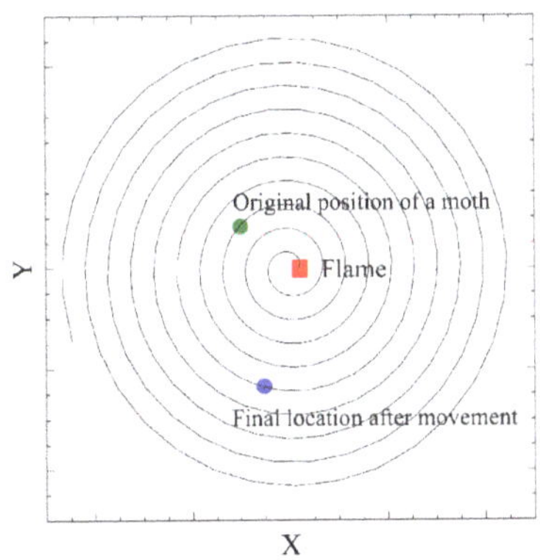

Figure 9. An example of the use the spiral function.

3.4.3. Offspring Computing

The new locations of the moths are used as the representation of the distance between permutations. That is, now the new coordinates of the moths represent the decomposition (factorization) of the distance between permutations. That is how, in this research, both approaches hybridize to improve the performance of the proposed algorithm.

The way that the distances between permutations return to the original dimension, i.e., n, is used in the algorithm of [54]. Through some steps, explained with an example below, the research of these authors offers the possibility of finding the permutation that corresponds to each offspring.

Pseudocode 1. [54]'s Procedure Example

```
n ← number of positions (permutation length)
Let a sample moth location vector V = (2, 0, 1)
therefore n := 4
insert n in 0 → Inverse permutation vector = (4, _, _, _)
for j:=n − 1 to 1
insert j := 3 in V3·1 → Inverse permutation vector = (4, 3, _, _)
insert j := 2 in V2·0 → Inverse permutation vector = (2, 4, 3, _)
insert j := 1 in V1·2 → Inverse permutation vector = (2, 4, 1, 3)
```

According to the sample moth location vector shown above, the corresponding inverse permutation vector is (2, 4, 1, 3). Finally, each permutation vector is obtained by inverting and composing the consensus ranking. An inverse permutation exists when every number and the number of the place that it occupies are exchanged. Thus, the inverse of the inverse permutation vector is (3, 1, 4, 2). If we take as the central ranking (1, 2, 3, 4) as example, then, the composition of the previous result with the central ranking is (3, 1, 4, 2) as the final vector.

3.5. Replacement

The offspring should be evaluated to obtain their fitness. Finally, the replacement process used in this study is a binary tournament between the parents and the offspring.

All the stages of the proposed algorithm have been defined. The loop is performed going back to step–quay crane matrix computing. In addition, the central ranking should be updated with the new population after the replacement process. All of this is within a number of the generations. In this research, 100 generations were used. This is a fixed parameter. Then, the QCEDA framework is provided below

Pseudocode 2. QCEDA Framework

```
D0 ← Generate M individuals
t := 1
Do
    FitD(t−1) ← Evaluate individuals (fitness)
    q(t−1) ← q matrix computing from D(t−1)
    Dq(t) ← Sampling from q(t−1)
    σ0 ← Central ranking computing from D(t−1)
    K(t−1) ← Distance computing from D(t−1) and σ0
    M(t−1) ← Set moths from K(t−1)
    FitM(t−1) ← Set fitness from FitD(t−1)
    M(t−1) ← Sorting moths
    f ← Flames computing
    F(t−1) ← Set flames from M(t−1)
    FitF(t−1) ← Set fitness from FitM(t−1)
    M(t) ← Moth movement computing (genotype)
    Ds(t) ← Offspring computing from M(t)
    FitD(t) ← Evaluate individuals from Ds(t) and Dq(t)
    D(t) ← Replacement by binary tournament
    t := t + 1
Until (stopping criterion is met)
```

Additionally, a flow chart is detailed in Figure 10 to illustrate the overall process.

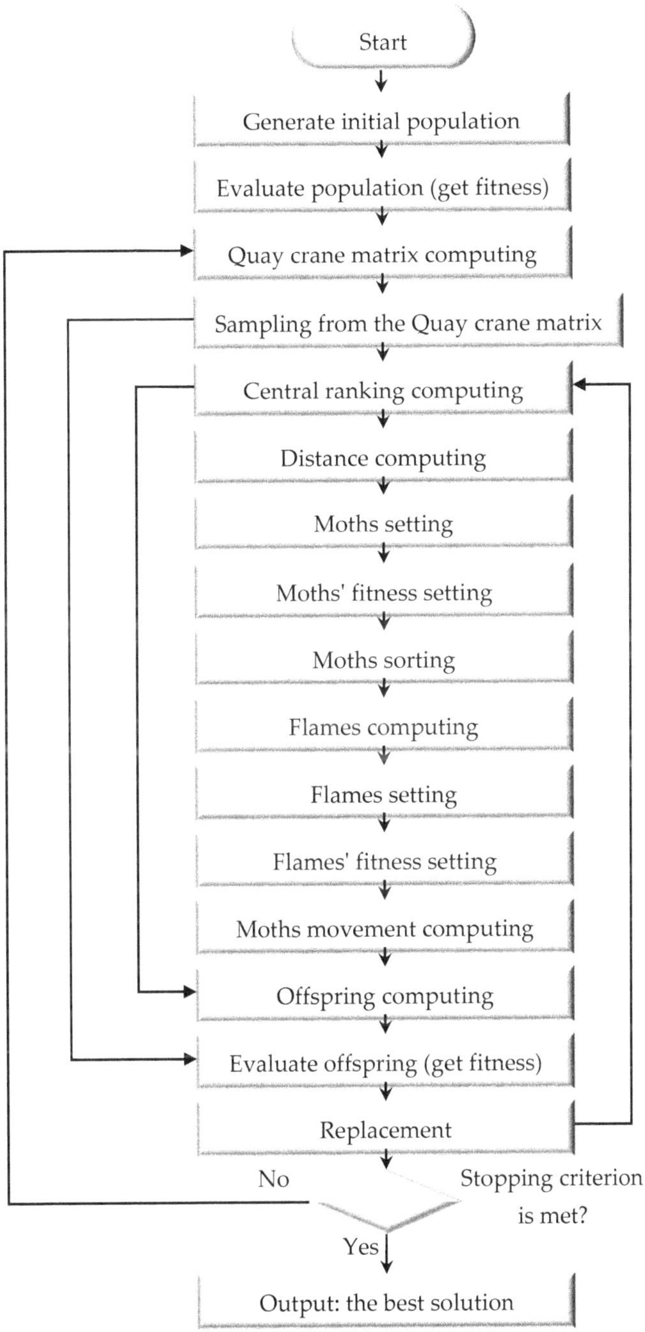

Figure 10. The core of the QCEDA approach.

4. Results and Comparison

4.1. Comparison with Standard Benchmarking Datasets

The set of instances of [55] are utilized in the comparison between different EDAs. The input data (instances) were already generated by [55], and these instances are available online. Ninety instances were used in the comparison. The comparison mentioned helps to determine the importance of hybridization for the performance of the proposed EDA. All the details of the instances are depicted below.

Basics of the Input Data

Headings present general information, such as
- The safety margin between the quay cranes is two ship bays
- The travel time of a QC between two adjacent bays is one

The body of the instance includes
- The number of containers (tasks)
- The starting position of each quay crane
- The problem number
- For each task
  - The location of the task,
  - The type of the task, i.e., deck or hold
  - The time required to perform the task
  - The type of operation, i.e., loading or discharging

All the trails were performed in a Lenovo™ ideapad 330, AMD A9-9425 Radeon R5, 3.10 GHz, 8 GB of RAM, Windows® 10 for 64 bits. C++ language is preferred to make all the comparisons. A total of 30 trials were run for all the dataset. The relative percentage increase (RPI) details how to compare the performance of the algorithms. This is the metric used for comparison in this research.

$$\text{RPI}(c_{\text{i}}) = (c_{\text{i}} - c^{+})/c^{+} \tag{14}$$

where c_i is the average wait time executed in the *i*th replication, and c^+ is the best average wait time found for each instance used in this study. The reason to utilize the RPI is to compare two quantities while taking into account the "sizes" of the things being compared. The comparison is expressed as a ratio and is a unitless number. By multiplying these ratios by 100, they can be expressed as percentages [56].

The performance of the QCEDA, through the experimental results, is presented in Table 9. The results of the QCEDA scheme are concentrated between [0, 0.03], over the best result found for all the instances. Therefore, the QCEDA is an outstanding technique to address the quay crane scheduling problem.

Table 9. Performance of the QCEDA using [55]'s instances.

		Intervals		
Layouts	**Trials**	[0, 0.03)	[0.03, 0.06)	[0.06, or more)
90	2700	1933	713	54
		71%	26%	3%

4.2. Comparison with Pure EDAs

The EDAs considered in the comparison are the MIMIC, the COMIT and the BOA. These EDAs are considered as pure EDAs. These EDAs utilize complex probability models and their performance is based on such models. Finally, the QCEDA, the proposed approach, is a hybrid EDA.

The experimental results distribution, for each interval, is shown in Table 10. The QCEDA results are comparatively concentrated, in the range of [0, 0.03], whereas the other algorithms results are concentrated in the range of [0.06, or more].

Table 10. Distribution of the results between pure EDAs and the QCEDA approach.

	Intervals		
Algorithms	[0, 0.03)	[0.03, 0.06)	[0.06, or more)
MIMIC	0	0	2700
COMIT	0	0	2700
BOA	0	0	2700
QCEDA for the QCSP	1933	713	54

Figure 11 indicates the results obtained by the algorithms for the QCSP. Through box and whisper charts, the dispersion of the values obtained, for the RPI, is appreciated. As the complexity of the probability model increases in the pure EDAs, the results improve. However, the QCEDA scheme outperforms all the previous results. The dispersion results of the QCEDA scheme is less than the other algorithms after all the trails were run, i.e., no matter how many times we run the instances, the performance of the QCEDA achieved the best value more times than the other algorithms. This means that the solutions found by the QCEDA are more concentrated around the best value than those by the other algorithms.

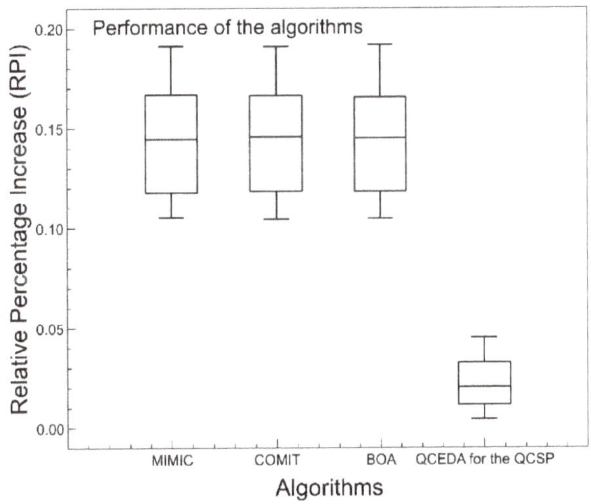

Figure 11. Comparative results for the QCSP.

4.3. Comparison with Multi-Objective Algorithms

The QCEDA is evaluated against other multi-objective algorithms to enhance its novelty. Two benchmark algorithms are used in the comparative, the algorithm from [57], named NSGA, and from [58], named NSGA-II. These algorithms are well known in the literature related to the multi-objective approach. The code of these algorithms is available in the Kanpur genetic algorithms laboratory web. For the multi-objective approach, two antagonistic objectives are analyzed: the average waiting time and the makespan. This means that both fitnesses are computed for each solution. The objectives were utilized as input parameters for finding the area obtained from the nondominated solutions, located in the first layer of the Pareto-front, i.e., the best Pareto-front from each algorithm. This is the new dependent variable for the experiment. All the details are depicted below.

The Best Pareto-Front Process

```
Generate initial population
t :=1
Do
For each member of population
        Get the average waiting time
        Get the makespan
Identify the nondominated solutions from the initial population
Compute the area from the nondominated solutions from the initial population
Continue with the rest of the QCEDA steps to get offspring
For each offspring
        Get the average waiting time
        Get the makespan
Identify the nondominated solutions from the offspring
Compute the area from the nondominated solutions from the offspring
Save the best Pareto-front area between parents and offspring
t :=t + 1
Until (stopping criterion is met)
```

Then, the RPI is adapted to use the same Equation (14), where c^* is the best area found by any of the algorithm configurations, and c_i is the area obtained from the best Pareto-front obtained in the i-th replication by a given algorithm configuration.

Although the probability model, the GMD proposed, has been competitive for finding suitable offspring, the best Pareto-front contributes to enhance the results when the nondominated solutions are coupled with the GMD process.

The experimental results distribution is depicted in Table 11. From the table, it is possible to identify that the overall results of the QCEDA approach are competitive. Based on the results, the QCEDA approach can efficiently find the closest solutions to the best solution, 2008 times in the first interval, whereas the other algorithms results are far from this amount.

Table 11. Distribution of the results between multi-objective algorithms and the QCEDA approach.

	Intervals		
Algorithms	[0, 0.03)	[0.03, 0.06)	[0.06, or more)
NSGA	160	21	3029
NSGA-II	159	25	3026
QCEDA for the QCSP	2008	791	411

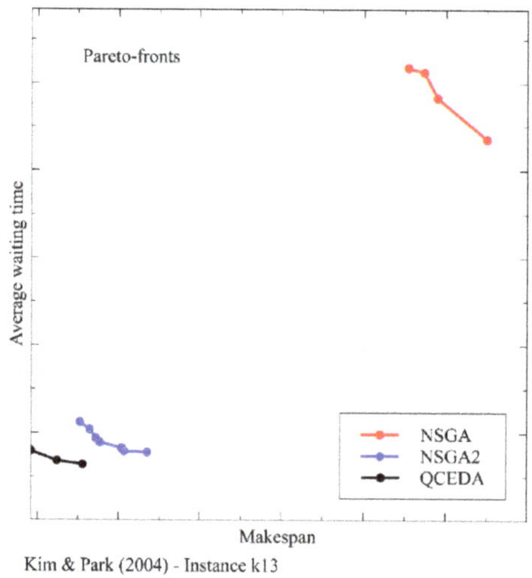

In addition, the Pareto-fronts obtained from each algorithm are shown at the bottom of the Table 11, for the instance k13.

Figure 12 includes the performances of the NSGA, the NSGA-II, and the QCEDA schemes. The QCEDA is efficient to find the best solutions. The dispersion of the QCEDA is almost the same as the other algorithms. However, the solutions found by the QCEDA are more concentrated around the best value than other algorithms, i.e., the average of the solutions of the QCEDA converge better than other approaches to the best-found value.

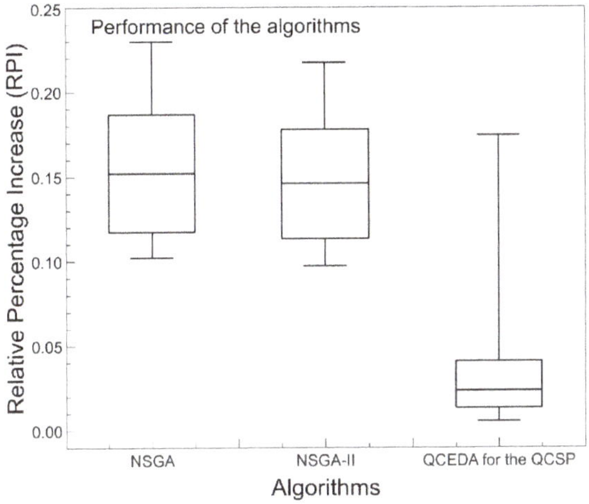

Figure 12. Performance of the multi-objective algorithms.

4.4. Comparison with Recent Algorithms

Four current schemes are used against the QCEDA. The [51] algorithm, the [50] algorithm, the [52] algorithm, and the [49] algorithm. The author has implemented all the recent evolutionary algorithms used in the comparison section. The implementation was made by following the indications of the respective papers, and the objective function is the same for all the comparison process, i.e., the average waiting time.

The experimental results distribution is detailed in Table 12. From the table, it is possible to identify that the overall results of the QCEDA scheme is, again, competitive. According to the results, the QCEDA algorithm can efficiently find the closest solutions to the optimal, more times in the range of [0, 0.03], than any another algorithm used in the comparison.

Although there are no optimal results reported in the literature regarding the average waiting time, it is possible to show the optimality gap for the makespan. These results are provided in the bottom of the Table 12 for the all instances used in the comparison.

Figure 13 depicts the results of the algorithms. All the results were outperformed by the QCEDA. Again, the dispersion of the QCEDA is less than other algorithms (between 0.01 and 0.05); this means that the solutions found by the QCEDA are more concentrated around the best value than other algorithms, i.e., the average of solutions of the QCEDA converges better than other approaches to the best found value. Furthermore, based on the results, it is possible to justify the proposal of a hybrid EDA, not only as a reason for publication; it also avoids the necessity of requiring building complex probability models. In addition, results help to determine what is most useful, i.e., the development of complex probability models or the hybridization of an EDA with other methods to obtain the same or better solutions. This justifies the reason for performing this research; there exists a wide field of development to determine which methods are more suitable to obtain the better performance of an EDA. Based on the results obtained in this study, it is more efficient to produce new and better offspring through such hybridization.

Table 12. Distribution of the results between recent algorithms and the QCEDA approach.

Intervals			
Algorithms	[0, 0.03)	[0.03, 0.06)	[0.06, or more)
Chung & Choy [51]	0	0	2700
Kaveshgar et al. [50]	0	0	2700
Wang et al. [52]	0	0	2700
Chung & Chan [49]	0	0	2700
QCEDA for the QCSP	155	55	2490

Optimality Gap				
Chung & Choy [51]	Kaveshgar et al. [50]	Wang et al. [52]	Chung & Chan [49]	QCEDA for the QCSP
2.73%	2.68%	2.76%	2.63%	2.48%

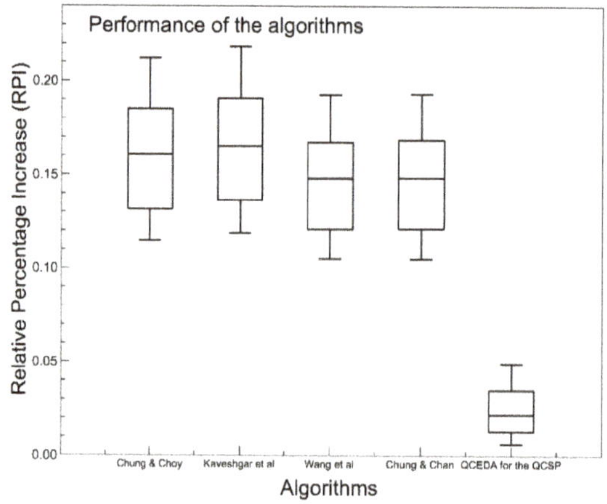

Figure 13. Comparative results.

4.5. Computational Cost Results

Finally, Figure 14 details the computational cost results of the QCEDA scheme. As we can see, it is competitive with the other algorithms used in the comparison.

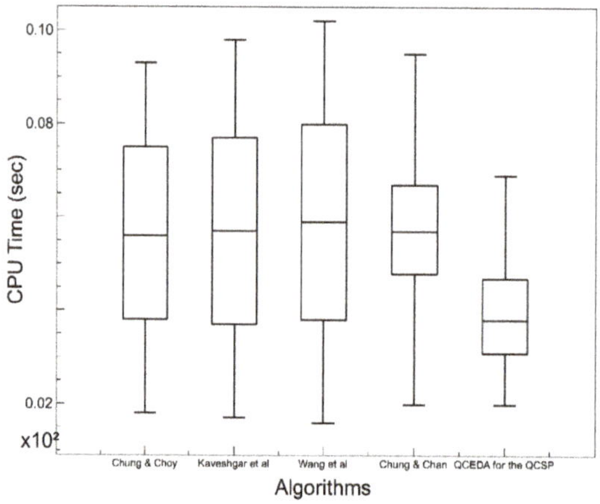

Figure 14. Computational cost results.

4.6. Convergence Patterns

Figure 15 presents the convergence patterns of the QCEDA scheme.

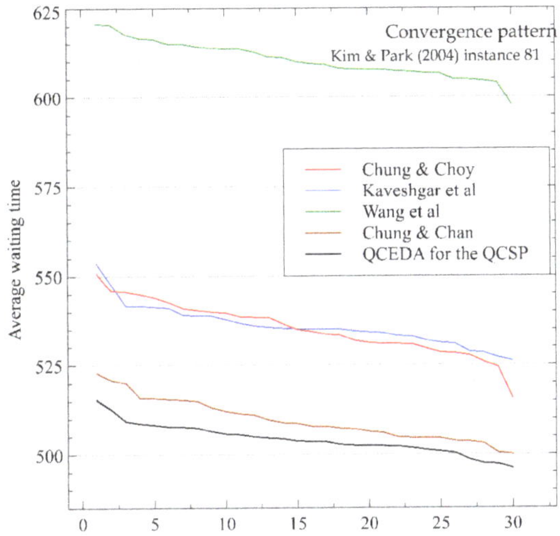

Figure 15. Convergence patterns.

5. Discussion, Conclusions and Future Research

The contributions of this article are

- To propose a hybrid EDA, not only a reason for publication; this also avoids the necessity of requiring building complex probability models.
- To determine what is most useful, i.e., the development of complex probability models or the hybridization of an EDA with other methods to obtain the same or better solutions.
- As a reason for doing this research, there exists a wide field of development to determine which methods are more suitable to obtain a better performance of the EDA.
- The research motivation is to show how the bioinspired techniques help to reduce the deficiencies of an EDA.
- Based on the results obtained in this study, it is more efficient to produce new and better offspring through such hybridization.

Based on the results detailed above, it is possible to conclude that the hybrid EDAs have a better performance, or equal in effectiveness, than the pure EDAs. It can be established that, for combinatorial problems, the estimation of the high order interactions can be omitted if developing a competitive algorithm is the objective. The proposed hybridization, with the Mallows—Moth-flame approach, is more useful to obtain the same or better solutions. The dispersion results of the QCEDA scheme is always less than the other algorithms used in the comparison section. This means that the solutions found by the QCEDA are more concentrated around the best value than other algorithms, i.e., the average of the solutions of the QCEDA converge better to the best found value than other approaches.

Furthermore, based on the results, a viable alternative to build better hybrid EDAs is to use already defined probability models for the permutations field. In this research, the proposed model is able to adapt itself to the structure of the representation of the members of the population. The main goal is achieved, i.e., to model the solution space from another perspective. The Mallows distribution falls in this approach, but there could be more distributions for this purpose.

Disadvantages of the proposed scheme

- The main disadvantage of the proposed QCEDA is to learn and program different procedures to obtain the distance, and factorization of distance, between solutions using a permutation based representation.
- Only the nondominated solutions located in the first layer of the Pareto-front are used in this research to make comparisons between algorithms. Other metrics should be considered in order to show the weaknesses and/or drawbacks of the proposed scheme.

As future research.

- New and more comparisons should be considered, using the QCEDA, in future work.
- Other combinatorial problems should be considered.
- In addition, a substitution for the Mallows distribution and/or to hybridize with other methods, should be considered to enhance the proposed scheme.
- Furthermore, diversity mechanisms should be considered to improve the proposed scheme.
- Finally, the application of the proposed scheme should consider the solution of real world problems in future work.
- The QCEDA should attend to more and different restrictions.

Funding: This research received no external funding.

Data Availability Statement: The datasets used and/or analyzed during the current study are available from the corresponding author on reasonable request.

Acknowledgments: Special thanks to all the reviewers.

Conflicts of Interest: The authors declare no conflict of interest. The funders had no role in the design of the study; in the collection, analyses, or interpretation of data; in the writing of the manuscript or in the decision to publish the results.

References

1. Wang, Y.; Wang, S. Deploying, scheduling, and sequencing heterogeneous vessels in a liner container shipping route. *Transp. Res. Part E Logist. Transp. Rev.* **2021**, *151*, 102365. [CrossRef]
2. Al-Refaie, A.; Abedalqader, H. Optimal berth scheduling and sequencing under unexpected events. *J. Oper. Res. Soc.* **2020**, 1–15. [CrossRef]
3. Malekahmadi, A.; Alinaghian, M.; Hejazi, S.R.; Saidipour, M.A.A. Integrated continuous berth allocation and quay crane assignment and scheduling problem with time-dependent physical constraints in container terminals. *Comput. Ind. Eng.* **2020**, *147*, 106672. [CrossRef]
4. Kavoosi, M.; Dulebenets, M.A.; Abioye, O.F.; Pasha, J.; Wang, H.; Chi, H. An augmented self-adaptive parameter control in evolutionary computation: A case study for the berth scheduling problem. *Adv. Eng. Inform.* **2019**, *42*, 100972. [CrossRef]
5. Dulebenets, M.A. A novel memetic algorithm with a deterministic parameter control for efficient berth scheduling at marine container terminals. *Marit. Bus. Rev.* **2017**, *2*, 302–330. [CrossRef]
6. Kim, Y.; Park, K.; Ko, J. A symbiotic evolutionary algorithm for the integration of process planning and job shop scheduling. *Comput. Oper. Res.* **2003**, *30*, 1151–1171. [CrossRef]
7. Rossi, A.; Dini, G. Flexible job-shop scheduling with routing flexibility and separable setup times using ant colony optimisation method. *Robot. Comput.-Integr. Manuf.* **2007**, *23*, 503–516. [CrossRef]
8. Dauod, H.; Li, D.; Yoon, S.; Srihari, K. Multi-objective optimization of the order scheduling problem in mail-order pharmacy automation systems. *Int. J. Adv. Manuf. Technol.* **2016**. [CrossRef]
9. Yue, L.; Guan, Z.; Saif, U.; Zhang, F.; Wang, H. Hybrid Pareto artificial bee colony algorithm for multi objective single machine group scheduling problem with sequence dependent setup times and learning effects. *Springerplus* **2016**, *5*. [CrossRef]
10. Huang, S.; Tian, N.; Wang, Y.; Ji, Z. Multi objective flexible job shop scheduling problem using modified discrete particle swarm optimization. *Springerplus* **2016**, *5*. [CrossRef]
11. Xu, H.; Bao, Z.; Zhang, T. Solving dual flexible job-shop scheduling problem using a Bat Algorithm. *Adv. Prod. Eng. Manag.* **2017**, *12*, 5–16. [CrossRef]
12. Prins, C. A simple and effective evolutionary algorithm for the vehicle routing problem. *Comput. Oper. Res.* **2004**, *31*, 1985–2002. [CrossRef]
13. Gao, J.; Sun, L.; Gen, M. A hybrid genetic and variable neighborhood descent algorithm for flexible job shop scheduling problems. *Comput. Oper. Res.* **2008**, *35*, 2892–2907. [CrossRef]

14. Zeng, Q.; Yang, Z. Integrating simulation and optimization to schedule loading operations in container terminals. *Comput. Oper. Res.* **2009**, *36*, 1935–1944. [CrossRef]
15. Lee, L.; Chew, E.; Tan, K.; Wang, Y. Vehicle dispatching algorithms for container transshipment hubs. *OR Spectr.* **2010**, *32*, 663–685. [CrossRef]
16. Schittekat, P.; Kinable, J.; Sörensen, K.; Sevaux, M.; Spieksma, F. A metaheuristic for the school bus routing problem with bus stop selection. *Eur. J. Oper. Res.* **2013**, *229*, 518–528. [CrossRef]
17. Li, X.; Gao, L. An effective hybrid genetic algorithm and tabu search for flexible jobshop scheduling problem. *Int. J. Prod. Econ.* **2016**, *174*, 93–110. [CrossRef]
18. Jarboui, B.; Eddaly, M.; Siarry, P. An estimation of distribution algorithm for minimizing the total flowtime in permutation flowshop scheduling problems. *Comput. Oper. Res.* **2009**, *36*, 2638–2646. [CrossRef]
19. Chen, S.; Chen, M.; Chang, P.; Zhang, Q.; Chen, Y. Guidelines for developing effective Estimation of Distribution Algorithms in solving simple machine scheduling problems. *Expert Syst. Appl.* **2010**, *37*, 6441–6451. [CrossRef]
20. Pan, Q.; Ruiz, R. An estimation of distribution algorithm for lot-streaming flow shop problems with setup times. *Omega* **2012**, *40*, 166–180. [CrossRef]
21. Chen, Y.; Chen, M.; Chang, P.; Chen, S. Extended artificial chromosomes genetic algorithm for permutation flowshop scheduling problems. *Comput. Ind. Eng.* **2012**, *62*, 536–545. [CrossRef]
22. Wang, L.; Wang, S.; Xu, Y.; Zhou, G.; Liu, M. A bi-population based estimation of distribution algorithm for the flexible job-shop scheduling problem. *Comput. Ind. Eng.* **2012**, *62*, 917–926. [CrossRef]
23. Mirjalili, S. Moth-flame optimization algorithm: A novel nature-inspired heuristic paradigm. *Knowl.-Based Syst.* **2015**, *89*, 228–249. [CrossRef]
24. Ceberio, J.; Irurozki, E.; Mendiburu, A.; Lozano, J. A Distance-Based Ranking Model Estimation of Distribution Algorithm for the Flowshop Scheduling Problem. *IEEE Trans. Evol. Comput.* **2014**, *18*, 286–300. [CrossRef]
25. Mallows, C. Nonnull ranking models. *Biometrika* **1957**, *44*, 114–130. [CrossRef]
26. Fligner, M.; Verducci, J. Distance based ranking models. *J. R. Stat. Soc.* **1986**, *48*, 359–369. [CrossRef]
27. Fligner, M.; Verducci, J. Multistage ranking models. *J. Am. Stat. Assoc.* **1988**, *83*, 892–901. [CrossRef]
28. Mühlenbein, H. The equation for response to selection and its use for prediction. *Evol. Comput.* **1997**, *5*, 303–346. [CrossRef] [PubMed]
29. De Bonet, J.; Isbell, C.; Viola, P. MIMIC: Finding Optima by Estimation Probability Densities. In Proceedings of the 1997 Conference on Advances in Neural Information Processing Systems, Denver, CO, USA, 2–6 December 1997; pp. 424–430.
30. Baluja, S.; Davies, S. Using Optimal Dependency-Trees for Combinatorial Optimization: Learning the Structure of the Search Space. In Proceedings of the Fourteenth International Conference on Machine Learning, San Francisco, CA, USA, 8–12 July 1997; Fisher, D., Ed.; pp. 30–38.
31. Pelikan, M.; Goldberg, D.; Cantú-Paz, E. *Linkage Problem, Distribution Estimation, and Bayesian Networks*; The University of Illinois, Genetic Algorithm Laboratory: Urbana-Champaign, IL, USA, 1998.
32. Pérez-Rodríguez, R.; Jöns, S.; Hernández-Aguirre, A.; Ochoa, C. Simulation optimization for a flexible jobshop scheduling problem using an estimation of distribution algorithm. *Int. J. Adv. Manuf. Technol.* **2014**, 3–21. [CrossRef]
33. Ceberio, J.; Irurozki, E.; Mendiburu, A.; Lozano, J. A review on estimation of distribution algorithms in permutation-based combinatorial optimization problems. *Prog. Artif. Intell.* **2012**, *1*, 103–117. [CrossRef]
34. Garg, H. A hybrid GA-GSA algorithm for optimizing the performance of an industrial system by utilizing uncertain data. In *Handbook of Research on Artificial Intelligence Techniques and Algorithms*; Vasant, P., Ed.; IGI Global: Hershey, PA, USA, 2014. [CrossRef]
35. Garg, H. A hybrid PSO-GA algorithm for constrained optimization problems. *Appl. Math. Comput.* **2016**, *274*, 292–305. [CrossRef]
36. Garg, H. A hybrid GSA-GA algorithm for constrained optimization problems. *Inf. Sci.* **2019**, *478*, 499–523. [CrossRef]
37. Peña, J.; Robles, V.; Larrañaga, P.; Herves, V.; Rosales, F.; Pérez, M. GA-EDA: Hybrid Evolutionary Algorithm Using Genetic and Estimation of Distribution Algorithms. In *IEA/AIE 2004, LNAI 3029*; Orchard, R., Yang, C., Ali, M., Eds.; Springer: Berlin/Heidelberg, Germany, 2004; pp. 361–371.
38. Zhang, Q.; Sun, J.; Tsang, E.; Ford, J. Estimation of distribution algorithm with 2-opt local search for the quadratic assignment problem. *Stud. Fuzziness Soft Comput.* **2006**, *192*, 281–292.
39. Liu, H.; Gao, L.; Pan, Q. A hybrid particle swarm optimization with estimation of distribution algorithm for solving permutation flowshop scheduling problem. *Expert Syst. Appl.* **2011**, *38*, 4348–4360. [CrossRef]
40. Fang, C.; Kolisch, R.; Wang, L.; Mu, C. An estimation of distribution algorithm and new computational results for the stochastic resource-constrained project scheduling problem. *Flex. Serv. Manuf. J.* **2015**. [CrossRef]
41. Wang, K.; Huang, Y.; Qin, H. A fuzzy logic-based hybrid estimation of distribution algorithm for distributed permutation flowshop scheduling problems under machine breakdown. *J. Oper. Res. Soc.* **2016**, *67*, 68–82. [CrossRef]
42. Pérez-Rodríguez, R.; Hernández-Aguirre, A.; Cruz, I. An estimation of distribution algorithm coupled with the generalized Mallows distribution for a school bus routing problem with bus stop selection. *Rev. Iberoam. Autom. E Inform. Ind.* **2017**, *14*, 288–298. [CrossRef]
43. Pérez-Rodríguez, R.; Hernández-Aguirre, A. A hybrid estimation of distribution algorithm for flexible job-shop scheduling problems with process plan flexibility. *Appl. Intell.* **2018**. [CrossRef]

44. Pérez-Rodríguez, R.; Hernández-Aguirre, A. A hybrid estimation of distribution algorithm for the vehicle routing problem with time windows. *Comput. Ind. Eng.* **2019**, *130*, 75–96. [CrossRef]
45. Abou Kasm, O.; Diabat, A. The quay crane scheduling problem with non-crossing and safety clearance constraints: An exact solution approach. *Comput. Oper. Res.* **2019**, *107*, 189–199. [CrossRef]
46. Sun, D.; Tang, L.; Baldacci, R. A benders decomposition-based framework for solving quay crane scheduling problems. *Eur. J. Oper. Res.* **2019**, *273*, 504–515. [CrossRef]
47. Chen, J.H.; Lee, D.H.; Goh, M. An effective mathematical formulation for the unidirectional cluster-based quay crane scheduling problem. *Eur. J. Oper. Res.* **2014**, *232*, 198–208. [CrossRef]
48. Bierwirth, C.; Meisel, F. A fast heuristic for quay crane scheduling with interference constraints. *J. Sched.* **2009**, *12*, 345–360. [CrossRef]
49. Chung, S.-H.; Chan, T.-S. A workload balancing genetic algorithm for the quay crane scheduling problem. *Int. J. Prod. Res.* **2013**, *51*, 4820–4834. [CrossRef]
50. Kaveshgar, N.; Huynh, N.; Rahimian, S.-K. An efficient genetic algorithm for solving the quay crane scheduling problem. *Expert Syst. Appl.* **2012**, *39*, 13108–13117. [CrossRef]
51. Chung, S.-H.; Choy, K.-L. A modified genetic algorithm for quay crane scheduling operations. *Expert Syst. Appl.* **2012**, *39*, 4213–4221. [CrossRef]
52. Wang, Y.; Chen, Y.; Wang, K. A case study of genetic algorithms for quay crane scheduling. In *Opportunities and Challenges for Next-Generation Applied Intelligence*; Chien, B., Hong, T., Eds.; Studies in Computational Intelligence; Springer: Berlin/Heidelberg, Germany, 2009; p. 214.
53. Borda, J. Memoire sur les elections au scrutin. *Histoire de l'Academie Royale des Science* **1781**, *102*, 657–665.
54. Meilă, M.; Phadnis, K.; Patterson, A.; Bilmes, J. Consensus ranking under the exponential model. *arXiv* **2007**, arXiv:1206.5265.
55. Kim, K.; Park, Y. A crane scheduling method for port container terminals. *Eur. J. Oper. Res.* **2004**, *156*, 752–768. [CrossRef]
56. Törnqvist, L.; Vartia, P.; Vartia, Y. How Should Relative Changes Be Measured? *Am. Stat.* **1985**, *39*, 43–46. [CrossRef]
57. Srinivas, N.; Deb, K. Muiltiobjective optimization using nondominated sorting in genetic algorithms. *Evol. Comput.* **1994**, *2*, 221–248. [CrossRef]
58. Deb, K.; Pratap, A.; Agarwal, S.; Meyarivan, T. A fast and elitist multiobjective genetic algorithm: NSGA-II. *IEEE Trans. Evol. Comput.* **2002**, *6*, 182–197. [CrossRef]

Mathematical and Computational Applications

MDPI

Article

Time to Critical Condition in Emergency Services

Pedro A. Pury

Facultad de Matemática, Astronomía, Física y Computación, Universidad Nacional de Córdoba, Ciudad Universitaria, Córdoba X5000HUA, Argentina; pedro.pury@unc.edu.ar

Abstract: Providing uninterrupted response service is of paramount importance for emergency medical services, regardless of the operating scenario. Thus, reliable estimates of the time to the critical condition, under which there will be no available servers to respond to the next incoming call, become very useful measures of the system's performance. In this contribution, we develop a key performance indicator by providing an explicit formula for the average time to the shortage condition. Our analytical expression for this average time is a function of the number of parallel servers and the inter-arrival and service times. We assume exponential distributions of times in our analytical expression, but for evaluating the mean first-passage time to the critical condition under more realistic scenarios, we validate our result through exhaustive simulations with lognormal service time distributions. For this task, we have implemented a simulator in *R*. Our results indicate that our analytical formula is an acceptable approximation under any situation of practical interest.

Keywords: first-passage time; Markov chain; queueing theory; simulation; OR in health services; KPI

MSC: [2010] 90B22; 60K25

Citation: Pury, P.A. Time to Critical Condition in Emergency Services. *Math. Comput. Appl.* **2021**, *26*, 70. https://doi.org/10.3390/mca26040070

Received: 14 July 2021
Accepted: 28 September 2021
Published: 30 September 2021

1. Introduction

The problem of assigning resources to respond to a stochastic demand is a ubiquitous topic in operational research. The trade-off between service quality and operational efficiency is a crucial aspect of the Emergency Medical Services (EMS), where the lives of patients depend on the timeliness of care. Thus, the development of Key Performance Indicators (KPIs) to objectively quantify the performance across the operational, clinical and financial departments is a current demand of an industry that is becoming increasingly data-driven. KPIs are the basic tools for planners, and the nature of each KPI selects a particular feature of the system and determines its data gathering strategy.

Among the most intuitive and used operational KPIs in EMS are the successive times involved in the service cycle: call reception, patient triage, dispatch, ambulance turnout, travel from the base to the emergency site, paramedic care, eventual transfer of the patient to a hospital and return of the ambulance to its base. Response time (the interval between the reception of an emergency call and the arrival of a paramedic at the scene of the event) is a common operational metric of EMS, and it is considered a good indication of the quality offered by the service [1]. One reason for its popularity as a KPI resides in the fact that it is directly quantifiable and easily understood by the public and policy makers. Additionally, the EMS industry has the goal of providing care within eight minutes for cardiac arrest [2] and major trauma [3]. However, there is evidence that exceeding that response time criterion does not affect patient survival after a traumatic impact injury [4,5]. Moreover, solutions that only focus on shortening the response time are cost prohibitive and put the safety of patients, attendant crew and the public at risk [6]. A rational approach to the ambulance business process should simultaneously consider multiple metrics and operational trade-off between administrator-oriented and patient-centered KPIs [7].

One of the most important aspects of emergency medical management is avoiding the oversaturation of the system. Therefore, in this work, we consider the First-Passage

Time (FPT) [8] to the critical condition, under which there will be not available servers to respond to the next incoming call. Criticality prediction is of special interest for the quality of the medical service (response time off-target) as well as for the financial management of the service, given that, as we will see, the critical condition strongly depends on the number L of ambulances simultaneously in service and because queueing a call may involve transferring it to another EMS. Any system operating under fixed conditions with a given number of servers and a First-Come First-Served (FCFS) discipline is a discrete one-dimensional stochastic process over the occupation states of servers. Therefore, the first-passage time to the state of oversaturation, in which there are no available servers, is finite. The question is how long is that time. Thus, the mean first-passage time (MFPT) becomes a relevant key performance indicator for the operational condition in EMS.

In urban emergency services logistics, there are two distinct fields: capacity planning and location analysis. Both fields are interrelated in the districting problem or how the region should be partitioned into areas of primary responsibility (districts) [9,10]. Here, MFPT can be applied to an operational subarea or district, preferably intended to be independently served by a subset of ambulances (intradistrict dispatches). In this case, MFPT gives us the average time to request an ambulance from another operational zone to answer the next emergency call when the primary equipment is busy (interdistrict dispatches). MFPT can also be a useful KPI in the decision-making process of emergency departments [11,12], where MFPT provides the average time to a shortage of intensive therapy beds when the FCFS discipline is used after the patient's triage.

Queueing theory has been widely applied in health care in the last 70 years [13]. Since Larson's seminal article [9], quite a few queueing models have been developed to incorporate the intrinsic probabilistic nature of urban EMS derived from the Poisson nature of the call arrival process and the variability in service times. Multiple queueing systems have been developed that respond with different emphasis on the KPIs selected in each case. In this contribution, we use a birth–death process to properly analyse the dependence of MFPT on the number of servers and the rest of the system's operational parameters. The birth–death process is basic to queueing models involving exponential inter-arrival and service time distributions. In the Kendall–Lee notation, we have $M/M/L/FCFS/\infty/\infty$ [8]. Thus, our analytical model is based only on two average times: T_C, the mean inter-arrival time, and T_S, the mean service time of a single server, that is, the time it takes for an ambulance to complete a trip from the instant a call is assigned until the release of this server. Several analytical results are well known in operational research under that assumption [8].

However, experimental evidence from an emergency service indicates that service time distributions are well fitted by lognormal distributions [12,14–16]. Hence, our objective is also to numerically evaluate the deviation between analytical and simulated results for MFPT. Thus, we present an R-simulator for a system of L servers in parallel with general distributions for inter-arrival and service times and FCFS discipline: $GI/G/L/FCFS$ [8]. This tool allows the user to calculate the key performance indicators of direct interest in the industry beyond the known analytical results, which are limited to exponential distributions. Particularly, we show our work on Mean First-Passage Time (MFPT) to system critical condition.

In this way, the motivation of our work is two-fold. On the one hand, we provide an explicit analytical expression for the MFPT to the critical condition, and, on the other hand, under more realistic conditions, we analyse the validity of our assumptions through exhaustive simulations. In Section 2, we provide our analytical expression for MFPT and explore the generic nature of the method, postponing detailed mathematical derivations to the appendices. Additionally, in that section, we describe the simulation framework for experimentation. Section 3 deals with the numerical results, and last, in Section 4, we discuss the importance of our contribution.

2. Model and Simulator

In this section, we develop an analytical closed-form solution for the MFPT and present a simulation framework based on discrete events.

2.1. Markov Chain Model for Servers in Parallel

We consider a stochastic continuous-time birth–death process [8] that describes the time evolution of the occupation state of a set of L servers in parallel. Changes in the state of the Markov chain imply the release of a server or putting one into action (if at least one is available). The state with occupation n corresponds to n received calls not completely served yet. Thus, when $n = 0$, all the servers are free, and there are neither trips in process nor calls in queue. For $0 < n \leq L$, there are no waiting calls, and n servers are in course of action. In an equivalent way, we can say that n calls are simultaneously being served. Particularly, when $n = L$, the system is saturated, that is, all servers are occupied. Even though there is not any call in the waiting queue, all servers have been assigned to calls, and consequently, there are not any servers available to process the next eventual incoming call. For $n > L$, the system is oversaturated, and there are $n - L$ calls in the waiting queue. At any time, the system can change its state of occupation between its nearest neighbours. Therefore, we denote the transition rate from the state n to $n+1$ by ω_n^+, whereas the transition rate toward the lower occupation state ($n \to n-1$) is ω_n^-. We assume that the time between calls is an exponential random variable with the mean number of calls per unit time $\lambda = 1/T_C$. On the other hand, the service time is also an exponential variable with the rate or mean number of services per unit time and per server $\mu = 1/T_S$. Thus, random call arrival times and service times consumed in each trip are generated from continuous time distributions. T_C and T_S are the average inter-arrival and service time, respectively. In our particular problem with L servers, the transition probabilities rates of the birth–death process are defined by [17]

$$
\begin{aligned}
\omega_n^+ &= \lambda \ \forall n\,, \\
\omega_n^- &= \begin{cases} n\,\mu & \text{for } n \leq L\,, \\ L\,\mu & \text{for } n \geq L\,. \end{cases}
\end{aligned}
\tag{1}
$$

In this manner, ω_n^+ results constant and only ω_n^- depends on the state of the system. Then, all the experimental information needed to characterize our theoretical model are the average times T_C and T_S.

For fixed values of T_C and T_S, the Markov process always reaches the state $L+1$, that is, the critical condition in which the incoming call could not be served. Therefore, we focus our interest in the first-passage time (FPT) to the state $L+1$. That is the time needed by the system to reach the critical situation (first call is derived to the waiting queue) given an initial state without queue ($0 \leq n \leq L$). Following previous experience [18], the MFPT, $T(n)$, from the initial state $n = 0, \ldots, L$, can be written as

$$
\begin{aligned}
T(0) &= T_C\left(L+1+\sum_{k=0}^{L-1}\frac{\gamma^{-k}}{k!}\sum_{i=k+1}^{L} i!\,\gamma^i\right), \\
T(1) &= T(0) - T_C\,, \\
T(n) &= T(0) - T_C\left(n+\sum_{k=0}^{n-2}\frac{\gamma^{-k}}{k!}\sum_{i=k+1}^{n-1} i!\,\gamma^i\right) \quad \text{for } 2 \leq n \leq L\,,
\end{aligned}
\tag{2}
$$

where the parameter $\gamma = \mu/\lambda = T_C/T_S$ is the inverse of Erlang's rate [8]. The derivation and mathematical details of Equation (2) are worked out in Ref. [18] and Appendix A. In this way, knowing γ and T_C, the expressions of Equation (2) can be numerically evaluated in a very direct way.

The average involved at this stage is over realizations of the stochastic process. Under actual operating conditions, where the dispatcher knows the system stress in real-time, Equation (2) makes it possible to predict the MFPT to respond accordingly. However, if we want to predict the MFPT under prospective stress conditions, we request for a quantity independent of the initial state in order to define a performance measure. Therefore, we need an average over $n = 0, \ldots, L$. For this purpose, we define

$$< T >= \sum_{n=0}^{L} P(n)\, T(n)\,, \tag{3}$$

where $P(n)$ is the probability of residence in the state n. To perform this calculation, we need to know the conditional probability of being at state n at time t given the initial state m, $P(n|m)(t)$. In order to simplify the problem, we propose to calculate $P(n)$ in the steady state regime, which is independent of the initial condition [8]: $P(n) = \lim_{t\to\infty} P(n|m)(t)$. Moreover, given that we are interested in the FPT to the critical condition (first jump to state $L+1$), we will approximate $P(n)$ by working with the finite Markov chain with reflecting boundaries at sites 0 and L. It is important to note that the steady state probabilities of the finite chain are approximately the same as the residence probabilities of our unbounded chain, since in the lapse before FPT there are no calls in the queue. Under these assumptions, we obtain

$$P(n) = \frac{1}{S}\frac{\gamma^{-n}}{n!}\,, \quad \text{for } 0 \le n \le L\,, \tag{4}$$

where S is given by the normalization condition, $S = \sum_{n=0}^{L} \frac{\gamma^{-n}}{n!}$. The mathematical derivation of these expressions is relegated to the Appendix B. The truncated Poisson distribution given in Equation (4) is known as the Erlang B-formula, and it has been proven that this equilibrium distribution of the number of occupied servers is independent of the form of the service time distribution [19]. Moreover, the Erlang B-formula is also valid for heterogeneous servers, provided however, that all servers have equal mean service times T_S [20].

Equation (3) plus Equations (2) and (4) provide us with a closed form expression for calculating the MFPT averaged on initial states.

2.2. Simulation Framework

To compare and analyse the prediction of Equations (2) and (4) with realistic situations, we have developed a flexible discrete-event simulator (DES) [21] with a process-oriented approach that implements several features inherent to EMS management.

The architecture of our simulator is outlined in Figure 1. The input parameters configure the statistical distributions and set the number of servers (L). The proposed simulator consists of three main modules. The first module, called `simserveRs`, is the simulator kernel. Each thread of the simulation is triggered with a new call arrival. Using a pseudo-random number generator (RNG), it draws a call time, summing an exponential random value to the time of the previous call, and compares it with the release times of busy servers. Then, the kernel iteratively puts the older calls in the queue in service, whereas the release times are less than the last call time (FCFS discipline) [22]. Afterward, if there are no available servers, the incoming call is derived to the waiting queue; otherwise, the call is dispatched to a server, picking it at random among the free ones. At last, `simserveRs` assigns a service time drawn from the desired distribution. Thus, at each event, the simulation engine updates the system state composed by the servers and the queue.

The module `simcritical` launches `simserveRs` with the wanted initial condition and stops the execution when the first call is derived to the queue. Then, the average given by Equation (3) is calculated, and the aggregation over simulations is performed. The last module reckons the MFPT from each initial condition using the analytical expressions given by Equation (2).

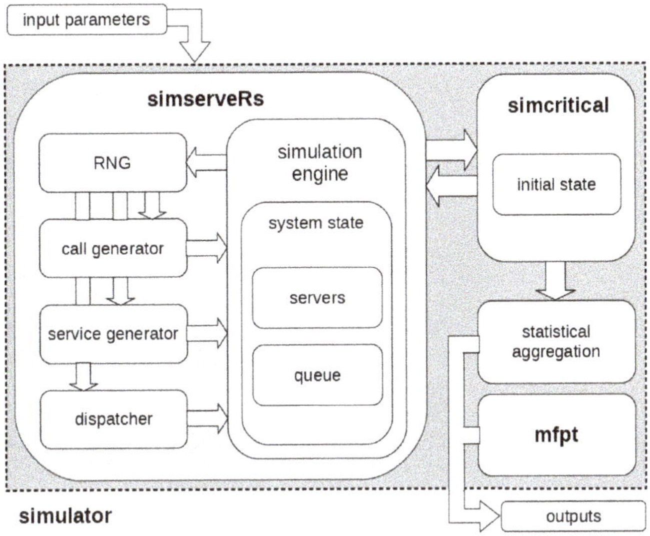

Figure 1. Architecture scheme of the discrete-event simulator.

The software is implemented in *R*, and it is available as open source (see details in Ref. [23]). The simulator can be simply adapted to incorporate most of the features of an actual EMS operator, such as disaggregating the service time into its components (e.g., preparation of ambulance at base, transit time, attention time and transit time to hospital) and implementing dispatching policies with distance or traffic time criteria. Additionally, the extensions of the simulator to any kind of distributions for inter-arrival times (e.g., Erlang) and service times (e.g., gamma or lognormal) are direct. Our simulator was developed in the second half of 2017 for a project related to process optimization in EMS management. The comparison between our simulation outputs and the historical data from an EMS operator showed a satisfactory statistical agreement in several operating scenarios. Over time, several generic DES frameworks for queueing systems have been developed, which deliver such functionalities and implement more efficient methods (see Ref. [24] and references therein). However, for the practical reason of evaluating the results of Section 2.1, the open source version of our simulator becomes an appropriate tool.

3. Results

In Figure 2, we show the non-linear behaviour of $< T >$ as a function of T_C and T_S and its strong dependence on L, as they are derived from Equations (2)–(4).

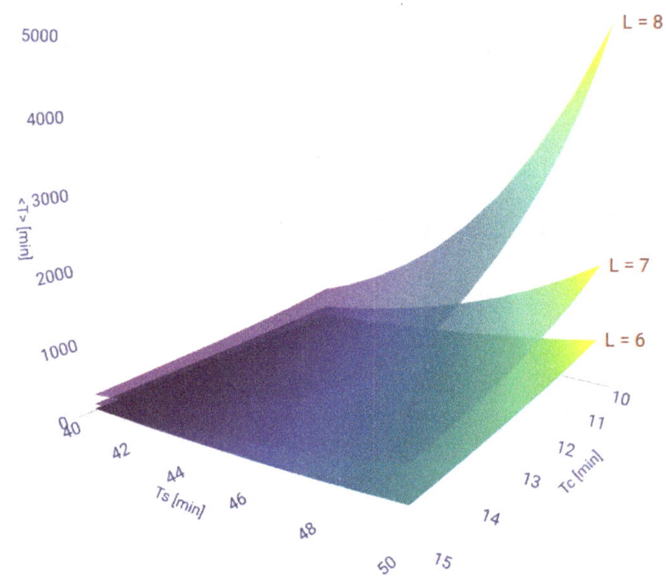

Figure 2. Average MFPT as function of T_C and T_S for three values of L.

The non-linear nature of $< T >$ is given by the powers of γ in Equation (2) and implies a sensitive dependence of the time to the critical condition on its variables. Thus, on the scale of the figure, variations in the order of one minute in T_C or T_S may represent variations of tens to thousands of minutes in $< T >$. The value of L determines the number of terms in Equations (2) and (3). Therefore, adding or removing a server, with the same values of T_C and T_S, can make a substantial change in the value of the average MFPT, as can be seen in Figure 2. The sensitivity of our problem on its parameters is very difficult to grasp intuitively and to predict from past experiences in practice.

The main reason for using the birth and death queueing model $(M/M/L)$ is the fact that we can derive the closed-form result given in Section 2. However, the analytical prediction of any performance measure needs to be contrasted with numerical outputs from more realistic scenarios. As an illustration, we take values of inter-arrival and service times measured by the EMS *Sistema de Urgencias del Rosafe* (URG) [25] in Córdoba, Argentina. URG is one of several private EMS operators in a city of about 1.39 million inhabitants. In 2016, URG operated a fleet of nine ambulances that were usually stationed in predefined parking spots scattered throughout the metropolitan zone. The operating scenario distinguishes several daily time bands within which the mean values T_C and T_S are relatively constant, although these, in turn, show seasonal changes throughout the year.

In Figure 3, we show histograms of real data corresponding to 2568 calls received by URG between May 1 and October 31, 2016, late evening (20:00 to 23:00 h). In this time period, we found good stationary statistics in the data, and no critical condition was reported in which there were no ambulances available to serve a call. The service time value measures the time elapsed from dispatch until the ambulance is released. Very low service times correspond to situations that were quickly resolved at scenes close to the ambulance base, whereas the distribution tail involves complex cases with the transfer of the patient to a hospital.

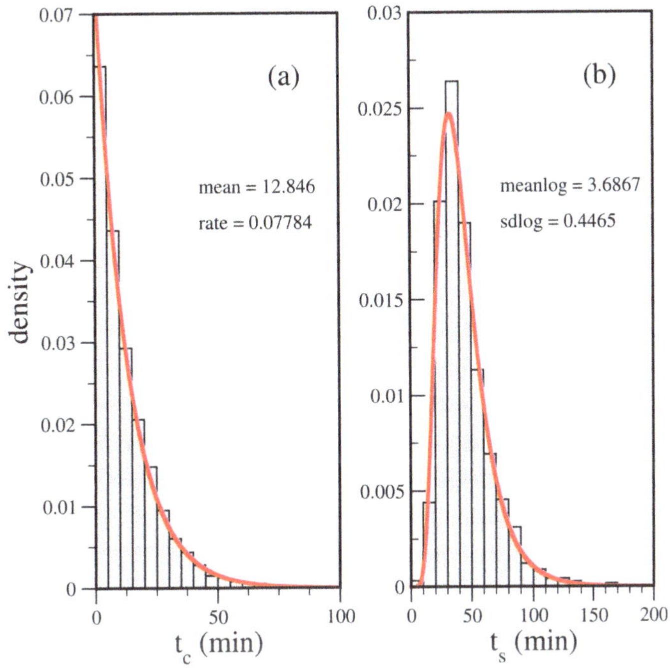

Figure 3. Histograms of real data corresponding to 2568 calls: (**a**) inter-arrival times; (**b**) service times. Solid lines are the best fits: (**a**) exponential; (**b**) lognormal. The insets show the fitting parameters.

From the figure, we can see that the inter-arrival times fit perfectly with an exponential distribution ($T_C = 12.85$ min, goodness-of-fit tests: Kolmogorov–Smirnov p-value $= 0.73$, Cramer–von Mises p-value $= 0.75$), but the best fit for service times is with a lognormal distribution ($T_S = 44.1$ min, goodness-of-fit tests: Kolmogorov–Smirnov p-value $= 0.017$, Cramer–von Mises p-value $= 0.052$). Thus, in this case, it is apparent that the main limitation of our analytical model is the assumption that the distribution of service times is exponential.

The input traffic to a call centre is a nonstationary Poisson process [7,14,26]. However, the arrival rate function $\lambda(t)$ is roughly constant over periods of a few hours [11,14,16,27]. Lognormal distributions for service times have already been reported in the literature [12,14–16]. The fact that the distribution of the sum of a few independent, but not necessarily identical, lognormal random variables could be approximated by a lognormal distribution [28] may explain the experimental findings when we only use two fitting parameters.

In Table 1, we show a comparison between analytical results given by Equation (2) and simulated MFPT from each initial state of a system with seven servers to the critical condition. The simulations were performed using the fitted distributions in Figure 3 and also using an exponential distribution for service times with a mean value equal to that of the fitted distribution in the right panel.

Table 1. Comparison among analytic and simulated values of MFPT (in minutes) for a system with seven servers. The simulated values are reported with its standard error for 10,000 simulations. In both cases, the mean service time is $T_S = 44.1$ min, whereas, in the first case, the probability distribution of service time is *exponential*, and in the second case, it is *lognormal* with `meanlog` $= 3.6867$ and `sdlog` $= 0.4465$. The percentage errors are that of the simulated values with respect to the analytical predictions.

$T_C = 5$ min					
Initial State	**Exponential**			**Lognormal**	
	Analytic	**Simulated**	**ε%**	**Simulated**	**ε%**
0	67.6	67.2 ± 0.4	0.6	49.8 ± 0.3	35.7
1	62.6	62.2 ± 0.4	0.5	44.8 ± 0.3	39.7
2	57.1	56.7 ± 0.4	0.7	39.8 ± 0.3	43.5
3	50.8	50.5 ± 0.4	0.6	34.7 ± 0.3	46.4
4	43.7	43.4 ± 0.4	0.7	29.4 ± 0.2	48.6
5	35.4	35.1 ± 0.3	0.9	23.7 ± 0.2	49.4
6	25.8	25.4 ± 0.3	1.6	17.1 ± 0.2	50.9
7	14.2	14.1 ± 0.2	0.7	9.3 ± 0.2	52.7
$T_C = 10$ min					
Initial State	**Exponential**			**Lognormal**	
	Analytic	**Simulated**	**ε%**	**Simulated**	**ε%**
0	315.2	310.9 ± 2.5	1.4	244.2 ± 2.1	29.1
1	305.2	300.7 ± 2.5	1.5	234.2 ± 2.1	30.3
2	292.9	288.4 ± 2.5	1.6	223.9 ± 2.1	30.8
3	277.3	272.7 ± 2.5	1.7	212.0 ± 2.1	30.8
4	256.7	252.3 ± 2.5	1.7	197.7 ± 2.1	29.8
5	228.1	224.3 ± 2.5	1.7	176.9 ± 2.0	28.9
6	185.6	182.1 ± 2.4	1.9	145.6 ± 2.0	27.5
7	117.7	115.2 ± 2.0	2.2	92.1 ± 1.7	27.8
$T_C = 15$ min					
Initial State	**Exponential**			**Lognormal**	
	Analytic	**Simulated**	**ε%**	**Simulated**	**ε%**
0	1378.4	1380 ± 13	0.1	1201 ± 12	14.8
1	1363.4	1365 ± 13	0.1	1186 ± 12	15.0
2	1343.3	1345 ± 13	0.1	1169 ± 12	14.9
3	1314.6	1317 ± 13	0.2	1147 ± 12	14.6
4	1270.4	1272 ± 13	0.1	1114 ± 12	14.0
5	1195.1	1198 ± 13	0.2	1056 ± 12	13.2
6	1052.2	1058 ± 12	0.5	937 ± 12	12.3
7	745.4	751 ± 12	0.7	670 ± 11	11.3

The analytical predictions agree perfectly with the simulations for the exponentially distributed service times in all cases. For the lognormal distribution of service times, the deviations between the prediction and simulation are about 50% in the worst situations. However, these cases are of little practical interest. Given a fixed number of servers, low values of T_C imply very low MFPT values that are incompatible with EMS response times, while situations with very high values of MFPT are often related with idle infrastructure, which are also avoided in practice.

In order to study the differences in the values of FPT under different service time distributions, in Figure 4, we superimpose two histograms of simulated values for a system with seven servers. Both cases correspond to an initial condition with four occupied servers and the same exponential inter-arrival time distribution, but we use two different service time distributions with the same mean value. The FPT distribution with exponential service times is broader than the lognormal counterpart. Therefore, the MFPT under these conditions is shorter for the lognormal service time distribution (also see entry for initial state equal to 4 in Table 1).

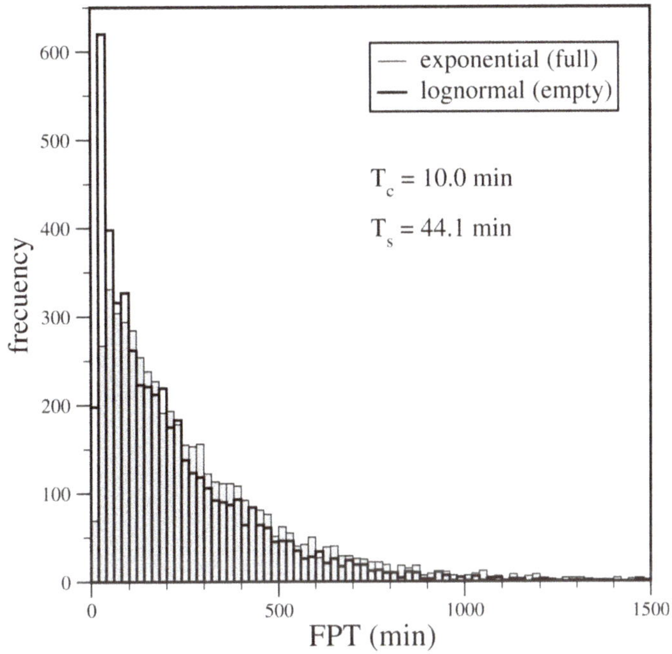

Figure 4. Histograms of FPT to a critical condition based on 5000 simulations from an initial state with 4 of 7 servers occupied. We compare two service time distributions with the same value of T_S: exponential (parameter $= 1/T_S$) and lognormal (`meanlog` = 3.6867 and `sdlog` = 0.4465).

Now, we investigate what happens with the residence probabilities in the long time limit. The analytical result given by the Erlang formula has been proven for systems without queue or $M/G/L$ blocking systems [19,20]. That is, if every server is busy when a call arrives, the call is lost; however, it is instructive to analyse the situation of a system with queue. The probability of residence in the queue, $P(n > L)$, is equivalent to the probability absorbed in the site $L + 1$ of the Markov chain. From Equation (A11), we can see that the probabilities in Equation (4) are the renormalized residence probabilities of a system with a queue in the interval $[0, L]$. To find out if this is also the case for lognormal service time distributions, we run our simulator 10^7 min in order to visit each state several times. In the first column of Table 2, we show the analytic result of Equations (A11)–(A13) and compare this prediction with the simulated values using four different service time distributions: an exponential and three different lognormal service time distributions—all of these with the same value $T_S = 44.1$ min ($T_C = 10$ min in all cases). The value `sdlog` = 0.25 corresponds to the almost symmetric case of the lognormal distribution, and `sdlog` = 1 corresponds to the more asymmetric situation (see Figure 6b). The exponential distribution is the case completely asymmetric, where the distribution does not have a maximum value. For comparison, we also introduce the middle value `sdlog`= 0.625.

Table 2. Comparison among analytic and simulated long-run probabilities of residence in each state of a system with seven servers, where q denotes the state with queued calls. All simulations correspond with a simulated running time of 1×10^7 min. In all cases, $T_C = 10$ min and $T_S = 44.1$ min, but the parameters in of the lognormal distributions are given by the pairs (`meanlog`, `sdlog`): (a) $(3.75521, 0.25)$, (b) $(3.59115, 0.625)$ and (c) $(3.28646, 1.0)$. See Figure 6b.

State	Analytic	Exponential	Lognorm (a)	lognorm (b)	lognorm (c)
0	0.012	0.011	0.011	0.011	0.012
1	0.051	0.050	0.049	0.050	0.052
2	0.112	0.112	0.109	0.111	0.113
3	0.165	0.165	0.164	0.164	0.165
4	0.182	0.182	0.184	0.182	0.181
5	0.160	0.160	0.167	0.163	0.158
6	0.118	0.118	0.127	0.122	0.115
7	0.074	0.074	0.085	0.081	0.075
q	0.126	0.127	0.104	0.114	0.130

The difference among analytical and simulated values are less than 5% in all states with the exception of the queue. The analysed case with queue is a worse situation than the finite chain with both reflecting ends, used in the derivation of Equation (4), because of the probability of residence in the queue may be greater than the saturated state. Therefore, we find it is valid to use the analytic result of Equation (4) for taking the average in Equation (3) and also in the simulation of MFPT with lognormal service time distributions in a system with queue.

In Figure 5, we show the plots of MFPT averaged over the initial conditions, $< T >$, as a function of the mean inter-arrival time T_C using the probabilities given by Equation (4). We have sketched a characteristic situation for the service time, and we considered several numbers of servers $L = 5, \ldots, 9$. The curves clearly show the non-linear behaviour of $< T >$ and allow us to evaluate the quality of the fit for the simulated situations achieved with our analytical expression.

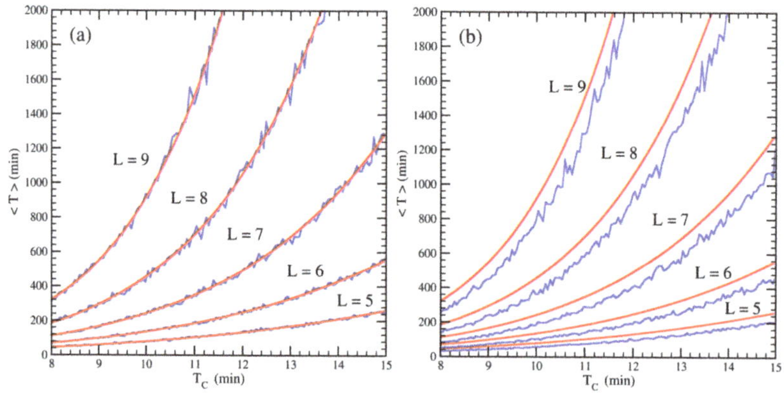

Figure 5. Analytic (red) and simulated (blue) average MFPT with (**a**) exponential and (**b**) lognormal distributed service times (`meanlog` $= 3.6867$ and `sdlog` $= 0.4465$). In both cases, $T_S = 44.1$ min. Simulated values for each value of T_C correspond to an average based on 1000 executions.

Again, in the left panel, we find excellent agreement among the analytical predictions and the simulations for exponentially distributed service times. In contrast, in the right panel, for lognormal distributed service times, we see that the analytical curves always run over the corresponding simulation. This fact has just been seen in Figure 4, where the FPT distribution has a longer tail in the case of service times distributed exponentially with respect to lognormal service times. Thus, the mean value of the FPT distribution (MFPT) is higher for exponential service times. Therefore, the analytic model underestimates the number of necessary servers under a given stress condition. Drawing a horizontal line in Figure 5b, when we move in the direction of increasing demand (that is, lowering T_C), we first cross the blue (simulated) curves in all the considered cases. This is an important fact to keep in mind if we want to use the analytical model to find the best number of servers in a particular operational scenario. However, the discrepancies observed between the simulations and the closed form expression for the average MFPT are not significant enough to cause a considerable effect in the estimation of the optimal number of servers, given the separation among the curves for different values of L. Thus, for example, from Figure 5b, we can see that to obtain an average MFPT of 6 hs, given $T_C = 14$ min, six servers are needed, whereas for $T_C = 10$, eight servers are needed under the same service conditions.

We also analyse the effect of asymmetry in the service time distribution on average MFPT. For this purpose, we simulate the average MFPT for a whole family of service time distribution with fixed T_S but changing the parameter `sdlog` in the interval $[0.25, 1]$, where the minimum value corresponds to the almost symmetric case and the maximum corresponds to the more asymmetric situation. The average MFPT as a function of the `sdlog` parameter is shown in Figure 6.

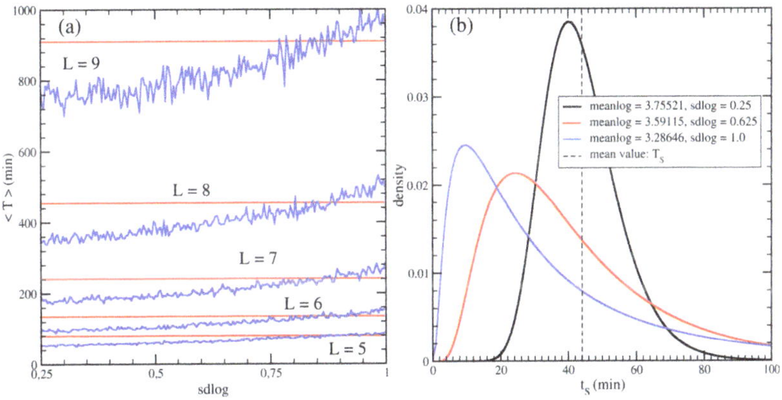

Figure 6. (**a**) Analytic (red) and simulated (blue) average MFPT with lognormal distributed service times with fixed $T_S = 44.1$ min (`sdlog` $\in [0.25, 1]$ and `meanlog` $= \ln(T_S) -$ `sdlog`$^2/2$) and $T_C = 10$ min. Simulated points for each value of `sdlog` are averages based on 1000 runs. (**b**) Sketches of lognormal densities corresponding to the extreme values (black and blue) and to the middle value (red) of `sdlog` in panel (**a**).

In all cases, the analytical prediction gives an acceptable approximation for the simulated data.

Finally, in the left panel of Figure 7, we plot the probability that one or more servers are available at the instant of a emergency call, $P(n < L)$ [27], as a function of the time between calls for a fixed mean service time. We are using Equation (4) for the calculation of this probability. In the right panel, we plot the average MFPT as a function of this availability probability for the same values of T_C given in the left panel. We can observe a very strong sensitivity of the average MFPT to the availability probability for high values of system availability. Thus, only controlling the availability of the system is not enough to assure a long enough time to the critical condition. A dispatcher observing a high availability probability value might conclude that the system is running unreasonably idle; however, the time to critical condition may be shorter than the expectation. More interesting, however, is that the function $< T >$ vs. $P(n < L)$ is practically independent of the number of servers.

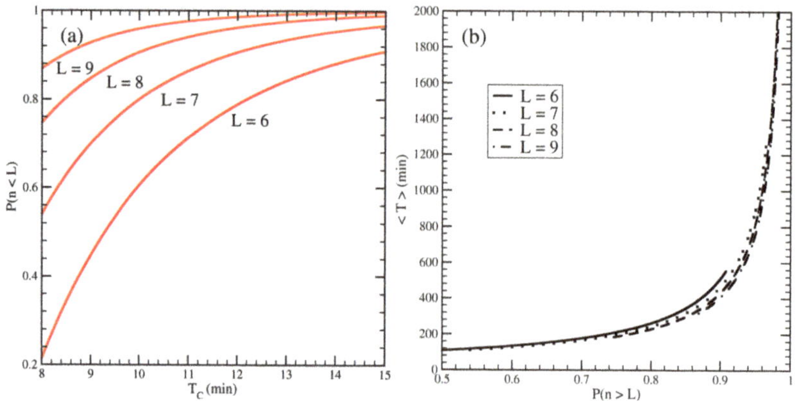

Figure 7. (**a**) The probability of one or more servers are free vs T_C. (**b**) Average MFPT as a function of the probability of servers' availability for $T_C \in [8, 15]$ min. In both panels, $T_S = 44.1$ min.

4. Concluding Remarks

MFPT is a useful KPI that allows estimating the running operative lapse of a system, under a given stress condition, before a service disruption. In this work, we have presented a closed-form expression to calculate the MFPT for a system of servers in parallel, and we also provide a simulation framework for the MFPT. Our formula, based on a birth–death process, only uses the average time between demands and the average service time. Our results make it possible to predict the MFPT given the stress of the system at a particular moment or to analyse the servers shortage time under generic operating conditions by averaging over the initial states of the system. The main limitation of our results, as is often with analytical exact results in queueing theory, is the assumption of an exponential distribution for service times. The impact of this limiting assumption is confronted with results of simulations using more realistic service distributions. Our results indicate that our analytical formula is an acceptable approximation under practical situations. Interesting potential future work may be to consider the implementation of accurate approximations for the $M/G/L$ problem [29] to the MFPT calculation. In addition, our simulation scheme allows us to evaluate the MFPT in any $GI/G/L/FCFS$ server configuration.

Funding: This work was partially supported by the *Sistema de Urgencias del Rosafe SA*, Córdoba, Argentina and by grant 33620180101258CB from Secretaría de Ciencia y Tecnología de la Universidad Nacional de Córdoba, Argentina.

Conflicts of Interest: The author declare no conflict of interest. The funders had no role in the design of the study; in the collection, analyses, or interpretation of data; in the writing of the manuscript, or in the decision to publish the results.

Abbreviations

The following abbreviations are used in this manuscript:

DES	Discrete-Event Simulator
EMS	Emergency Medical Services
FCFS	First-Come First-Served
FPT	First-Passage Time
KPI	Key Performance Indicator
MFPT	Mean First-Passage Time
RNG	pseudo-Random Number Generator
URG	Sistema de Urgencias del Rosafe

Appendix A. MFPT

For a birth–death process with asymmetric and site-dependent transition probabilities ($w_k^+ \neq w_k^-$), the analytical expression for the MFPT with a reflecting boundary condition at origin and an absorbing one at $L+1$ is given by

$$\begin{aligned} T(0) &= \sum_{k=0}^{L} \frac{1}{w_k^+} + \sum_{k=0}^{L-1} \frac{1}{w_k^+} \sum_{i=k+1}^{L} \prod_{j=k+1}^{i} \frac{w_j^-}{w_j^+}, \\ T(1) &= T(0) - \frac{1}{w_0^+}, \\ T(n) &= T(0) - \sum_{k=0}^{n-1} \frac{1}{w_k^+} - \sum_{k=0}^{n-2} \frac{1}{w_k^+} \sum_{i=k+1}^{n-1} \prod_{j=k+1}^{i} \frac{w_j^-}{w_j^+} \quad \text{for } 2 \leq n \leq L. \end{aligned} \tag{A1}$$

For details and derivation of Equation (A1), see Section 6 in Ref. [18]. In our model with L servers, using Equation (1) and the parameter γ defined in the text, we can recast the products in Equation (A1) as

$$\prod_{j=k+1}^{i} \frac{w_j^-}{w_j^+} = \gamma^{i-k} \prod_{j=k+1}^{i} j = \gamma^{i-k} \frac{i!}{k!}. \tag{A2}$$

Thus, we can also recast the sums in Equation (A1) as

$$\sum_{i=k+1}^{n-1} \prod_{j=k+1}^{i} \frac{w_j^-}{w_j^+} = \frac{\gamma^{-k}}{k!} \sum_{i=k+1}^{n-1} i!\, \gamma^i, \tag{A3}$$

and

$$\sum_{k=0}^{n-2} \frac{1}{w_k^+} \sum_{i=k+1}^{n-1} \prod_{j=k+1}^{i} \frac{w_j^-}{w_j^+} = \frac{1}{\lambda} \sum_{k=0}^{n-2} \frac{\gamma^{-k}}{k!} \sum_{i=k+1}^{n-1} i!\, \gamma^i. \tag{A4}$$

Replacing the last expression in Equation (A1), we obtain in a direct manner Equation (2) in Section 2.1.

Appendix B. Steady State

Following Ref. [8], we can construct the steady state for the problem of a Markov chain with a reflecting boundary condition at the origin. In the long-run, the time-independent probability of residence at state n, π_n, must satisfy

$$\begin{aligned} \omega_1^- \pi_1 - \omega_0^+ \pi_0 &= 0, \\ \omega_{n+1}^- \pi_{n+1} + \omega_{n-1}^+ \pi_{n-1} - (\omega_n^+ + \omega_n^-) \pi_n &= 0, \quad \text{for } n \geq 1. \end{aligned} \tag{A5}$$

Thus, we can prove by induction that

$$\pi_n = \frac{\omega_{n-1}^+ \cdots \omega_0^+}{\omega_n^- \cdots \omega_1^-} \pi_0 = \prod_{j=1}^{n} \frac{\omega_{j-1}^+}{\omega_j^-} \pi_0, \text{ for } n \geq 1. \tag{A6}$$

From the normalization condition, $\sum_{n=0}^{\infty} \pi_n = 1$, results $\pi_0 = 1/S_\pi$, where

$$S_\pi = 1 + \sum_{n=1}^{\infty} \prod_{j=1}^{n} \frac{\omega_{j-1}^+}{\omega_j^-}. \tag{A7}$$

The existence of the steady state is then determined by the convergence of the series in Equation (A7).

For the model given by Equation (1), the products in Equation (A6) and (A7) can be written as,

$$\prod_{j=1}^{n} \frac{\omega_{j-1}^{+}}{\omega_{j}^{-}} = \begin{cases} \left(\frac{\lambda}{\mu}\right)^{n} \prod_{j=1}^{n} \frac{1}{j} = \frac{\gamma^{-n}}{n!}, & \text{for } 0 \leq n \leq L, \\ \left(\frac{\lambda}{\mu}\right)^{n} \prod_{j=1}^{L} \frac{1}{j} \prod_{j=L+1}^{n} \frac{1}{L} = \frac{\gamma^{-n}}{L!} \frac{1}{L^{n-L}}, & \text{for } n \geq L. \end{cases} \tag{A8}$$

Substituting Equation (A8) into Equation (A7), we obtain

$$S_{\pi} = \sum_{n=0}^{L} \frac{\gamma^{-n}}{n!} + \frac{L^{L}}{L!} \sum_{n=L+1}^{\infty} (L\,\gamma)^{-n}. \tag{A9}$$

The convergence of the series in the last expression only occurs if $L\,\gamma > 1$. In this case,

$$\sum_{n=L+1}^{\infty} (L\,\gamma)^{-n} = \frac{(L\,\gamma)^{-L}}{L\,\gamma - 1}. \tag{A10}$$

Substituting Equations (A8)–(A10) into Equation (A6), we obtain

$$\pi_{n} = \frac{1}{S_{\pi}} \begin{cases} \frac{\gamma^{-n}}{n!}, & \text{for } 0 \leq n \leq L, \\ \frac{L^{L}}{L!} (L\,\gamma)^{-n}, & \text{for } n \geq L, \end{cases} \tag{A11}$$

where

$$S_{\pi} = \sum_{n=0}^{L} \frac{\gamma^{-n}}{n!} + \frac{\gamma^{-L}}{L!\,(L\gamma - 1)}. \tag{A12}$$

In this manner, the long-term probability of having the system *with calls in the waiting queue* results in

$$\pi_{q} = \sum_{n=L+1}^{\infty} \pi_{n} = \frac{\gamma^{-L}}{S_{\pi}\, L!\,(L\gamma - 1)}. \tag{A13}$$

The last expression is also is known as Erlang C-formula.

We now consider the case of a *finite* Markov chain with reflecting boundaries at the origin and at site L. The time-independent probabilities of residence, $P(n)$, in the each state n satisfy Equation (A5) but are supplemented with the additional reflecting condition at L,

$$\begin{aligned} \omega_{1}^{-} P(1) - \omega_{0}^{+} P(0) &= 0, \\ \omega_{n+1}^{-} P(n+1) + \omega_{n-1}^{+} P(n-1) - (\omega_{n}^{+} + \omega_{n}^{-}) P(n) &= 0, \quad \text{for } n \geq 1, \\ \omega_{L-1}^{+} P(L-1) - \omega_{L}^{-} P(L) &= 0. \end{aligned} \tag{A14}$$

Following the above procedure, we find the first line of Equation (A8) again, which directly leads to Equation (4) in Section 2.1, where S is the normalization on the interval $[0, L]$.

References

1. Dos Santos Cabral, E.L.; Castro, W.R.S.; de Medeiros Florentino, D.R.; de Araújo Viana, D.; da Costa Junior, J.F.; de Souza, R.P.; Rêgo, A.C.M.; Araújo-Filho, I.; Medeiros, A.C. Response time in the emergency services. Systematic review. *Acta Cirúrgica Brasileira* **2018**, *33*, 1110–1121. [CrossRef] [PubMed]
2. Eisenberg, M.; Bergner, L.; Hallstrom, A. Cardiac Resuscitation in the Community: Importance of Rapid Provision and Implications for Program Planning. *J. Am. Med. Assoc.* **1979**, *241*, 1905–1907. [CrossRef]
3. Feero, S.; Hedges, J.R.; Simmons, E.; Irwin, L. Does out-of-hospital EMS time affect trauma survival? *Am. J. Emerg. Med.* **1995**, *13*, 133–135. [CrossRef]

4. Pons, P.T.; Markovchick, V.J. Eight minutes or less: Does the ambulance response time guideline impact trauma patient outcome? *J. Emerg. Med.* **2002**, *23*, 43–48. [CrossRef]
5. McCoy, C.E.; Menchine, M.; Sampson, S.; Anderson, C.; Kahn, C. Emergency Medical Services Out-of-Hospital Scene and Transport Times and Their Association with Mortality in Trauma Patients Presenting to an Urban Level I Trauma Center. *Ann. Emerg. Med.* **2013**, *61*, 167–174. [CrossRef] [PubMed]
6. Al-Shaqsi, S.Z.K. Response time as a sole performance indicator in EMS: Pitfalls and solutions. *Open Access Emergercy Med.* **2010**, *2*, 1–6. [CrossRef]
7. Singer, M.; Donoso, P. Assessing an ambulance service with queuing theory. *Comput. Oper. Res.* **2008**, *35*, 2549–2560. [CrossRef]
8. Winston, W.L. *Operations Research: Applications and Algorithms*, 4th ed.; Duxbury Press: Belmont, CA, USA, 2003; Chapter 20.
9. Larson, R.C. A hypercube queuing model for facility location and redistricting in urban emergency services. *Comput. Oper. Res.* **1974**, *1*, 67–95. [CrossRef]
10. Takeda, R.A.; Widmer, J.A.; Morabito, R. Analysis of ambulance decentralization in an urban emergency medical service using the hypercube queueing model. *Comput. Oper. Res.* **2007**, *34*, 727–741. [CrossRef]
11. Uriarte, A.G.; Zúñiga, E.R.; Moris, M.U.; Ng, A.H.C. How can decision makers be supported in the improvement of an emergency department? A simulation, optimization and data mining approach. *Oper. Res. Health Care* **2017**, *15*, 102–122. [CrossRef]
12. Cildoz, M.; Ibarra, A.; Mallor, F. Accumulating priority queues versus pure priority queues for managing patients in emergency departments. *Oper. Res. Health Care* **2019**, *23*, 100224. [CrossRef]
13. Lakshmi, C.; Sivakumar, A.I. Application of queueing theory in health care: A literature review. *Oper. Res. Health Care* **2013**, *2*, 25–39. [CrossRef]
14. Brown, L.; Gans, N.; Mandelbaum, A.; Sakov, A.; Shen, H.; Zeltyn, S.; Zhao, L. Statistical Analysis of a Telephone Call Center. *J. Am. Stat. Assoc.* **2005**, *100*, 36–50. [CrossRef]
15. Ingolfsson, A.; Budge, S.; Erkut, E. Optimal ambulance location with random delays and travel times. *Health Care Manag. Sci.* **2008**, *11*, 262–274. [CrossRef]
16. Van Buuren, M.; Kommer, G.J.; van der Mei, R.; Bhulai, S. A simulation model for emergency medical services call centers. In Proceedings of the 2015 Winter Simulation Conference (WSC), Huntington Beach, CA, USA, 6–9 December 2015; pp. 844–855. [CrossRef]
17. Cox, D.R.; Smith, W.L. *Queues*; Methuen: London, UK, 1961.
18. Pury, P.A.; Cáceres, M.O. Mean first-passage and residence times of random walks on asymmetric disordered chains. *J. Phys. A Math. Gen.* **2003**, *36*, 2695–2706. [CrossRef]
19. Takacs, L. On Erlang's Formula. *Ann. Math. Stat.* **1969**, *40*, 71–78. [CrossRef]
20. Fakinos, D. The $M/G/k$ Blocking System with Heterogeneous Servers. *J. Oper. Res. Soc.* **1980**, *31*, 919–927. [CrossRef]
21. Banks, J.; Carson II, J.S.; Nelson, B.L.; Nicol, D.M. *Discrete-Event System Simulation*, 5th ed.; Pearson Education Limited: Harlow, UK, 2014.
22. Krivulin, N.K. A recursive equations based representation for the $G/G/m$ queue. *Appl. Math. Lett.* **1994**, *7*, 73–77. [CrossRef]
23. Pury, P.A. simserveRs: R Simulator for Parallel Servers with FCFS Queueing Discipline. Available online: https://bitbucket.org/realhubot/simservers/ (accessed on 22 August 2021).
24. Ucar, I.; Smeets, B.; Azcorra, A. simmer: Discrete-Event Simulation for R. *J. Stat. Softw.* **2019**, *90*, 1–30. [CrossRef]
25. Sistema de Urgencias del Rosafe SA. Available online: https://www.urg.com.ar/ (accessed on 22 August 2021).
26. Sze, D.Y. OR Practice—A Queueing Model for Telephone Operator Staffing. *Oper. Res.* **1984**, *32*, 229–249. [CrossRef]
27. Hall, W.K. Management science approaches to the determination of urban ambulance requirements. *Socio-Econ. Plan. Sci.* **1971**, *5*, 491–499. [CrossRef]
28. Wu, J.; Mehta, N.B.; Zhang, J. Flexible lognormal sum approximation method. In Proceedings of the IEEE Global Telecommunications Conference (GLOBECOM '05), St. Louis, MO, USA, 28 November–2 December 2005; Volume 6, pp. 3413–3417. [CrossRef]
29. Hokstad, P. Approximations for the $M/G/m$ Queue. *Oper. Res.* **1978**, *26*, 510–523. [CrossRef]

Mathematical and Computational Applications

MDPI

Article

Analysis and Detection of Erosion in Wind Turbine Blades

Josué Enríquez Zárate [1,2], María de los Ángeles Gómez López [3], Javier Alberto Carmona Troyo [4] and Leonardo Trujillo [4,*]

1 AP Engineering, Oaxaca 70110, Mexico; jenriquezza@ap-engineering.com.mx
2 Tecnológico Nacional de México/IT del Valle de Etla, Oaxaca 68230, Mexico
3 Tecnológico Nacional de México/IT de Tuxtla Gutiérrez, Tuxtla Gutiérrez 29050, Mexico; 16119325angy@gmail.com
4 Tecnológico Nacional de México/IT de Tijuana, Tijuana 22414, Mexico; javier.carmona@itlp.edu.mx
* Correspondence: leonardo.trujillo@tectijuana.edu.mx

Abstract: This paper studies erosion at the tip of wind turbine blades by considering aerodynamic analysis, modal analysis and predictive machine learning modeling. Erosion can be caused by several factors and can affect different parts of the blade, reducing its dynamic performance and useful life. The ability to detect and quantify erosion on a blade is an important predictive maintenance task for wind turbines that can have broad repercussions in terms of avoiding serious damage, improving power efficiency and reducing downtimes. This study considers both sides of the leading edge of the blade (top and bottom), evaluating the mechanical imbalance caused by the material loss that induces variations of the power coefficient resulting in a loss in efficiency. The QBlade software is used in our analysis and load calculations are preformed by using blade element momentum theory. Numerical results show the performance of a blade based on the relationship between mechanical damage and aerodynamic behavior, which are then validated on a physical model. Moreover, two machine learning (ML) problems are posed to automatically detect the location of erosion (top of the edge, bottom or both) and to determine erosion levels (from 8% to 18%) present in the blade. The first problem is solved using classification models, while the second is solved using ML regression, achieving accurate results. ML pipelines are automatically designed by using an AutoML system with little human intervention, achieving highly accurate results. This work makes several contributions by developing ML models to both detect the presence and location of erosion on a blade, estimating its level and applying AutoML for the first time in this domain.

Keywords: wind energy; wind turbine blades; erosion; modal analysis; aerodynamic analysis; AutoML

Citation: Enríquez Zárate, J.; Gómez López, M.d.l.A.; Carmona Troyo, J.A.; Trujillo, L. Analysis and Detection of Erosion in Wind Turbine Blades. *Math. Comput. Appl.* **2022**, 27, 5. https://doi.org/10.3390/mca27010005

Academic Editor: Guillermo Valencia-Palomo

Received: 30 December 2021
Accepted: 11 January 2022
Published: 13 January 2022

1. Introduction

Wind energy offers an important supply of electricity without pollution problems presented by conventional forms of energy. There is a global interest for the development and use of alternative energy sources, including geothermal, photovoltaic, hydroelectric, tidal wave, biomass and others [1]. Unlike other technologies, wind farms have a very low impact on their environment, which has resulted in increased use worldwide, with some countries obtaining as much as 20 percent of their energy from the wind [2].

Wind turbines can be categorized, for example, based on their rotation axis, which can be vertical or horizontal [3]. Horizontal axis turbines are the most common and can be classified according to the rotation of the rotor with respect to the tower. These machines are composed of a foundation, a tower, a rotor, nacelle with power train and the blades. The blades are one of the most important components, if not the most, since they are in charge of collecting the energy from the wind, converting the linear movement of the wind into a rotary movement of the rotor. This energy is transmitted to the hub, from the hub it proceeds to a mechanical transmission system and from there it proceeds to the generator that transforms it into electrical energy.

Blades can suffer different types of failures due to a variety of phenomena [4], such as the following: bending, twisting, cracks and erosion. Detecting these types of faults, particularly at early stages, is important for avoiding catastrophic failures, reducing down times and taking corrective actions in a timely manner [5]. In particular, automatically detecting erosion at the leading edge of the blade tip presents a challenge because it is not trivial to properly measure erosion without direct access to the blade [4]. Typically, fault detection in wind turbine blades has been carried out by visual means (https://energy.sandia.gov/programs/renewable-energy/wind-power/, accessed on 29 December 2021), but it is also possible to use different sensor schemes implemented via a SCADA system [5,6]. Afterward, these data can be analyzed by computational models, such as those derived by means of machine learning (ML), which have been found to be of great utility in the detection of damages in a variety of complex scenarios [7–9]. One approach of note, for example, is the use of sound to detect such damage [10]. Moreover, to implement ML methods, sufficient data are required to model blades with and without erosion; however, extracting these data from the field or in controlled scenarios can be a complex task to perform [11]; thus, one possible alternative is to exploit simulation software such as QBlade [12].

The goal of this work is two fold. First, it presents a detailed analysis of the effects that erosion has on wind turbine blades, considering modal and numerical analysis, with respect to the physical stress caused by erosion on the blades and the power-generating capacity of a wind turbine. Second, this work presents a methodology to construct ML models that can detect the presence and location of erosion in a blade and measure the amount of erosion as well. The paper makes two important contributions. First, we show for the first time that AutoML can be successfully applied in this problem domain, which has not been considered before for this problem. Second, we show that it is possible to detect the presence of erosion on the blade, determine its location and predict the amount of damage caused by erosion.

The remainder of this paper is organized as follows. Section 2 presents wind turbine blades and the aerodynamic, modal and numerical analysis of the blades studied in this work. Section 3 deals with the detection of erosion with ML methods, including an overview of related works and our experimental approach and results. Finally, conclusions and future work are presented in Section 4.

2. Wind Turbine Blades

During regular operation of a wind turbine blade, air mixed with sand and water droplets will cause severe erosion, which can produce a loss of fatigue resistance of the blade's surface and cause heavy damage to the blade material, resulting in a loss in electrical potential [13]. Before analyzing the effect of erosion on a wind turbine blade, we first consider the aerodynamic and structural response of the blade caused by erosion.

2.1. Aerodynamic Analysis

Generally, a wind turbine blade is divided in three sections: tip, mid span and root (see Figure 1). These sections are exposed to potential structural damage during normal operation, such as cracks, wrinkles, delamination, debonding and erosion. In particular, this paper is focused on the erosion problem around to the leading edge of the blades. This section presents an analysis of erosion on the blades, identifying the negative effects that erosion has on the power-generating capacity based on power coefficient C_p as well as the power generated by the wind turbine P_G.

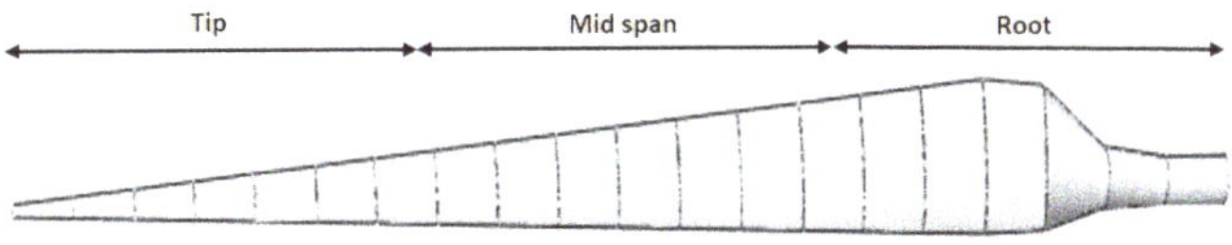

Figure 1. Sections of a wind turbine blade.

In order to analyze erosion in a wind turbine blade it is necessary to identify the development of several dynamic parameters, such as the drag coefficient C_D, lift coefficient C_L and frequency ω_n. For aerodynamic analysis, this study uses the Theory of Momentum (TM), Blade Element Momentum (BEM), Geometry of Blade (GB) and the evaluation of the performance of the wind rotor, considering a case study of a 1 Kw Wind Turbine Generator (WTG). In order to obtain a design as close to the optimum as possible, several aerodynamic surfaces were evaluated to identify suitable conditions of aerodynamic lift based on the pressure gradient of aerodynamics airfoils. The design parameters of the blade and the technical specification of the WTG are provided in Table 1.

Table 1. Design parameters of the 1 Kw WTG.

Parameter	Value	Variables	Units
Rated power	1000	P_{nom}	W
Number of blades	3	B	[-]
Rated speed	8	u_{nom}	m/s
Speed of rotation	295.63	Ω	rpm
Gearbox efficiency	0.95	$\eta_{gearbox}$	[-]
Generator efficiency	0.95	η_{gen}	[-]
Wind density	1.185	ρ	kg/m^3
Dynamic Viscosity	1.78×10^{-5}	μ	Pa×s

For this design, we consider a power coefficient $C_p = 0.40$ based on the Betz limit [14], which allows calculating the length (theoretical) of the blade, which results in $R = 1.628$ [m], and the relationship between the speed and the blade tip in nominal conditions resulting in a value of $\lambda_{nom} \approx 8$. In order to obtain the Reynolds number of the blade, we separate the blade into three sections. We consider an average airfoil in every section of the blade, resulting in the following Reynolds' numbers. For the root section, the average chord was 0.14 m and Reynolds was 147,000; the average chord of the body is 0.10 m and Reynolds was 198,000, while the average chord of the tip is 0.023 m and Reynolds was 78,000. Finally, BEM parameters used to simulate the blade design were obtained using the method of the *Viterna* extrapolation [15], which include the following: aspect ratio of 10, lift coefficient adjustment of 0.7 and the number of elements is seven.

Subsequently, from the previous parameters, it is necessary to evaluate several aerodynamics airfoils from databases such as NACA, NREL and WORTMANN. In this work, we evaluated 49 airfoils using the open source *QBlade* software with the goal of selecting the best and most stable design. The chosen airfoils were FX 63-137 and E216 based on their lift C_L and drag C_D coefficients, both of which influence power coefficient C_p of the turbine.

The designed blade, considering the position of the selected airfoils, chords, spin and Reynolds numbers calculated using QBlade, is show in Figure 2a. The evaluation of aerodynamic performance of the blade using QBlade was carried out using the following parameters: 1 kW of power, an input speed of 3 m/s, cutting speed of 15 m/s, rotor speed of 296 rpm and a density of 1.185 kg/m^3. In particular, the work considers the wind conditions of *La Ventosa, Oaxaca, Mexico* in the Isthmus of Tehuantepec for the simulations, because this geographical zone is located between the Pacific Ocean and the Gulf of Mexico, which is a perfect location for a wind farm [16]. The wind farms are placed near the Pacific Ocean, within a range of 20 and 60 km. In this zone, wind conditions are considered as good (class 4) and excellent (class 7). With a wind speed of 8 m/s, the power generated was 1.38 kW, and the power coefficient was $C_p = 0.4197$ with a simulation of 1.4 s.

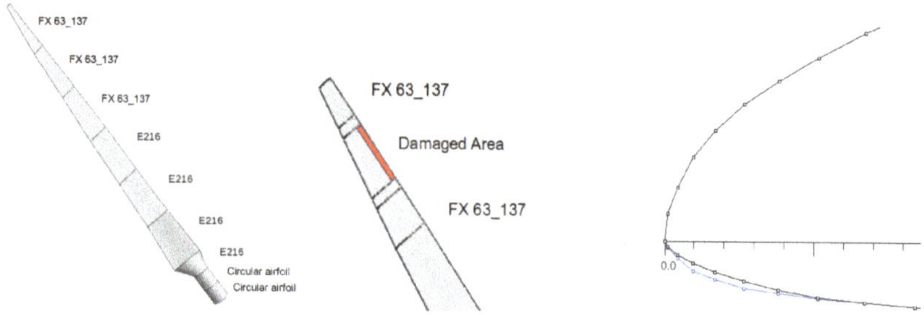

(a) (b) (c)

Figure 2. (**a**) Wind turbine blade design. (**b**) Area of erosion damage. (**c**) Blue line represents the clean Airfoil FX 63_137, while the dark line is the eroded blade considering case BI_1.

2.2. Modal Analysis

Structural analysis characterizes the behavior of the blade in the presence of erosion. In particular, this section presents vibration analysis using simulated erosion in the leading edge of the blade tip to compare a clean model with an eroded blade. The material, or mass, loss is expected to cause a variation in the vibration response of the blade, which directly affects the wind rotor and other subsystems of the turbine, such as the gearbox, main shaft and the generator due to the coupling between them. Furthermore, structural analysis is carried out to analyze the response frequency of the blade through modal analysis using a finite element software. This is performed by considering both a clean blade and an eroded version. For structural analysis, the mechanical and physical proprieties of the blade materials are described in Table 2b. Furthermore, dynamic analysis is also necessary in order to obtain natural frequencies and modal shapes of the blade, which was carried out using the following mesh parameters: quadratic order element, uniform size function, medium center of relevance, fast transition, low smoothness, 732 elements and 1768 nodes.

Simulated erosion was carried out with QBlade considering the tip of the blade, which should effect drag and lift coefficients, as well as the power coefficient and the electric power generated by the turbine. Erosion analysis consist of modifying the geometry of the blade to represent damage on the surface of the tip. The damage due to erosion reduces the original mass of the blade, which is represented in terms of a percentage in profile reduction, as shown in Figure 2b. For this analysis, three different depths of erosion were considered for the bottom edge of the profile, denoted as percentages by $BI_1 = 10\%$, $BI_2 = 18\%$ and $BI_3 = 25\%$. In each case, these percentages represents a reduction in profile coordinates relative to the clean blade. This is shown in Figure 2c, where the black line represents the clean profile, and the blue line represents the eroded profile for the BI_1 case.

Modal analysis is performed to evaluate the effects of different levels of erosion on the mechanical structure of the blade through their frequency response. The complete results are summarized in Table 3, where four modal frequencies are evaluated for each case of erosion depth compared with the numerical response of the clean blade.

Table 2. (a) Wind turbine blade design. (b) Blades properties.

(a)

Position	Airfoil	Chord
0	Circular Airfoil	0.07
0.077	Circular Airfoil	0.07
0.077	Circular Airfoil	0.07
0.228	E216	0.14
0.46	E216	0.12
0.693	E216	0.11
0.926	E216	0.1
1.158	FX 63-137	0.08
1.391	FX 63-137	0.06
1.6323	FX 63-137	0.023

(b)

Property	Value	Units
Density	2000	(Kg/m^3)
Orthotropic elasticity		
Young's Module in x	50,000	MPa
Young's Module in y	8000	MPa
Young's Module in z	8000	MPa
Poisson's ratio xy	0.3	-
Poisson's ratio yz	0.4	-
Poisson's ratio xz	0.3	-
Stiffness module xy	5000	MPa
Stiffness module yz	3846.2	MPa
Stiffness module xz	5000	MPa

Table 3. Comparative analysis of modal frequencies of the eroded and the clean blades.

Mode	BI_1	BI_2	BI_3	Clean
		(Hz)		
1	3.8954	3.2521	3.9509	3.9887
2	15.631	12.874	14.903	14.317
3	23.60	22.268	23.411	22.469
4	42.867	35.699	38.514	38.207

2.3. Numerical Analysis

The drag and lift coefficients, respectively, C_D and C_L, are compared for the three levels of erosion and the clean blade in Figure 3a. The change in the original design of the blade causes a variation in the power coefficient C_P and the electric power generated P_G.

In order to complete analysis, the lift coefficient C_L is compared with respect to the leading edge of the blade, which is denoted by α using QBlade, as shown in Figure 3b. Parameter α represents the angle of inclination of the blade with respect to the direction of the wind force. This angle can change in a range between −20 to 20 degrees. In this study, the simulation tests were carried out considering α with values of 10 and 20 degrees. For the three erosion cases considered here, BI_1, BI_2 and BI_3, α = 10 degrees. Figure 3b shows the behavior of C_L relative to our three erosion levels. The percentage of erosion on the leading edge of the blade impacts its aerodynamic response, particularly the lift and drag coefficients (C_L) and (C_D). Figure 3a,b show both coefficients reducing their performance when erosion is increased, this considering a particular angle of attack of 10 degrees. The power coefficient (C_p) of the wind rotor is affected by the presence of erosion as well, minimizing the electricity production potential (P_G), shown in Figure 3c,d.

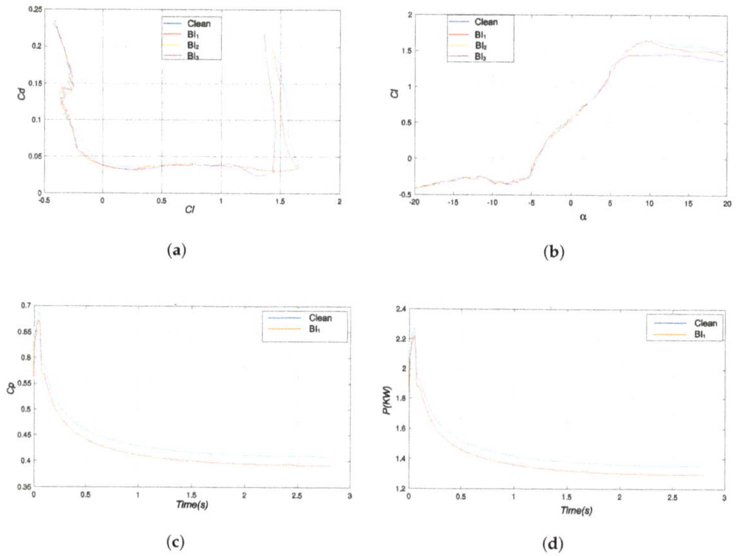

Figure 3. (**a**) Relationship between the drag coefficient (C_D) and the lift coefficient (C_L). (**b**) Lift coefficient C_L Vs angle of attack α. (**c**) Power coefficient C_p for BI_1 and the clean wind turbine blade. (**d**) Power output P_G for BI_1 and the clean blade.

Table 4 presents comparative results of the drag coefficient C_d and the lift coefficient C_l for the three levels of erosion, with respect to the clean blade. Both coefficients are reduced when the depth of the damage is increased on the surface of the blade, affecting aerodynamic performance.

Table 4. Comparison of the leading edge in three different erosion's depth for the drag coefficient C_d and the lift coefficient C_l. Last two columns show the percentage difference relative to the clean blade (C_d(%) and C_l(%)).

Blade	α	C_d	C_l	C_d(%)	C_l(%)
Clean	10	0.036	1.66	-	-
BI_1	10	0.036	1.64	0%	1.3%
BI_2	10	0.03	1.46	16.7%	13%
BI_3	10	0.028	1.44	22.3%	14%

Finally, the behavior of the power coefficient C_P considering BI_1 is shown in Figure 3c, where the blue line represents the clean blade and the red line represents the eroded blade. It is possible to observe the variation of C_P, which decreases to $C_P = 0.39$, while $C_P = 0.41$ for the clean blade. The power output generated P_G is also affected by erosion, which decreases to around $P_G = 1.30$ kW with respect to $P_G = 1.36$ kW for the clean blade, as shown in Figure 3d.

2.4. Experimental Modal Analysis

In order to validate our simulated model of erosion, the following tests were carried out using a physical model. The blade was manufactured with a mixture of fiberglass, catalyst and resin. To cause the different levels of erosion on the leading edge of the blade, a Dremel tool was used, which uses a disc to gently sand and abrade the area of interest. Finally, to measure the percentage of erosion, a Vernier was used, measuring depth and thickness at the worn area. The experimental test of the blade was carried out in the Laboratory of Vibration of the Department of Mechatronics Division in the CINVESTAV–IPN. To evaluate the levels of erosion discussed in the numerical results, it was necessary the use an electromagnetic shaker model ET-139 (manufactured by Labworks©) to excite the blade, which is mounted over a mechanical rail supported by a mechanism of ball bearing that allows limited horizontal displacement. The electromagnetic shaker is controlled via a linear power amplifier (manufactured by Labworks©) model PA-138. The response in terms of acceleration and force applied by the electromagnetic shaker to the blade is monitored by using an impedance head placed at the stinger of the shaker connected via a cable to the data acquisition system (manufactured by Klister LabAmp©). The frequency response of the system is fed back using accelerometers model 8640A (manufactured by Klister©) mounted at the tip section of the blade. The data acquisition system is connected via USB to an external graphical interface implemented in MATLAB/Simulink©for the analysis of experimental data captured with a Sensoray©card. A comparative study of the experimental and Finite Element Method (FEM) response (simulation) of a wind turbine blade of 1 Kw of electrical power of the first three vibration modes is summarized Table 5. Results indicate that our simulated model behaves consistently with the real-world tests. The variations in the frequency response of the blade is due to the loss of material on the leading edge of the tip section, which can compromise the stability and balance of the wind rotor.

Table 5. Modal analysis of the simulated (FEM) and physical (experimental) blade under different erosion conditions; values given in Hz.

Mode	FEM	Experimental
	Level of Erosion BI_1	
1	3.89	3.45
2	15.6	14.6
3	23.6	29.3
	Level of Erosion BI_2	
1	3.25	3.5
2	12.8	14.9
3	22.2	29.2
	Level of Erosion BI_3	
1	3.95	3.49
2	14.9	14.8
3	23.4	29.0

3. Erosion Detection with Machine Learning

This section deals with the automatic detection of erosion in wind turbine blades. The goal is to detect where erosion has occurred using a classification model, considering the three different cases of where erosion can appear on the blade tip: on the bottom edge, the top edge or both. Moreover, we predict, using regression models, the amount of erosion on the blade. Both tasks are posed as supervised learning problems and solved using ML algorithms by performing feature extraction on the power and vibration response of the blade through QBlade simulation. However, before presenting our proposal, we briefly survey related works in this domain.

3.1. Related Work

Several works have applied ML towards the detection of different types of problems in wind turbines. For instance, static and dynamic regression models have previously been used to detect failures in wind turbines based on vibration analysis [5]. Another example is [17], which presents an approach to predict when preventive maintenance should be performed, focusing on the remaining useful life of a wind turbine before a failure occurs and diagnosing the type of failure. The proposal involves low implementation costs because it is based solely on information collected from the very common SCADA system. A recent example also includes forecasting of wind speed assessment using satellite data and ML [18], specifically a neural network.

In [11], the authors classify the occurrence of different types of failures in blades, using a piezoelectric accelerometer to measure the vibration of the blade. That work considered five types of damage to the top of the leading edge of the blades, namely bending, cracks, looseness, pitch, twist and erosion. To classify the signals time-domain, feature extraction is performed on the vibration signals, focusing on different types of summary statistics. In a more recent study by the same authors [19], they also use vibration signals, histogram features and ML to monitor the condition of wind turbine blades, in this case using *lazy* classifiers.

ML has also been applied to maintenance management of blades in [20]. The work is based on the detection of delamination, a common structural problem that can generate large costs. Continuous monitoring of turbines is the focus of [21], using real data from a SCADA system to predict damage to the structure and blades of wind turbines. The authors present two models for this: the first is the use of multilayer neural network, and the second is adaptive networks with a fuzzy inference system. The proposal is to monitor the power curve signal, achieving good precision. Sound analysis, a unique approach, has also been used for fault detection, extracting descriptive feature of acoustic waves and detecting damage using common ML methods [10]. A related work can be found in [22], where ML is used to estimate turbine energy yield losses due to erosion on the leading edge of the blade.

In general, few works deal specifically with erosion, and those that do not focus on a detailed analysis of this type of failure.

3.2. Data Set

The dataset was generated with QBlade and the procedure outlined in Section 2.2. A total of 100 blades were simulated for each type of erosion of the blade tip (bottom edge, top edge or both), producing a dataset with 300 samples, similar to [11]. This work assumes that all blades used in a real-world setting will have a certain amount of erosion in at least one of the edges of the tip. Therefore, we do not consider the case in which the blade is completely clean. It must be stated that, while we are relying on simulated data, it has been shown that simulated results of wind turbine blade performance are reliable predictors of on site behavior [22,23].

In order to simulate an eroded blade, each of the contour points of the blade profile was perturbed, adding displacements within the range of 8% to 18%. The same seven contour points on the tip of the blade were modified to model different levels of erosion. Half of the samples were generated using uniform grid sampling, while the other half of samples were generated with random values within the specified range. For example, for the lower edge cases, 8% of erosion was removed from the Y-axis coordinate value; subsequently, the percentage of erosion increased by 0.2% up to 18% to generate 50 samples. For the remaining 50 samples, the amount of erosion was determined randomly by using a uniform distribution $U[8, 18]$. Random samples were used to simulate a rugged surface on the blade, which can be caused by random events such as contact with insects or large sand particles. Our approach is justified since roughness on a blade is often simulated with a random surface [24,25].

Each blade was simulated with QBlade using the settings in Section 2.1. The acceleration response at the blade tip was obtained, using the QFEM tool for structural design and modal analysis of each blade. The NREL FAST tool [26] was used to carry out analysis of the dynamic response of wind turbines. The vibration of the acceleration signal at the tip of the blade is selected as output. The simulation parameters are as follows: time step of 0.1, 3 blades, a rotor speed of 296 rpm and air density of 1.225 k/m^3. The wind fields are specified in Table 6 by mostly using the same simulation values used to determine power output. The difference is simulation time and air density.

Table 6. Wind field and wind simulation parameters.

Windfield Parameter	Value
Time (s)	60
Timesteps	100
Point per direction	20
Simulation Parameter	**Value**
Rotor Radius (m)	30
Hub Height (m)	60
Mean Wind Speed (m/s)	13
Measurement Height (m)	10
Turbulence Intesity (%)	10
Roughness Length (m)	1.00×10^{-2}

3.3. Feature Extraction

After obtaining the samples of both power and acceleration for the different blades, we proceeded to perform feature extraction. For this work, feature detection was carried out in the time domain given the success of such measures in similar work [11] and in the analysis of other complex signals [27,28].

Let $\xi(t) \in \mathbb{R}^T$ denote the vector containing a time series from a single signal, and T denotes the number of samples in ξ. A feature of $\xi(t)$ is denoted by x, while the matrix $\mathbf{X} = [\mathbf{x}_1, .., \mathbf{x}_F]$ contains all features from all samples, $\mathbf{x}_i$ is the vector of a single feature, and F is the total number of features extracted.

The feature extracted from the signals include six statistical descriptors, namely mean, median, maximum, minimum, sum, standard deviation, variance and kurtosis. Moreover, we also extract the following:

- Power: $P_\xi = \frac{1}{T}\Sigma_{-\infty}^{\infty}|\xi(t)|^2$;
- First difference: $\delta_\xi = \frac{1}{T-1}\Sigma_{t-1}^{T-1}|\xi(t+1) - \xi(t)|$;
- Normalized first difference: $\bar{\delta_\xi} = \frac{\delta_\xi}{\sigma_\xi}$;
- Second difference: $\gamma_\xi = \frac{1}{T-2}\Sigma_{t-1}^{T-2}|\xi(t+2) - \xi(t)|$;
- Normalized second difference: $\bar{\gamma_\xi} = \frac{\gamma_\xi}{\sigma_\xi}$.

Power $P\xi$ measures the strength of the signal or the energy consumed per unit of time. The first and second differences show the changes of a signal in time. The normalized first difference is also known as the Normalized Length Density and is used to quantify the self-similarities contained in a signal. Additionally, we also extract what is referred to as Hjorth features [29]. These include the following: Activity, Mobility and Complexity. The Activity feature represents the variance of the signal and is computed by the following.

$$A_\xi = \frac{\Sigma_{t=1}^{T}(\xi(t) - \mu_\xi)^2}{T} .$$

The Mobility feature is defined by the standard deviation of the slope of the EEG signal using as reference the standard deviation of the amplitude expressed as the following ratio by time unit.

$$M_\xi = \sqrt{\frac{var(\dot{\xi}(t))}{var(\xi(t))}} .$$

The Complexity feature measures the signal's variation using a smooth curve as reference provided by the following.

$$C_{\xi} = \frac{M(\dot{\xi}(t))}{M(\xi(t))} .$$

Another time domain feature is the Non-Stationary Index (NSI) [30]. Signal ξ is divided into segments, and their respective μ_i is computed. NSI is defined as the standard deviation of the segments' μ_i. When NSI is high, the signal is considered to be "less stationary."

The last feature includes Higher Order Crossings (HOC) [31]. The feature describes the oscillatory nature of signal counting the number of sign changes over multiple variants of the signal. A total of 10 distinct featur HOC features were extracted.

In total, this work considers 27 time domain features to characterize the signals of interest extracted from the wind turbine blade.

Classification and Regression Problems

The above feature extraction process produces a total of 27 time domain features for each signal. These features are used to pose three classifications by using the following: (1) the features from the power signal; and (2) adding the features from the acceleration signal. An ML model will learn to use these features to determine what edge of the blade is affected by erosion.

Moreover, the same feature set will be used to generate a regression model to estimate the exact amount of erosion. In this scenario, the objective is to predict the level of erosion, which ranges from 8 to 18 percent. In this case, the location of the erosion (top, bottom or both) is not taken into account, and the percentage of erosion is the target of the learning process.

3.4. *Auto Machine Learning with H2O-DAI*

AutoML is an approach for automating the design, tuning, implementation and evaluation of complete ML pipelines. The goal is to simplify the manner in which ML models are tested and evaluated such that the process by which the models are generated provide a comprehensive evaluation of the best possible approach to solve a given problem. In this proposal, we use H2O-DAI, which stands for H2O Driverless AI, which offers a very simple user interface and a comprehensive set of tools to perform AutoML [32]. For instance, it makes a choice from a set of state-of-the-art models, such as XGBoost [33], Generalized Linear Models [34] and Deep Learning [35].

There are basically four tuning hyperparameters that are used to configure the AutoML process of H2O-DAI; these include the following. Accuracy refers to the amount of effort to find the best possible pipeline in the range (1–10); it is set to 7 in these experiments. Time controls the duration of the search process, it is set to 2 in our experiments. Interpretability controls the amount of feature engineering performed by the AutoML system. In this case, since a diverse set of features is already being used, it is set to 8. Moreover, to evaluate performance, 6-fold cross validation was used. All experiments were carried out on an IBMP Power 8 Server for High-Performance Computing with two Power 8 processors and two NVIDIA Tesla P100 GPUs.

3.4.1. Classification Results

Summary of the results are presented as an average confusion matrix based on the classification achieved on the testing folds of the cross validation process. Results are shown in Table 7 when using the power output for feature extraction, where class labels are shown as Bottom, Top and Both for each of the three types of erosion. Other noteworthy classifier performance scores include (given as the average $\pm$ standard deviation over all the testing folds) the following: Area Under the Receiver Operating Characteristic Curve of 0.99 ± 0.001 and an F1-Score of 0.98 ± 0.008.

Table 7. Average confusion matrix using the power signal for feature extraction.

	Bottom	Top	Both	Error
Bottom	99%	1%	0	1%
Top	0	100%	0	0
Both	0	1%	99%	1%

H2O DAI converged to an XGBoost model for classification [36], using a total of seven input features, four of which are raw features from the 27 time domain features and three automatically engineered features. In particular, for this version of the problem, H2O DAI focused on statistical features, such as the variance, mean and median, but it also used the power feature.

Extending the feature set, incorporating the features computed on the acceleration signal produced optimal results, as shown in the confusion matrix of Table 8. In this case, H2O DAI also converged to an XGBoost model, using a total of 10 input features, including five automatically engineered features. It is notable that all of the features used in this case are features extracted from the acceleration signal, including the first differential, the NSI and the power features.

Table 8. Average confusion matrix using both the power signal and the acceleration signal for feature extraction.

	Bottom	Top	Both	Error
Bottom	100%	0	0	0%
Top	0	100%	0	0
Both	0	0	100%	0

3.4.2. Regression Results

The same configuration of H2O DAI is used, which was reported above for classification, with the exception that the scoring function is the root mean squared error (RMSE). Results are presented in Table 9, showing the average performance on the test sets of the 6-fold cross validation. H2O DAI was applied on three groups of features: power signal, acceleration signal and both, showing the mean absolute error (MAE), coefficient of determination R^2 and the root mean square percentage error loss (RMSPE). In all cases, H2O DAI converged to a Light Gradient Boosting Machine (Light GBM) [37]. Results show that using both signals for feature extraction produced a highly accurate model in terms of both R^2 and RMSPE.

Table 9. Regression results for H2O DAI estimating the percentage of erosion showing the average and standard deviation.

	Power	Acceleration	Both
MAE	0.002 (0.0007)	0.001 (0.0007)	0.0009 (0.0001)
R^2	0.98 (0.008)	0.98 (0.012)	0.99 (.0004)
RMSPE	2.8 (0.4)	1.9 (1.14)	0.97(0.14)

4. Concluding Remarks

This study presents an in-depth analysis of the aerodynamic and modal response of an eroded wind turbine blade. Efficiently and effectively detecting erosion on a blade can have substantial impacts in preventative and timely maintenance of wind turbines. Results show that it is possible to accurately determine where the erosion is present on the blade (top edge, bottom edge or both) and to estimate the level of erosion (between 8 and 18 percent). This is accomplished by analyzing the power signal of the wind turbine and the vibrations of the blade tip. A large set of time domain features was extracted, and the modeling process is carried out by using an AutoML system, namely H2o DAI. As such, this study represents the first contribution that tackles both the detection, localization and estimation of erosion level on the leading edge of a blade using ML. Moreover, this work is the first to apply AutoML in this domain. The process of designing the ML pipeline was carried out in an automatic fashion, without hampering performance and requiring very little human intervention in the design process. This could motivate further collaborative and multidisciplinary research between applied ML and wind energy maintenance and production.

The results presented in this work are consistent with those reported by [11,19], with the slight performance difference probably due to working with simulated data in our case, which is nonetheless a good predictor of real-world performance, as shown by [22,23] and partially validated by our experiments with a physical model.

Among both signals that were analyzed, the accelerometer readings seem to be more informative relative to the power signal, based both on the classification (erosion detection) and regression (erosion level estimation) problems, with small but consistent differences. Moreover, the best performance was achieved when both signals are used for feature extraction. It should be possible to use both models to automatically detect the presence and level of erosion in a properly instrumented wind turbine blade. Future work will focus on applying the same experimental procedure in a fully working prototype: first in a wind tunnel and then in the field.

Author Contributions: Conceptualization: J.E.Z. and L.T.; methodology: J.E.Z., M.d.l.Á.G.L. and J.A.C.T.; formal analysis and investigation: J.E.Z., M.d.l.Á.G.L. and J.A.C.T.; writing—original draft preparation: L.T. and J.A.C.T.; writing—review and editing: J.E.Z. and L.T.; resources: J.E.Z. and L.T.; supervision: J.E.Z. and L.T. All authors have read and agreed to the published version of the manuscript.

Funding: This research received no external funding.

Acknowledgments: The authors want to thank Daniel E. Hernández Morales and Luis A. Cárdenas Florido for their technical support in the development of this research, as well as AP ENGINEERING INNOVACIÓN TECNOLÓGICA EN ENERGÍAS S.A DE C.V. The authors also thank Gerardo Silva Navarro from the Laboratory of Vibration of the Department of Mechatronics Division in the CINVESTAV–IPN.

Conflicts of Interest: The authors declare that they have no conflicts of interest.

References

1. Kaldellis, J.K.; Zafirakis, D. The wind energy (r) evolution: A short review of a long history. *Renew. Energy* **2011**, *36*, 1887–1901. [CrossRef]
2. Hernández-Escobedo, Q.; Perea-Moreno, A.J.; Manzano-Agugliaro, F. Wind energy research in Mexico. *Renew. Energy* **2018**, *123*, 719–729. [CrossRef]
3. Spera, D.A. *Wind Turbine Technology: Fundamental Concepts in Wind Turbine Engineering*, 2nd ed.; ASME Press: New York, NY, USA, 2009.
4. Li, D.; Ho, S.C.M.; Song, G.; Ren, L.; Li, H. A review of damage detection methods for wind turbine blades. *Smart Mater. Struct.* **2015**, *24*, 033001. [CrossRef]
5. Oliveira, G.; Magalhães, F.; Cunha, Á.; Caetano, E. Vibration-based damage detection in a wind turbine using 1 year of data. *Struct. Control Health Monit.* **2018**, *25*, e2238. [CrossRef]
6. Blanco Martínez, A. Study and Design of Classification Algorithms for Diagnosis and Prognosis of Failures in Wind Turbines from SCADA Data. Ph.D. Thesis, Universitat de Vic-Universitat Central de Catalunya, Barcelona, Spain, 2018.
7. Eftekhar Azam, S.; Rageh, A.; Linzell, D. Damage detection in structural systems utilizing artificial neural networks and proper orthogonal decomposition. *Struct. Control Health Monit.* **2019**, *26*, e2288. [CrossRef]
8. Tang, Z.; Chen, Z.; Bao, Y.; Li, H. Convolutional neural network-based data anomaly detection method using multiple information for structural health monitoring. *Struct. Control Health Monit.* **2019**, *26*, e2296. [CrossRef]
9. Gu, J.; Gul, M.; Wu, X. Damage detection under varying temperature using artificial neural networks. *Struct. Control Health Monit.* **2017**, *24*, e1998. [CrossRef]
10. Krause, T.; Ostermann, J. Damage detection for wind turbine rotor blades using airborne sound. *Struct. Control Health Monit.* **2020**, *27*, e2520. [CrossRef]

11. Joshuva, A.; Sugumaran, V. Wind turbine blade fault diagnosis using vibration signals through decision tree algorithm. *Indian J. Sci. Technol.* **2016**, *9*, 1–7. [CrossRef]
12. Alaskari, M.; Abdullah, O.; Majeed, M.H. Analysis of wind turbine using QBlade software. *IOP Conf. Ser. Mater. Sci. Eng.* **2019**, *518*, 032020. [CrossRef]
13. Du, Y.; Chen, W. Numerical simulation of the wind turbine erosion. In Proceedings of the 3rd International Conference on Material, Mechanical and Manufacturing Engineering (IC3ME 2015), Guangzhou, China, 27–28 June 2015; Atlantis Press: Dordrecht, The Netherlands, 2015.
14. Vennell, R. Exceeding the Betz limit with tidal turbines. *Renew. Energy* **2013**, *55*, 277–285. [CrossRef]
15. Mahmuddin, F.; Klara, S.; Sitepu, H.; Hariyanto, S. Airfoil Lift and Drag Extrapolation with Viterna and Montgomerie Methods. *Energy Procedia* **2017**, *105*, 811–816. [CrossRef]
16. Elliott, D.; Schwartz, M.; Scott, G.; Haymes, S.; Heimiller, D.; George, R. *Wind Energy Resource Atlas of Oaxaca*; Technical Report; National Renewable Energy Lab. (NREL): Golden, CO, USA, 2003.
17. Zhao, Y.; Li, D.; Dong, A.; Kang, D.; Lv, Q.; Shang, L. Fault prediction and diagnosis of wind turbine generators using SCADA data. *Energies* **2017**, *10*, 1210. [CrossRef]
18. Majidi Nezhad, M.; Heydari, A.; Pirshayan, E.; Groppi, D.; Astiaso Garcia, D. A novel forecasting model for wind speed assessment using sentinel family satellites images and machine learning method. *Renew. Energy* **2021**, *179*, 2198–2211. [CrossRef]
19. Joshuva, A.; Sugumaran, V. A lazy learning approach for condition monitoring of wind turbine blade using vibration signals and histogram features. *Measurement* **2020**, *152*, 107295. [CrossRef]
20. Jimenez, A.; Gómez Muñoz, C.Q.; García Márquez, F.P. Machine Learning for Wind Turbine Blades Maintenance Management. *Energies* **2017**, *11*, 13. [CrossRef]
21. Morshedizadeh, M. Condition Monitoring of Wind Turbines Using Intelligent Machine Learning Techniques. Master's Thesis, University of Windsor, Windsor, ON, Canada, 2017.
22. Cappugi, L.; Castorrini, A.; Bonfiglioli, A.; Minisci, E.; Campobasso, M.S. Machine learning-enabled prediction of wind turbine energy yield losses due to general blade leading edge erosion. *Energy Convers. Manag.* **2021**, *245*, 114567. [CrossRef]
23. Wu, G.; Zhang, L.; Yang, K. Development and Validation of Aerodynamic Measurement on a Horizontal Axis Wind Turbine in the Field. *Appl. Sci.* **2019**, *9*, 482. [CrossRef]
24. Nikolov, I.; Madsen, C. Quantifying Wind Turbine Blade Surface Roughness using Sandpaper Grit Sizes: An Initial Exploration. In Proceedings of the 16th International Joint Conference on Computer Vision, Imaging and Computer Graphics Theory and Applications, Virtual Conference, 8–10 January 2021; SCITEPRESS—Science and Technology Publications: Setúbal, Portugal, 2021.
25. David, M. *Owner Reports-Airfoil Performance Degradation Due to Roughness and Leading-Edge Erosion, Data and Plots-Raw Data*; Technical Report; Pacific Northwest National Lab. (PNNL): Richland, WA, USA, 2020.
26. Jonkman, J.M. *Modeling of the UAE Wind Turbine for Refinement of FAST_AD*; Technical Report; National Renewable Energy Lab.: Golden, CO, USA, 2003.
27. Hernández, D.E.; Trujillo, L.; Z-Flores, E.; Villanueva, O.M.; Romo-Fewell, O. Detecting Epilepsy in EEG Signals Using Time, Frequency and Time-Frequency Domain Features. In *Computer Science and Engineering—Theory and Applications*; Sanchez, M.A., Aguilar, L., Castañón-Puga, M., Rodríguez-Díaz, A., Eds.; Springer International Publishing: Cham, Switzerland, 2018; pp. 167–182.
28. Jenke, R.; Peer, A.; Buss, M. Feature Extraction and Selection for Emotion Recognition from EEG. *IEEE Trans. Affect. Comput.* **2014**, *5*, 327–339. [CrossRef]
29. Hjorth, B. EEG analysis based on time domain properties. *Electroencephalogr. Clin. Neurophysiol.* **1970**, *29*, 306–310. [CrossRef]
30. Hausdorff, J.; Lertratanakul, A.; Cudkowicz, M.E.; Peterson, A.L.; Kaliton, D.; Goldberger, A.L. Dynamic markers of altered gait rhythm in amyotrophic lateral sclerosis. *J. Appl. Physiol.* **2000**, *88*, 2045–2053. [CrossRef] [PubMed]
31. Petrantonakis, P.C.; Hadjileontiadis, L.J. Emotion Recognition From EEG Using Higher Order Crossings. *IEEE Trans. Inf. Technol. Biomed.* **2010**, *14*, 186–197. [CrossRef] [PubMed]
32. Hall, P.; Kurka, M.; Bartz, A. Using H_2O Driverless AI. Mountain View, CA, USA, 2018. Available online: http://docs.h2o.ai (accessed on 29 December 2021).
33. Zheng, H.; Yuan, J.; Chen, L. Short-term load forecasting using EMD-LSTM neural networks with a Xgboost algorithm for feature importance evaluation. *Energies* **2017**, *10*, 1168. [CrossRef]
34. Freebury, G.; Musial, W. Determining equivalent damage loading for full-scale wind turbine blade fatigue tests. In Proceedings of the 2000 Asme Wind Energy Symposium, Reno, NV, USA, 10–13 January 2000; p. 50.
35. Stetco, A.; Dinmohammadi, F.; Zhao, X.; Robu, V.; Flynn, D.; Barnes, M.; Keane, J.; Nenadic, G. Machine learning methods for wind turbine condition monitoring: A review. *Renew. Energy* **2019**, *133*, 620–635. [CrossRef]
36. Chen, T.; Guestrin, C. XGBoost: A Scalable Tree Boosting System. In Proceedings of the 22nd ACM SIGKDD International Conference on Knowledge Discovery and Data Mining, San Francisco, CA, USA, 13–17 August 2016; Association for Computing Machinery: New York, NY, USA, 2016; pp. 785–794.
37. Ke, G.; Meng, Q.; Finley, T.; Wang, T.; Chen, W.; Ma, W.; Ye, Q.; Liu, T.Y. Lightgbm: A highly efficient gradient boosting decision tree. *Adv. Neural Inf. Process. Syst.* **2017**, *30*, 3146–3154.

Mathematical and Computational Applications

MDPI

Article

AutoML for Feature Selection and Model Tuning Applied to Fault Severity Diagnosis in Spur Gearboxes

Mariela Cerrada [1], Leonardo Trujillo [2,*], Daniel E. Hernández [2], Horacio A. Correa Zevallos [2], Jean Carlo Macancela [1], Diego Cabrera [1] and René Vinicio Sánchez [1]

[1] GIDTEC, Universidad Politécnica Salesiana, Cuenca 010105, Ecuador; mcerrada@ups.edu.ec (M.C.); jeanca.macancela@gmail.com (J.C.M.); dcabrera@ups.edu.ec (D.C.); rsanchezl@ups.edu.ec (R.V.S.)

[2] Tecnológico Nacional de México/IT de Tijuana, Tijuana 22414, Mexico; daniel.hernandezm@tectijuana.edu.mx (D.E.H.); horacio.correa17@tectijuana.edu.mx (H.A.C.Z.)

* Correspondence: leonardo.trujillo@tectijuana.edu.mx

Abstract: Gearboxes are widely used in industrial processes as mechanical power transmission systems. Then, gearbox failures can affect other parts of the system and produce economic loss. The early detection of the possible failure modes and their severity assessment in such devices is an important field of research. Data-driven approaches usually require an exhaustive development of pipelines including models' parameter optimization and feature selection. This paper takes advantage of the recent Auto Machine Learning (AutoML) tools to propose proper feature and model selection for three failure modes under different severity levels: broken tooth, pitting and crack. The performance of 64 statistical condition indicators (SCI) extracted from vibration signals under the three failure modes were analyzed by two AutoML systems, namely the H2O Driverless AI platform and TPOT, both of which include feature engineering and feature selection mechanisms. In both cases, the systems converged to different types of decision tree methods, with ensembles of XGBoost models preferred by H2O while TPOT generated different types of stacked models. The models produced by both systems achieved very high, and practically equivalent, performances on all problems. Both AutoML systems converged to pipelines that focus on very similar subsets of features across all problems, indicating that several problems in this domain can be solved by a rather small set of 10 common features, with accuracy up to 90%. This latter result is important in the research of useful feature selection for gearbox fault diagnosis.

Keywords: AutoML; feature selection; fault severity assessment; gearboxes; XGBoost classifiers

Citation: Cerrada, M.; Trujillo, L.; Hernández, D.E.; Correa Zevallos, H.A.; Macancela, J.C.; Cabrera, D.; Vinicio Sánchez, R. AutoML for Feature Selection and Model Tuning Applied to Fault Severity Diagnosis in Spur Gearboxes. *Math. Comput. Appl.* **2022**, 27, 6. https://doi.org/10.3390/mca27010006

Academic Editor: Nicholas Fantuzzi

Received: 1 December 2021
Accepted: 11 January 2022
Published: 13 January 2022

1. Introduction

Gearboxes are crucial devices in industrial processes, as they play an important role in power transmission. Then, fault detection and diagnosis in such devices are attracting growing interest in researches, with focus on fault severity assessment. In particular, when a fault is starting, the first stages of the failure mode are not easy to detect, in most cases, and the incipient fault is not advised until reaching severe stages that can cause damages to other devices, decrease the performance of the process, and produce economical losses [1,2].

Vibration signal is one of the most informative signals commonly used to determine the health condition of rotating machines [3]. Once a vibration signal is available, data-driven approaches can offer signal processing techniques to tackle the problem of fault detection and diagnosis by identifying certain signal characteristics in time, frequency or time-frequency domains, as proposed in [4] for gearboxes.

Particularly, gearboxes exhibit nonlinear and chaotic behavior [5], and these characteristics are enhanced in the presence of faults [6–8]. Then, the identification of informative characteristics in the vibration signal produced by faulty conditions is not easy to accomplish in gearboxes by using standard signal processing, and usually requires expert knowledge. Additionally, previous studies have visually shown that the vibration signal behavior is non-monotonic to the fault severity increment in helical gearboxes [9,10]; that is, the signal amplitude does not increase with the fault. Under this scenario, Machine Learning (ML) approaches can address properly the task of fault detection and diagnosis. Moreover, signal processing permits different characterization of the vibration signal useful for the application of ML-based approaches providing high-performance solutions, commonly being the supervised fault classification.

ML-based classifiers using the k-nearest neighbor method (KNN) [11–14], artificial neural networks (ANN) [15–18] and support vector machines (SVM) [19–23], are frequently developed to propose fault classification models. Random Forest (RF) and Decision Trees (DT) are also reported as fault classifiers for gearboxes because of their powerful performance in cases where only a few samples are available and high dimensional feature spaces [24–28].

The performance of a conventional ML-based fault classification model is highly dependent on the input feature quality, avoiding the well-known curse of dimensionality, and the selection of the best classification model. Particularly, feature selection is a stage that must be carefully accomplished after features extraction is performed on the vibration signal. Statistical condition indicators (SCI) serve as features extracted from the vibration signal in the time domain, some of them being closely related to the vibration analysis such as the root-mean-square, standard deviation, kurtosis, skewness, among others [29]. Other features are related to the biomedical field to analyze surface electromyography signals [30,31]. The availability of a large set of features makes both feature selection or the mining for new features a process that is not easily generalized.

Although the recent applications of Deep Learning (DL) models to fuse the stages of feature extraction and selection are being widely reported [9,32–38], the selection of fault-related features like SCI extracted from the raw vibration signal is still a field of interest, mainly due to the easy understanding of such SCI. Moreover, the necessity of developing the whole ML pipeline, including not only feature engineering and selection but also classification model selection, hyperparameter optimization and validation, is currently a challenge in ML system development, called Automated Machine Learning (AutoML) [39,40].

According to the case study and the related dataset, for example, fault classification of gearboxes under different failure mode severity or multi-fault scenarios, where different failure modes are combined, the ML-based approach requires the development of a new pipeline for each scenario, that is, feature engineering and model adjustment. In most cases, this process requires exhaustive training plans, including greedy searching on feature and parameter spaces which demands computational efforts and optimization algorithms to obtain a proper classification model as mentioned previously. This is why, the development of ML pipelines automatically is nowadays highly required in practical problems associated to high dimensional feature spaces and complex model requirements. The evaluation of the different computational tools providing this support is well received by the ML-based engineering applications community in helping to choose the proper model.

To face the automated development of ML pipelines systematically, this paper presents the application of two AutoML systems for fault severity classification, with evaluation and comparison from an empirical perspective. Particularly, the paper is focused on the feature selection stage, including feature engineering over SCI extracted from vibration signals related to the fault severity of three failure modes in gears, which are pitting, crack and broken tooth. In the following, SCI are named statistical features or features. The application of AutoML in the field of Prognosis and Health Management (PHM) is a more difficult challenge as the industrial equipment, particularly rotating machines, usually work under complex and time varying conditions of load and speed. Then, new contributions in this field are required.

The goal is the comparison of the informative capability between the original statistical features, through the performance evaluation of the ML classifiers proposed by the AutoML systems. The experimental framework uses the software tools *H2O Driverless AI* which is equipped with evolutionary algorithms to perform feature engineering and selection, and *TPOT*, which also uses an evolutionary search to build tree-based ML pipelines. Both tools are relatively simple and straightforward to use, offering basic and intuitive configurations for generating different types of pipelines using state-of-the-art techniques and implementations.

Results of the evaluation and comparison between H2O DAI and TPOT show that both platforms select common features regardless of the selected model. For some failure modes, TPOT reduces the feature space but increase it in others. Moreover, accuracy when using the whole feature space, without feature selection, remains very close for the pipelines created by both platforms. Regarding H2O DAI, the informative capability of ten time-domain vibration signal-related statistical features is highlighted, which are enough to obtain proper classification performance.

Moreover, not only are the best features identified for each failure mode, but a common set of features reports proper performance to classify all the failure modes. This is an important contribution in the field of fault severity classification in gearboxes, for which the search for a common set of features for several failure modes in the same machine are still under research.

The rest of the paper is organized as follows. Section 2 presents the background about the AutoML as an emerging area in ML, and the description of the failure modes in gearboxes under study in this paper. Section 3 discusses the previous works regarding the feature analysis for fault severity assessment in gearboxes by using vibration signals mainly, and recent applications of AutoML on this domain. Section 4 describes the test bed of the different cases study, the collection of the data set for each case and the corpus generations. Section 5 details the experimental framework and results by using the AutoML software *H2O Driverless AI*. Section 6 is devoted to the discussion, and finally Section 7 summarizes and concludes the paper.

2. Background

This paper presents an analysis of the applicability of AutoML in the automatic diagnosis of faults in rotating machinery, namely industrial gearboxes. Therefore, this section is intended to provide an overview of both domains, with the former covered in Section 2.1 and the latter discussed in Section 2.2.

2.1. Overview of AutoML

ML is everywhere now, with successful applications in diverse domains such as automatic programming [41] and the prediction of complex chemical processes [42], and the list of examples grows every day. This success, however, has created the need to simplify and accelerate the development of problem-specific ML deployments. Among the different strategies being taken, two paradigms are particularly promising: Transfer Learning [43–45] and AutoML [39,40,46–51]. The former entails using a model generated on a source task to solve a target task, simplifying learning on the target and making it more efficient. The latter, on the other hand, employs a search process to discover large model architectures [49–51] or to automatically derive complete ML pipelines [46–48]. Moreover, recent techniques are even hybridizing both methods for the construction of Deep Learning models [52].

While Transfer Learning opens up a wide range of possibilities in the domain of interest of the present work, such questions are left for future research. The focus of this paper is on AutoML. To understand AutoML, it is first necessary to review what a ML pipeline consists of. While different problems might require slight variations of the general case, a typical ML pipeline includes: (1) Feature Engineering; (2) Feature Selection; (3) Model/Algorithm selection; (4) Hyper-parameter optimization; and, finally, (5) Training and Evaluation [39,40]. In fact, some AutoML systems even facilitate model deployment for real-world use [48]. Each of the stages in a ML pipeline is a research area in and of itself, characterized by large combinatorial search spaces and highly complex and non-linear interactions. However, in general, it is not possible to determine the optimal approach for each stage based on the description or data of a particular ML task, much less the optimal configuration of a complete pipeline.

The general goal of AutoML is to simplify the design process of a ML pipeline, by automatically determining how to instantiate a ML pipeline based on a problem's training data and a user-defined objective or scoring function. A general ML pipeline is presented in Figure 1, highlighting the elements that are usually addressed by an AutoML system. One of the most widely used variants of AutoML is Neural Architecture Search (NAS), which deals with the automated design of neural networks with a search process. While NAS has a long history [53,54], it has recently had a resurgence thanks to the intrinsic difficulty of developing Deep Neural Networks manually [49–51]. However, NAS is very specific to Deep Networks, focusing primarily on the Model/Algorithm selection part of the ML pipeline. The most ambitious version of AutoML are systems that cover most of the elements of the ML pipeline. Such systems work under the assumption that it is necessary to consider the interactions of all the stages in a ML pipeline; that is, an optimal pipeline will exhibit synergy among most, if not all, of the stages. To do so, it is necessary to solve a complex optimization problem, which requires performing a search within all possible ML pipelines. Given the complexity of such a search space, this type of AutoML systems often employ heuristic or meta-heuristic algorithms, such as Evolutionary Algorithms (EAs) [55].

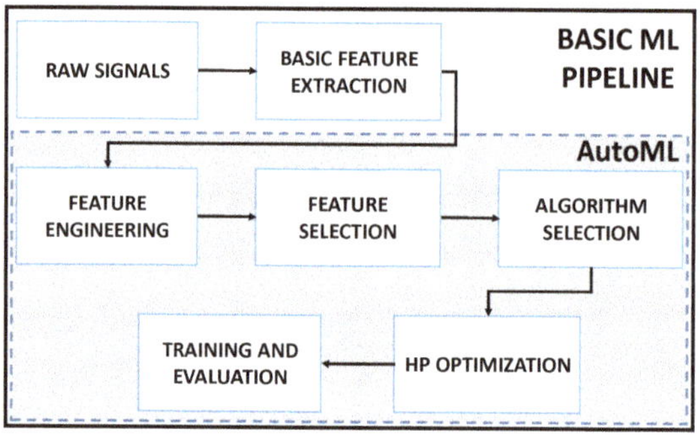

Figure 1. General ML pipeline and focus of an AutoML system.

Two such AutoML systems are TPOT [47] and H2O Driverless AI [48], with the former being an open-source academic software and the latter a commercial product specifically designed for industrial use (Driverless AI by H2O also offers academic versions of the software). TPOT is built on top of Scikit-learn [56], and while H2O is not, both share many of the underlying ML algorithms. TPOT uses genetic programming (GP), a form of EA, such that ML pipelines are coded using variable size tree structures. GP is the main search procedure in TPOT, indeed it can be seen as a basic GP system with a highly specialized primitive set and search space [55]. H2O Driverless AI, on the other hand, also uses EAs to perform Feature Engineering and Feature Selection, a task for which EAs are known to produce good results in difficult ML problems [45]. However, it employs a wider variety of mechanisms to generate the final pipeline, including Bayesian optimization for hyperparameter tuning and a large set of predefined engineered features from which the system can extend a problems feature space. In this work, we evaluate both systems to perform our experimental evaluation of AutoML because of their general simplicity and straightforward usage, where the user only needs to set very basic and intuitive configurations. Such is the goal of an AutoML system, to simplify the design process of a ML pipeline. Finally, it is important to highlight that the goal of this work is not specifically to compare the systems, even though their relative performances are contrasted. Our approach is to leverage the information produced by the ML pipelines produced by each system to better understand the feature selection problem in the domain of fault diagnosis in gearboxes, helped by this kind of AutoML systems.

2.2. Faults in Gearboxes

A gear is a toothed wheel usually attached to a rotating shaft. In a gear train, the teeth of one gear engage the teeth on the other gear. Gears and gear trains are important mechanical power transmission devices in industrial applications to produce speed and torque conversions from a rotating power source to another device. Gearboxes may work under constant or varying operation conditions of load and speed. According to the assembly of gear wheels, gear trains can be classified as a simple gear train, compound gear train, reverted gear train and planetary gear train [57]. In this work, the experimental test bed is a simple gear train assembled as a one-stage spur gearbox, with a number of teeth, Z1 and Z2, as shown in Figure 2.

Figure 2. One stage gearbox.

In heavy machinery, complex multistage gearboxes are used and the interaction of the gearbox elements with the working environment can drive the gears to faults that may cause significant economic losses. For that reason, fault diagnosis in gears has been the subject of intensive research [57].

One of the most dangerous damages in gears is the crack at the tooth root, that is caused because of significant changes in tooth stiffness [58]. A physical simulation of a man-made crack is shown in Figure 3a, where a different length and depth of crack throughout the base of the tooth were implemented. Tooth pitting is another common failure mode of a gearbox, which can appear at several levels according to pitted areas: slight micro pitting, macro pitting, macro pitting over 50%–100% of the gear tooth surface, and macro pitting over all the gear tooth surface [59]. Tooth pits can be simulated like circles of different diameter, depth and several number of holes over the tooth surface. A physical simulation of man-made pitting is shown in Figure 3b, where pits with different diameter and depth were implemented. Finally, another important research topic is the analysis of the effect of a broken tooth on the gearbox vibration, which is caused by an excessive impact load or unstable load. Under this fault, the size of the contact surface between the meshing teeth is decreased. In addition, the tooth becomes shorter and the contact length along the in volute profile of the damaged tooth is also decreased [60]. A physical simulation of the man-made broken tooth is shown in Figure 3c, where a different percentage of tooth loss was implemented. The simulated physical failure modes were implemented at the Vibrations Laboratory of the Universidad Politécnica Salesiana (UPS-Ecuador), and they are used in this work for generating the dataset of vibration signals.

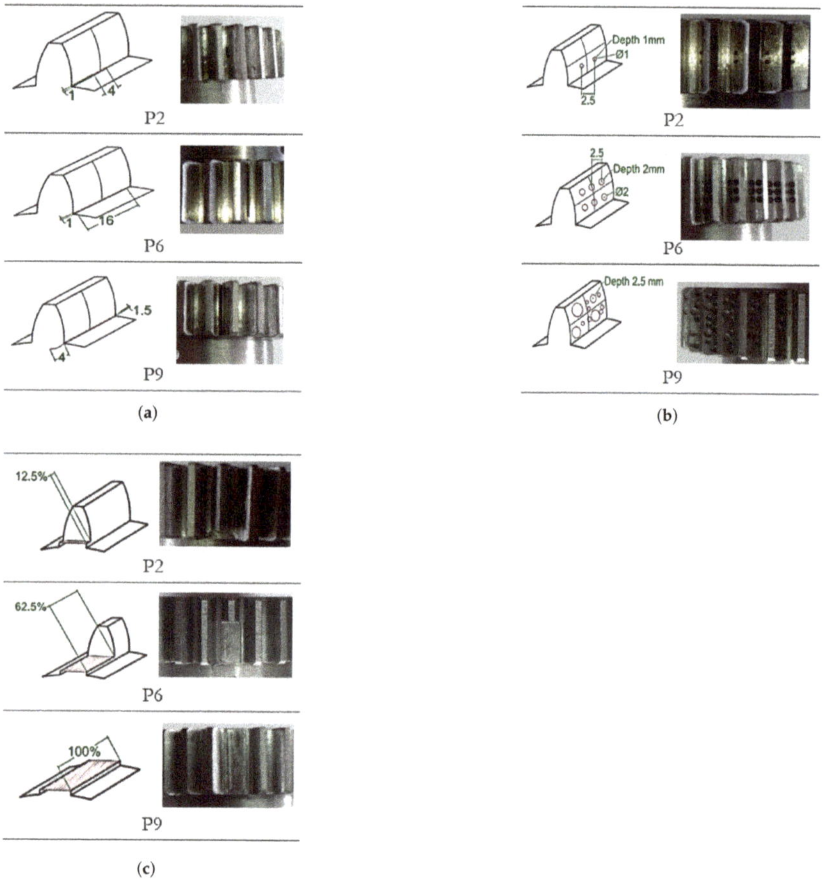

Figure 3. Details of the simulated faults for the three case studies. (**a**) Schemes and photography of crack levels. (**b**) Schemes and photography of pitting levels. (**c**) Schemes and photography of broken tooth levels.

3. Previous Work

Fault severity assessment in gears by using ML has been widely reported. This section is devoted to approaches using feature extraction by calculating statistical features, and focused on the problem of feature selection, and also recent approaches using artificial features extracted from DL models.

In [12], the identification of five different gear crack levels is performed by using features obtained from Wavelet Packet Decomposition (WPD). The first set of features is composed by 620 statistical features calculated from the wavelet coefficients at different levels, which are then reduced to seven significant principal components. KNN is used as a classifier and compared to other statistical models such as linear discriminant analysis, quadratic discriminant analysis, classification and regression trees, and naive Bayes classifier. Another approach to crack fault level identification is discussed in [13] by considering 25 features extracted from the time and frequency domains. A two-stage feature selection and weighting technique via Euclidean distance evaluation is proposed to select sensitive features and to weigh the selected features according to their sensitivity to each fault level. The Weighted K-Nearest Neighbor (WKNN) classifier is adopted to identify three gear crack levels under different loads and motor speeds.

Four different crack levels are detected in [61] by using statistical features and decision trees (DT). Similar work using DT and ARMA feature extraction from the vibration signals is presented in [62]. Ordinal rough approximation operators based on fuzzy covering and feature selection algorithms for ordinal classification are proposed in [63] for gear crack level identification. Finally, a DL approach using a Long Short-Term Memory (LSTM) based recurrent neural network is discussed in [64], to detect tooth crack growth at different stages, where the vibration signal is the input to the LSTM network. The LSTM prediction error is used as a measure of fault severity.

The detection of localized pitting damages in a worm gearbox by a vibration visualization method and ANNs is presented in [18]. Twelve statistical features are calculated from vibration signals in time and frequency domains for multi-class recognition, where each class is related to a severity level. In [29], a pitting severity assessment is tackled with supervised learning using an SVM-based ranking model that learns the ordinal information contained in the dataset. Thirty-four statistical features were calculated from vibration signals in the time and frequency domain, among others specifically designed for gearbox damage diagnosis. Three levels of damage were estimated. The approach in [65] uses Stacked Auto Encoders (SAE) for unsupervised feature extraction, feature reduction based on QR decomposition is then applied on the feature matrix provided by the SAE to obtain low dimensional data feeding an unsupervised K-means clustering.

Fault severity in broken tooth is also tackled by different approaches. The approach in [9] introduces Stacked Convolutional Autoencoders (SCAE) together with a Deep Convolutional Neural Network (DCNN) as a method for unsupervised hierarchical feature extraction for fault severity assessment in a helical gearbox. In that proposal, statistical features are not extracted, but the artificial ones are provided by the DCNN after training with initialization parameters given by an SCAE. These features feed a multilayer perceptron for classification. Another approach is provided in [38], where artificial features are provided by a CNN. The spectrogram of the vibration signal is used as the input to the CNN, the output of the last convolutional layer is connected to one softmax layer and finally, these outputs feed an SVM-based decision layer. An approach based on fuzzy transition is developed in [10] to predict the broken tooth severity in helical gearboxes. This is accomplished by two steps: a set of statistical features extracted from the vibrations signal are uses as input for a static fuzzy model to compute the weights of fuzzy transitions (WFT), and then a dynamic equation using WFT predicts the next degradation state of the rotating device. All the previous works focus on the same dataset related to ten severity damages of broken tooth in helical gearboxes.

Research on AutoML arises as a way to face the challenge of automating the Combined Algorithm Selection and Hyper-parameter tuning, called CASH by [66], and recent applications can be found. The review in [67] tackles the use of AutoML for developing smart devices to obtain auto generated embedded code ready to test and execute. An Automated Hyperparameter Search-Based Deep Learning Model for Highway Traffic Prediction is performed by AutoML in [68]. In the field of process monitoring, ref. [69] uses AutoML to optimize the parameters of a soft-sensor for monitoring the lysine fermentation process. The authors in [70] propose in the future research that the AutoML approach could be used for parameter optimization of a Variational Auto Encoder for nonlinear process monitoring.

However, ML-based fault diagnosis has just started to be analyzed under this approach with only a few works in the field of fault diagnosis for rotating machines. In [71] an approach using the Google DeepMind Team method called Neural Architecture Search (NAS) with reinforcement learning is proposed, as an AutoML tool for the automated search of a multiscale cascade CNN applied to the fault diagnosis of the gearbox data set of the PHM 2009 Data Challenge. In particular, two problems were addressed: fault detection and fault severity classification. Inspired by the NAS method, in [72] a neural network architecture automatic search method based on reinforcement learning is applied to rolling bearing fault diagnosis for two cases studies: the Case Western Reserve University (CWRU) bearing dataset and the locomotive bearing data set. Inspired by AutoML approaches, a self-optimizing module is proposed in [73] to dynamically adjust the knowledge base selection and parameter setting for training an Extreme Learning Machine based on physical knowledge. The self-optimizing module was applied on the CWRU bearing dataset.

4. Case Study

This section introduces the test bed and the experimental plan to acquire the database of vibration signals from three gearbox failure modes: pitting, crack and broken tooth. Next, details of the dataset are provided.

4.1. Experimental System and Data Acquisition

The real test bed available at the Universidad Politécnica Salesiana is shown in Figure 4, where realistic conditions of load and speed can be configured. The test bed is composed by a one stage gearbox (a) with spur gears, an induction motor (b) Siemens 1LA7 096-6YA60 2Hp 1200 rpm under variable speed controlled by a variable-frequency driver (c), and a magnetic brake system (d) for manual load control (e) via belt transmission (f). The four accelerometers (g) are IMI Sensor 603C01, 100 mV/g, placed in a vertical manner. The data acquisition cards (h) are National Instrument NI 9234 and cDAQ-9188, including anti-aliasing filtering with a sample frequency of 50 KS/s and, finally, the laptop with signal acquisition software (i) developed in Labview and Matlab.

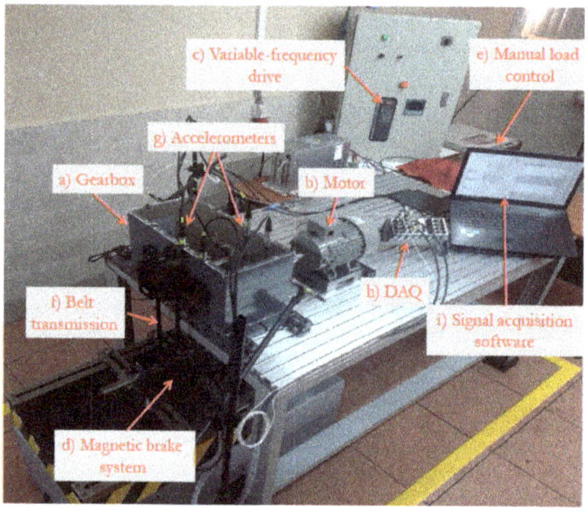

Figure 4. Test bed under realistic conditions.

The spur gears in the gearbox are made from steel E410, with a center distance of 90 mm, the ratio Z1/Z2 of the teeth number is 32/48, the module is 2.25 and the pressure angle is 20°. The vibration signals were collected with all four accelerometers, but this work only analyses the vibration signals collected from the accelerometer (A1) placed over Z1 in a vertical manner, as shown in Figure 5. The accelerometer (A1) provides the most informative vibration signal because it is close to the motor [74].

Three failure modes are considered, namely pitting, crack and broken tooth, each one configured on the gear Z1, as detailed in Table 1. For pitting, the severity level is related to the number of holes, their diameter and depth. For cracks, the severity level is related to the depth, width and length of the crack throughout the tooth. The broken tooth was implemented as a percentage of transverse breakage on the tooth. Vibrations signals were collected for nine severity levels for each failure mode, where P1 labels the Normal (N) or healthy state, and P2 to P9 label each fault severity level. An example of the vibration signals in the time domain of each failure mode in normal conditions, severity levels P2 and P9, is presented in Figure 6. These figures show the change of the vibration amplitude, but this change is not monotonic regarding the severity level. Details about these failure modes are given in Section 2.2.

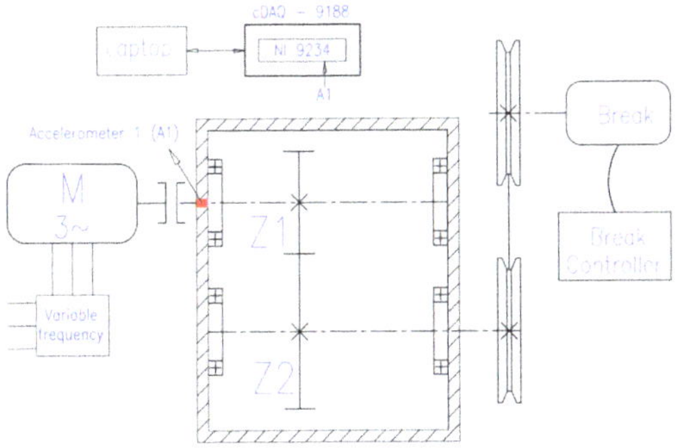

Figure 5. Schematic experimental setup.

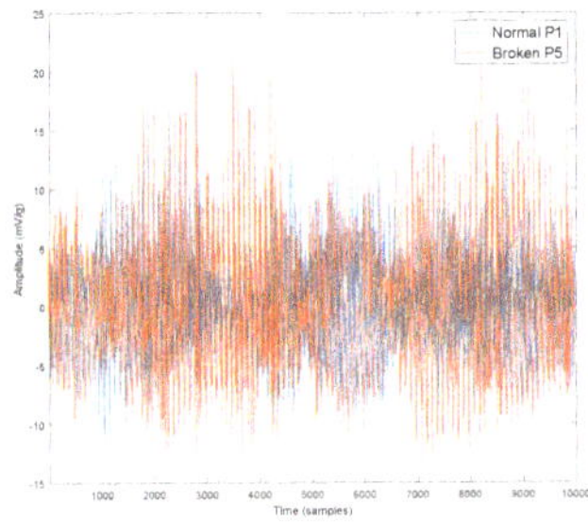

(**a**) Broken Tooth, P1 vs P5

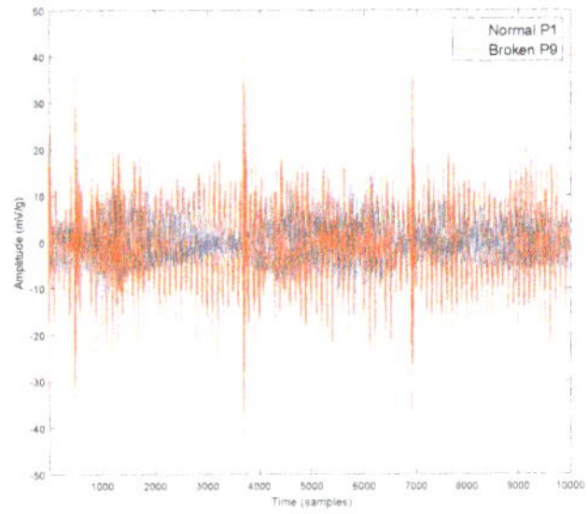

(**b**) Broken Tooth, P1 vs P9

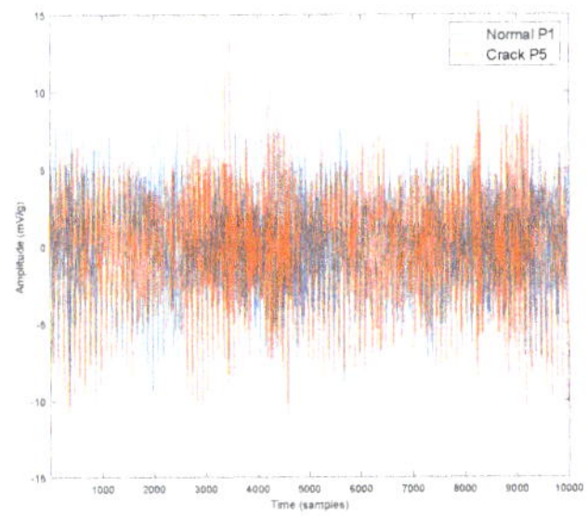

(**c**) Crack, P1 vs P5

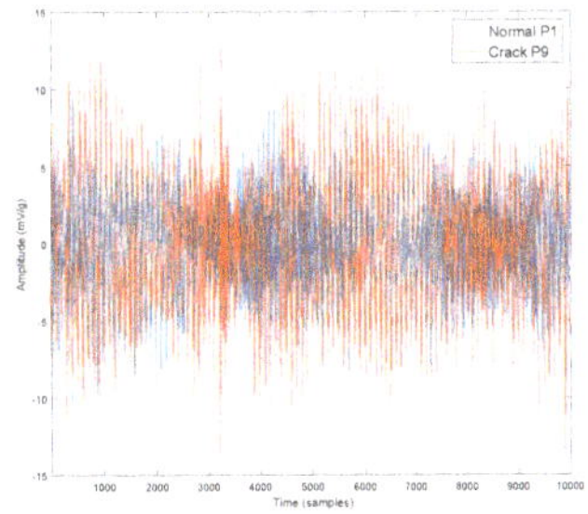

(**d**) Crack, P1 vs P9

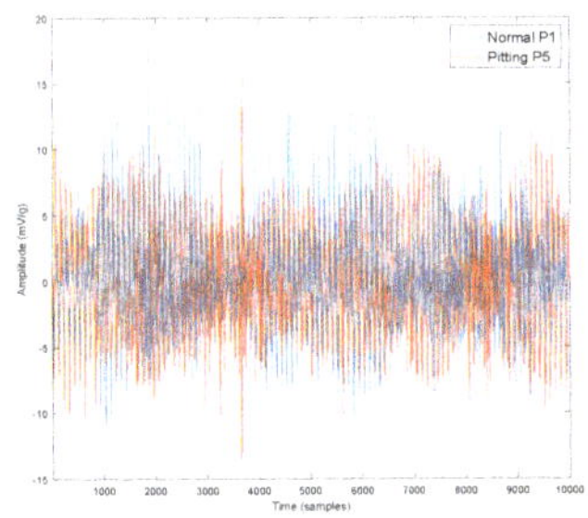

(**e**) Pitting, P1 vs P5

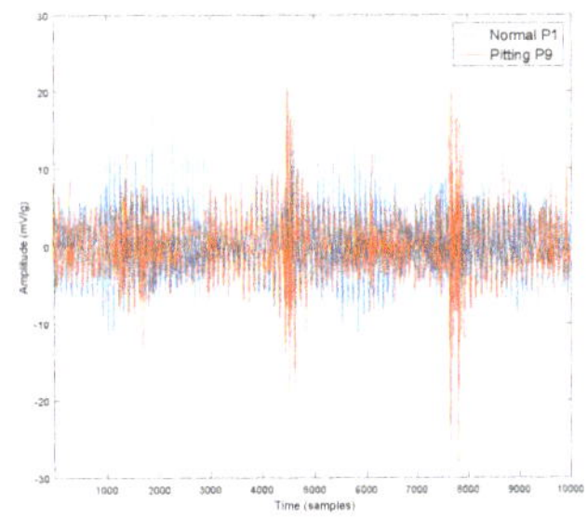

(**f**) Pitting, P1 vs P9

Figure 6. Samples of a vibration signal for simulated faults for the three case studies.

Table 1. Severity level characteristics of the three failure modes.

Label	Severity Level Pitting	Severity Level Crack	Severity Level Broken Tooth
P1	N	N	N
P2	Holes: 2 Diameter: 1 mm Depth: 1 mm 4.1%	Depth: 1 mm Width: 1 mm Length: 4 mm 4.9%	12.5%
P3	Holes: 2 Diameter: 1.5 mm Depth: 1.5 mm 7.3%	Depth: 1mm Width: 1 mm Length: 8 mm 9.8%	25.0%
P4	Holes: 4 Diameter: 1.5 mm Depth: 1.5 mm 14.7%	Depth: 1 mm Width: 1 mm Length: 10 mm 12.3%	37.5%
P5	Holes: 4 Diameter: 2 mm Depth: 2 mm 23.1%	Depth: 1mm Width: 1 mm Length: 12 mm 14.7%	50.0%
P6	Holes: 6 Diameter: 2 mm Depth: 2 mm 34.6%	Depth: 1 mm Width: 1 mm Length: 16 mm 19.7%	62.5%
P7	Holes: 6 Diameter: 2.5 mm Depth: 2.5 mm 49.91%	Depth: 1 mm Width: 1 mm Length: 20 mm 25%	62.5%
P8	Holes: 8 Diameter: 2.5 mm Depth: 2.5 mm 66.5%	Depth: 2 mm Width: 1.5 mm Length: along the tooth 50.0%	87.5%
P9	Holes: irregular Diameter: irregular Depth: 2.5 mm 83.1%	Depth: 4 mm Width: 1.5 mm Length: along the tooth 100%	100%

4.2. Dataset

The dataset of vibration signals were collected under the following conditions:

- The time length of each example is 10 s. Then, the signal is composed by 500,000 samples, according to the sample frequency of the DAQ card;
- The motor rotates at constant speeds of 180 rpm, 720 rpm and 960 rpm;
- The constant load was configured for no-load (0 N m), and loads generated with constant voltage application on the magnetic brake on 5 VDC (1.44 Nm) and 10 VDC (3.84 Nm);
- Each example configured under different speed and load were repeated 15 times.

Therefore, each severity level is composed by 135 examples and the whole dataset considering nine severity levels has 1215 examples. After that, 64 statistical features over the time domain of the vibration signal were extracted for each example to build the corpus matrix. The time domain offers a suitable condition indicator with easy interpretability. Some statistical features are mean, variance, standard deviation, kurtosis, skewness, energy, absolute mean, crest factor, norm entropy, Shannon entropy, sure entropy, among others. Detailed definitions of each condition indicator can be found in [29–31]. Then, the corpus is a 1215×64 matrix.

5. Experiments and Results

This section presents our general methodology to evaluate both the AutoML tools used in this paper, and the implementation details and the main findings. The methodology is presented in Figure 7, where processes (rounded squares), inputs and outputs (squares) are stated. In all of the work presented, it is important to always consider that there are three specific case studies that are analyzed, namely pitting, crack and broken tooth, each of the gearbox faults studied in this paper. The section is divided into two main subsections, each one devoted to one of the AutoML systems considered in this work.

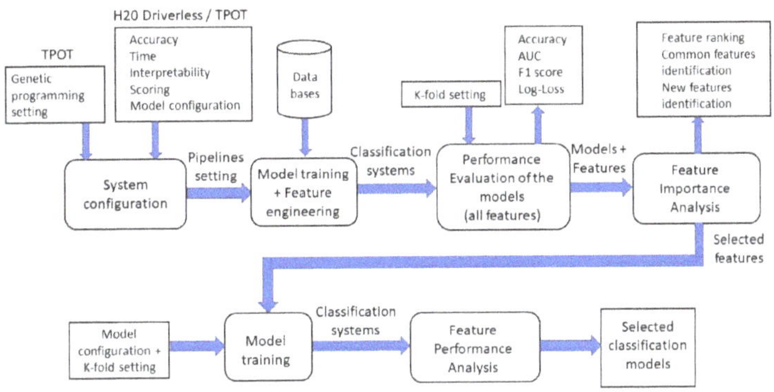

Figure 7. Methodological Framework.

5.1. *AutoML with H2O Driverless AI*

The experimental approach presented in this paper is focused on characterizing the performance of AutoML in each case study. This is done by analyzing H2O performance at different levels. First, we focus on basic generalization or testing performance, the most common approach towards evaluating ML solutions. Second, we evaluate the incidence of the Feature Engineering process on AutoML performance. It is widely understood that Feature Engineering is essential to obtain optimal performance with many ML algorithms [45]. Third, we evaluate the Feature Selection performed by AutoML, by analyzing the estimated feature importance provided by H2O and evaluating the impact that these features have on predictive accuracy. Finally, we will qualitatively compare in Section 6 the performance of the obtained AutoML results with those previously published in this domain The goal is to illustrate the differences and similarities of the AutoML pipelines relative to the standard development of ML pipelines.

5.1.1. System Configuration

H2O Driverless AI (DAI) offers an elementary user interface and control settings. In this work, we are using the version 1.6.5 of DAI, running on an IBM Power 8 HPC server, with two CPUs and 160 CPU cores, 512 GB RAM and two NVIDIA Tesla P100 GPUs with 16 GB RAM each. There are basically four elements that need to be configured by the user to perform the AutoML process, these are (We do not cover the complete setup process, only the most relevant aspects for our study; for a detailed description please see [48] and visit the online documentation):

1. **Accuracy [1–10]**: This setting controls the amount of search effort performed by the AutoML process to produce the most accurate pipeline possible, controlling the scope of the EA and the manner in which ensemble models are constructed. In all of our experiments, we set this to the highest value of 10;
2. **Time [1–10]**: This setting controls the duration of the search process and allows for early stopping using heuristics when it is set to low values. In all of our experiments, we set this to the highest value of 10;

3. **Interpretability [1–10]**: Guaranteeing that learned models are interpretable is one of the main open challenges in ML [75], which can be effected, for example, by model size [76]. DAI works under the assumption that model interpretability can be improved if the features used by the model are understandable to the domain expert, and if the relative number of features is kept as low as possible. This setting controls several factors, including the use of filtering features selection on the raw features, and, more importantly for our study, the amount of Feature Engineering methods used. In this work, we evaluate two extreme conditions for this setting, for each case study we perform two experiments, using a value of 1 and 10. A value of 10 filters out co-linear and uninformative feature, while also limiting the AutoML process to only use the original raw features of the problem data. On the other hand, the lower value uses all of the raw features and constructs a large set of new features using a variety of Feature Engineering methods;
4. **Scoring**: Depending on the type of problem, regression or classification, DAI offers a large variety of scoring functions that are to be optimized by the underlying search performed by the AutoML system, such as Classification Accuracy or Log-Loss for classification. In this work, we choose the Area Under the Receiver Operating Characteristic Curve (AUC), where optimal performance is achieved with a value of 1, and 0 otherwise. Since all case studies are multi-class problems, this measure is computed as a micro-average of the ROC curves for each class [77].

All of the other DAI settings are left to their default values, of which one in particular merits further explanation. As mentioned before, one of the tasks that an AutoML algorithm must determine is the ML model or algorithm to use for a given dataset. DAI offers the following options: XGBoost, Generalized Linear Models, LightGBM, Follow the Regularized Leader and RuleFIt Models. After initial exploratory experiments, we observed that under all our experimental conditions, and for each case study, DAI always chose XGBoost as the learning algorithm. Therefore, XGBoost is only considered in all the experiments reported below. XGBoost is an implementation of gradient boosted decision trees, that is highly efficient, particularly on GPU-enabled systems [78]. It is widely considered to be a state-of-the-art algorithm, that consistently outperforms most other techniques on a wide variety of domains. Finally, DAI does not only consider single models, but also attempts to build ensembles of several models, an approach that often outperforms single model approaches [79]. The ensemble of XGBoost models are linearly combined to produce the final output.

5.1.2. Evaluation of AutoML Pipelines

Given the three case studies (Pitting, Crack and Broken Tooth), and the two experimental configurations (Interpretability set to 1 and Interpretability set to 10), we have a total of six sets of experimental results. DAI performs 3-fold cross validation to compute the performance scores, and presents averages of performing this process 3 times. Table 2 summarizes the results for each experiment, focusing on the following performance metrics: Classification Accuracy, AUC, F1 Measure and Log-Loss, all of which are standard measures in the ML literature. These measures are given as averages over all test folds in the cross validation process, and the standard deviation is given in parentheses. The next two columns present the details on model size, given by the number of features used and the number of components in the final ensemble (In some cases the number of components in the final ensemble was larger than what we report in Table 2, but some models had a zero weight, so they were omitted from the final result). It is worthwhile to mention, once again, that while size is not equivalent to complexity or interpretability, in practice it often sufficiently commensurate to be used as a useful approximation [76]. The final column specifies the amount of computation time required for each experiment (in hours).

Table 2. Summary of experimental results for each problem (BT = Broken Tooth, Interpret = Interpretability and Acc = Accuracy). Testing performance is shown as averages and the standard deviation in parentheses. Model Size is given for the final ML pipeline found.

Configuration		Testing Performance				Model Size		
Problem	Interpret.	Acc.	AUC	F1	Log-Loss	Features	Ensemble	Time
Pitting	1	0.99(0.0025)	0.99(0.0002)	0.97(0.0116)	0.09(0.0185)	177	4	19
Pitting	10	0.99(0.0021)	0.99(0.0005)	0.96(0.0009)	0.12(0.0253)	27	1	8.5
Crack	1	0.99(0.0023)	0.99(0.0001)	0.98(0.0106)	0.08(0.0176)	131	4	18.5
Crack	10	0.99(0.0015)	0.99(0.0002)	0.98(0.0007)	0.08(0.0197)	25	2	14.8
BT	1	0.98(0.0027)	0.99(0.0003)	0.95(0.0124)	0.17(0.0161)	1019	3	19.5
BT	10	0.98(0.0040)	0.99(0.0002)	0.94(0.0180)	0.17(0.0556)	15	2	14.7

There are some notable results in Table 2. First, in terms of generalization performance, all of the ML pipelines produce very strong results. Second, the Interpretability setting does not seem to have a significant effect on performance, with almost identical results for both settings. Third, this setting does affect the number of features used and the running time of the experiments. It is clear that using an Interpretability setting of 1 does increase the number of features, in some cases quite drastically (Broken Tooth in particular), but it does not generate a similar performance increase. This phenomenon is well studied in the EA literature, and it is known as bloat [76]. While widely studied, and a variety of methods used to control it, it is obviously not addressed in this state-of-the-art AutoML system.

5.1.3. Analysis of Feature Importance

While feature engineering did not lead to substantial improvements in classification accuracy, however, the AutoML feature importance estimation and feature selection process are also evaluated. Table 3 presents a summary of the feature importance scores assigned by DAI to each of the features used in this study, which are given in the range of $[0,1]$, when setting $Interpretability = 10$. Features that are assigned a value below 0.003 are not used by the ML pipeline found in a particular problem, and those features with such a value in all three pipelines are omitted from the table. Notice that the number of features that meet these criteria for each problem is well below the total number of available features: 26 for pitting, 15 for Broken Tooth and 24 for crack.

Moreover, features that were used in the pipeline of all three problems (feature importance values greater than 0.003 in all pipelines) are highlighted in gray (nine features), these are: x_1, x_6, x_7, x_8 ,x_{27}, x_{35} x_{41}, x_{54}, and x_{61}. Features that are only used in at least two of the pipelines are in light gray (12 features), and features that are used once are in white (12 features). The final column of Table 3 presents the average Feature Importance score assigned by DAI to each feature, considering the score from the pipeline obtained from each problem. These results show that there is substantial agreement in which features are informative in this domain, irrespective of the specific type of fault.

Taking into consideration the results in Table 2, there are several observations that can be made regarding feature importance and the DAI results that did employ Feature Engineering (Interpretability = 1). These results show that feature engineering produces a larger set of informative features, compared to the original set of features. For instance, we can focus on the top 50 features determined by DAI, and analyze them based on their feature importance value to characterize their statistical distribution, as summarized in Table 4 (Note that for the experiments with Interpretability = 10 some features (in some cases most) did not achieve a feature importance value greater than 0.003, and were thus omitted from the analysis in Table 3. However, since DAI does not provide the feature importance value for these features, it was considered to be equal to 0.003 for this analysis). These results clearly show that the feature engineering process did produce a larger set of informative features, compared to the original set of raw features. We can also analyze the type of features used by DAI when setting Interpretability = 1, as summarized in Table 5, again focusing on the top 50 features for each problem. The table summarizes the percentage of features generated by different Feature Engineering heuristics; these are:

- Original: The original features in the dataset;
- Cluster Distance (CD): Uses a subset of features to cluster the samples, and uses the distance to a specific cluster as a new feature;
- Cluster Target Encoding (CTE): Also cluster the data, but computes the average value of the target feature of each cluster as a new feature;
- Interaction: Uses feature interactions as new features, based on simple arithmetic operations, namely addition, subtraction, division, and multiplication;
- Truncated SVD (TSVD): This heuristic trains a truncated SVD model on a subset of the original features, and uses the components of the SVD matrix as new features for the problem.

Table 3. Feature importance computed by DAI on each problem, for each of the features used in this study. Features not on this list had a zero value in all three problems. Dark gray rows indicate that a feature was used in all three problems, light gray in two, and white in at least one.

Feature	Pitting	Crack	Broken Tooth	Average
x_1	1	1	1	1
x_5	0.07	0	0	0.02
x_6	0.92	0.63	0.93	0.88
x_7	0.52	0.77	0.20	0.49
x_8	0.50	0.71	0.90	0.70
x_9	0	0.14	0.57	0.23
x_{10}	0.06	0	0	0.02
x_{11}	0	0.64	0	0.02
x_{12}	0.05	0.18	0	0.07
x_{13}	0.51	0	0.61	0.37
x_{15}	0.17	0.65	0	0.27
x_{16}	0.14	0	0	0.04
x_{17}	0.09	0	0	0.03
x_{19}	0.51	0.38	0	0.29
x_{20}	0.52	0.18	0	0.23
x_{21}	0.11	0.30	0	0.13
x_{22}	0.05	0	0.97	0.34
x_{24}	0.25	0	0	0.08
x_{26}	0	0.44	0	0.14
x_{27}	0.63	0.70	0.52	0.61
x_{28}	0.06	0.72	0	0.26
x_{30}	0	0	0.49	0.16
x_{33}	0	0	0.20	0.06
x_{35}	0.27	0.16	0.65	0.36
x_{39}	0.05	0.19	0	0.08
x_{41}	0.52	0.65	0.94	0.70
x_{42}	0.27	0.19	0	0.09
x_{45}	0	0.62	0	0.20
x_{54}	0.24	0.56	0.19	0.33
x_{58}	0.05	0	0.55	0.20
x_{61}	0.39	0.68	0.22	0.43
x_{62}	0	0.18	0	0.06
x_{63}	0	0.15	0	0.05
x_{64}	0.05	0.77	0	0.27

It appears that the original features and the CD and Interaction heuristics produce most of the informative features used by the DAI models. However, as shown in Table 2, these features did not lead to increased performance in the studied problem instances. Suggesting, once again, that these transformations are, in fact, reducing model interpretability without notably improving model accuracy.

Table 4. Feature importance values for the top 50 features found for each problem by DAI.

Configuration		Feature Importance				
Problem	**Interpret.**	**Min**	**Max**	**Median**	**Mean**	**Std**
Pitting	1	0.149	1	0.212	0.259	0.152
Pitting	10	0.114	1	0.183	0.233	0.147
Crack	1	0.281	1	0.414	0.462	0.180
Crack	10	0.003	1	0.052	0.166	0.248
BT	1	0.003	1	0.003	0.182	0.319
BT	10	0.003	1	0.062	0.237	0.301

Table 5. Percentage of types of features used by DAI on each problem, based on the 50 features with the highest feature importance scores.

Problem	Original	CD	CTE	Interaction	TSVD
Pitting	48%	32%	0%	18%	2%
Broken Tooth	32%	30%	0%	36%	2%
Crack	8%	52%	4%	36%	0%

5.1.4. Feature Importance and Classification Performance

To better understand how classification performance depends on the number of features used, two experiments were carried out. In both, an XGBoost classifier was evaluated on each problem using an increasing number of features, starting with the feature with the highest Feature Importance value and progressively adding the next highest and so on. In the first experiment, independent feature importance lists were used, as given by the DAI pipeline for each problem. For the second experiment, a common feature importance list was implemented, using the average importance value for each feature among the three problems. The values are presented in the Average column in Table 3.

This analysis allows us to characterize how the least important features impact classification accuracy. In these experiments, an 80%/20% training/testing split was carried out. Training was carried out using a Grid Search over hyperparameter space and 10-fold cross validation. Hyperparameter optimization considered the following hyperparameters and search ranges:

- Number of estimators (n_{est}): The number of weak-learners used by the XGBoost classifier. The search range is $\{50, 100, 150, 200\}$;
- Learning rate (lr): Size of each bootstrapping step, and it is critical to prevent overfitting. The search range is $\{0.01, 0.1, 0.2, 0.3\}$;
- Max Depth (d_{max}): Maximum depth of each weak-learner, which is represented as a decision tree. The search range is $[3, 10]$;
- Feature Subsampling (SS): Represents the fraction of features that are subsampled by a particular learner. The search range is $[0.1, 0.2]$;
- Gamma (γ): A regularization term that controls when a leaf is split in a weak-learner decision tree. The search range is $\{0, 0.1, 0.2\}$.

For the first experiment, the feature importance values for each problem are listed in Table 3. The results on the training and test set, for each number of features, are summarized in the first column of Figure 8, which shows the classification accuracy for each problem. What it is clear is that the highest accuracy is reached by all methods with 10 to 16 features, which is significant since it is less than the number of features used by the DAI pipelines from the Pitting and Crack problems. This suggest that further feature selection, at least in these two cases, can be beneficial. It is also important to note that this experiment is using a single XGBoost classifier, instead of the ensembles suggested by the DAI pipeline, but performance is nonetheless comparable. The second column of Figure 8 shows the optimal hyperparameter values found on each problem for each number of total features used, each one normalized to the range $[0, 1]$.

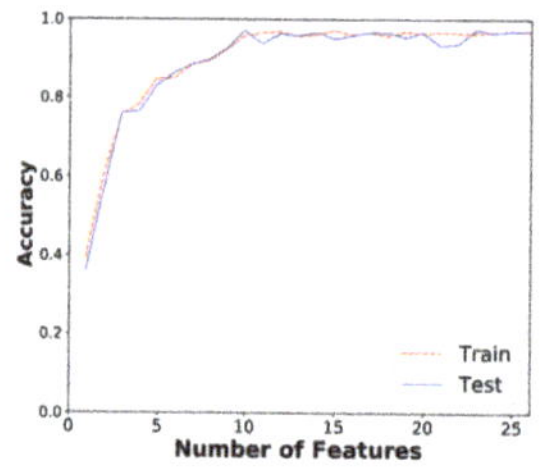

(**a**) Pitting: Accuracy

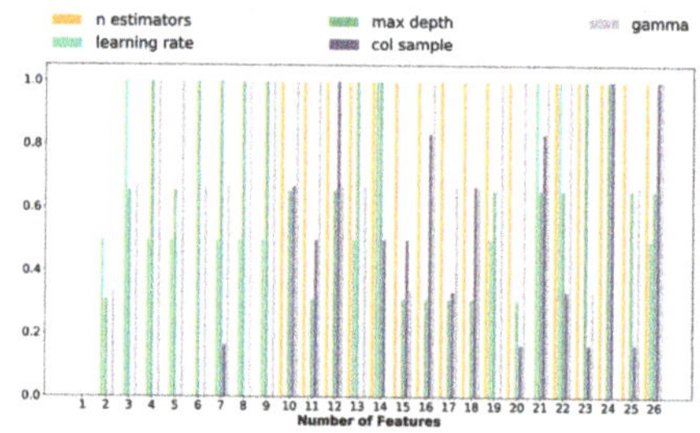

(**b**) Pitting: HP

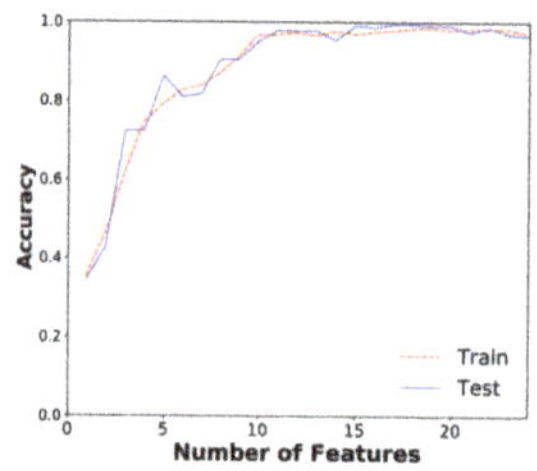

(**c**) Crack: Accuracy

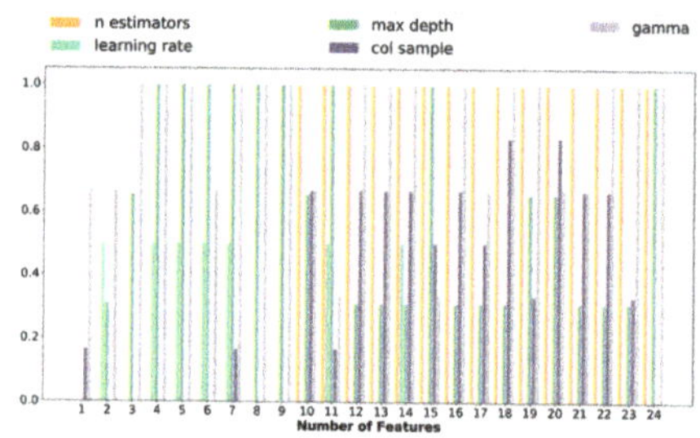

(**d**) Crack: HP

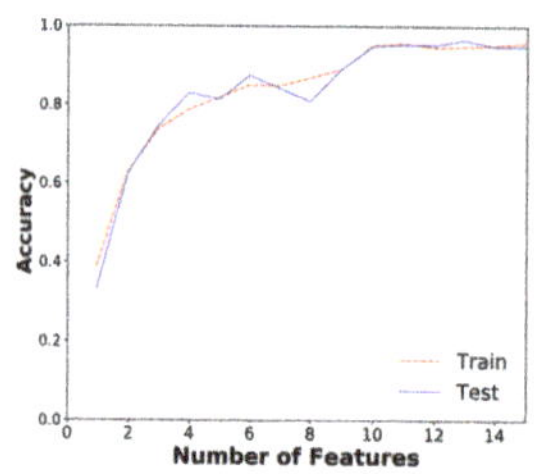

(**e**) Broken Tooth: Accuracy

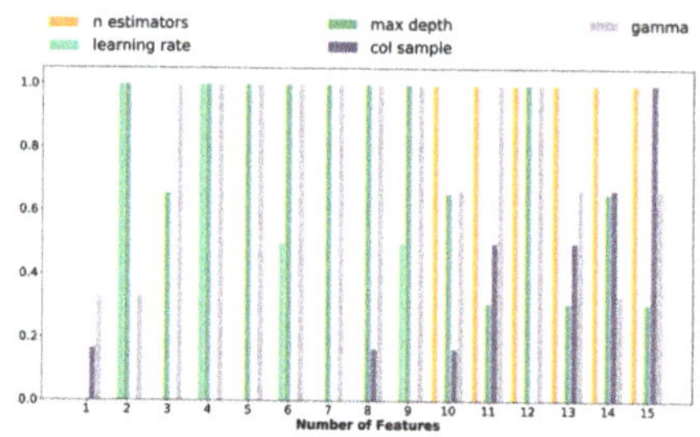

(**f**) Broken Tooth: HP

Figure 8. Plots show the impact of total features used by XGBoost classifier for each problem using individual feature lists. The left column (**a**,**c**,**e**) shows the impact on classification accuracy and the right column (**b**,**d**,**f**) shows the impact on hyperparameter optimization with Grid Search.

In the second experiment, as stated above, features were added using a common list of features based on the average feature importance values. Results are summarized in Figure 9 and several observations are pertinent when compared with the results in Figure 8. First, it is clear that performance does not decrease when using a common set of features, suggesting that the same set of features can be used to solve all three classification tasks. Second, hyperparameter optimization behaves very similar, across all problems and in both experimental setups. When using a few features, a large learning rate and deep trees are preferable, while more estimators, shallower trees and a smaller learning rate is better when using more features.

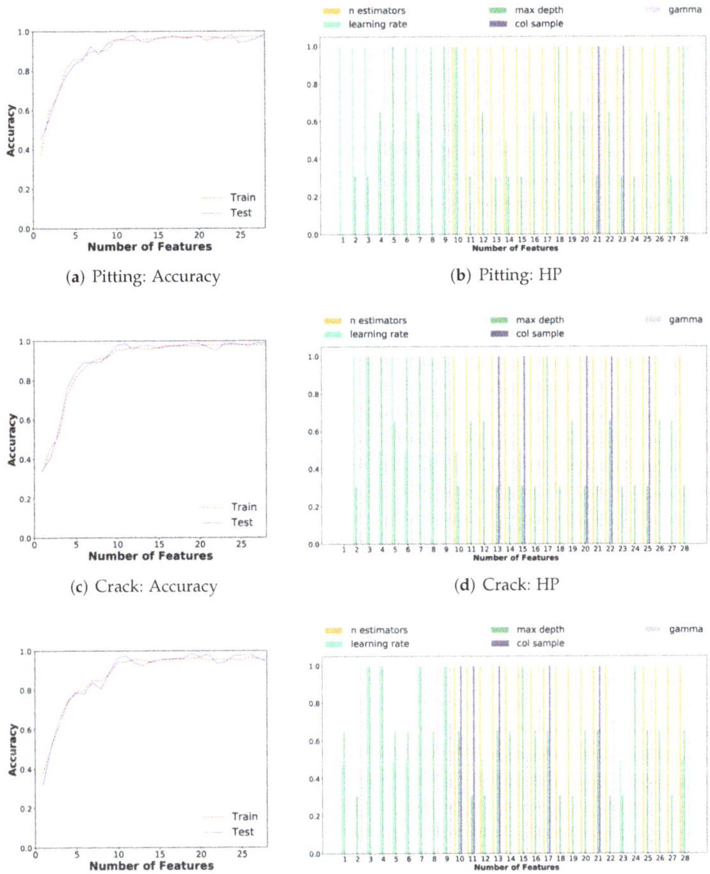

(a) Pitting: Accuracy (b) Pitting: HP

(c) Crack: Accuracy (d) Crack: HP

(e) Broken Tooth: Accuracy (f) Broken Tooth: HP

Figure 9. Plots show the impact of total features used by XGBoost classifier for each problem using a common feature list based on average feature importance. The left column (**a**,**c**,**e**) shows the impact on classification accuracy and the right column (**b**,**d**,**f**) shows the impact on hyperparameter optimization with Grid Search.

5.2. AutoML with TPOT

TPOT incorporates several similar basic elements as H2O does, which are feature transformation, feature engineering, feature selection, parameter optimization and modeling. However, it takes a different approach to do so, using a form of evolutionary search called Genetic Programming [47], and building on top of the well-known ML library Scikit-learn [56]. TPOT basically uses four types of pipeline elements, namely preprocessors, decomposition operators, feature selection and modeling. Preprocessors include various types of scaling techniques for input data, as well as basic feature engineering methods based on polynomial expansions of the input features. Decomposition operators include different forms of PCA, while feature selection techniques include recursive feature elimination (RFE) or filtering techniques such as SelectFwe, which is based on the family wise error of each feature [56]. For classification problems, modeling techniques include KNN, linear SVM, logistic regression, Gradient Boosting, Extra Tree classifiers and XGBoost, to name a few.

TPOT is configured to maximize classification accuracy as well as to minimize model complexity (given by the size of the pipeline). Since these two objectives can be potentially in conflict, TPOT employs a multi objective criteria to guide the search for the best ML pipeline for a given problem.

While TPOT does not include such a rich set of explicit feature engineering methods, it achieves the same type of results by employing stacking of models. In summary, stacking allows TOT to concatenate the output of several models in such a way that all the outputs from a particular model can be used as additional features to train another model which is further downstream in the pipeline. For instance, a Gradient Boosting model will produce m probabilistic outputs for each sample, where m is the number of classes; that is, the probability that a particular sample belongs to each class. This output vector can be concatenated with the original input features to form an extended feature set for another ML model.

5.2.1. TPOT Configuration and Results

The same datasets and cross validation procedure was used with TPOT as was done with H2O. However, the configuration of the system is notably different, with the underlying Genetic Programming algorithm requiring a set of specific hyperparameters to perform the run. For the presented experiments, we used the same configuration suggested in [47]. The average cross validation testing performance (accuracy) on each problem was: 0.99 for pitting, 0.98 for crack and 0.96 for Broken Tooth. Comparing these results with those presented in Table 2 shows that overall, the performance of the ML pipelines produced by both AutoML systems are quite similar. However, given the different approaches towards ML pipeline generation, the pipelines generated by TPOT are more heterogeneous than those produced by H2O.

TPOT Pitting Pipeline

For this problem, TPOT produced the most complex pipeline, which consisted of:

- First, an RFE feature selection step that reduced the feature space to hold the number of original features;
- A filtering of the features using SelectFwe, which left a total of 27 original features. It is of note that the number of features used in this pipeline is the same as those used by H2O on the same problem (see Table 2 for Pitting with Interpretability set to 10);
- Feature transformation by PCA, followed by a robust scaling;
- Finally, modeling was carried out by the Extra Tree classifier with 100 base learners.

TPOT Crack Pipeline

For this problem, TPOT produced the following pipeline:

- The pipeline stacked two models in this pipeline. The first model is a Multilayer Perceptron (MLP) with a learning rate of 1 and 100 hidden neurons;
- The outputs from the MLP were concatenated with the original feature set and used to train the second model, a Gradient Boosting classifier with learning rate 0.5, max depth of 7, minimum sample split of 19 and 100 base learners.

TPOT Broken Tooth Pipeline

For this problem, TPOT produced the following pipeline:

- The pipeline stacked two models in this pipeline. The first model is a Gradient Boosting classifier with learning rate 0.5, max depth of 4, minimum sample split of 10 and 100 base learners;
- The second model is a linear SVM classifier, with a squared hinge loss function, and L1 regularization.

It is noticeable that, unlike H2O, TPOT does not converge to XGBoost models, but does use a decision tree-based model in all three pipelines. Moreover, TPOT reduces the feature space on pitting, but increases the feature space with stacking on the other two problems.

5.2.2. Analysis of Feature Selection and Feature Importance

Unlike H2O, TPOT does not generate an explicit feature ranking or feature importance score, a notable and useful feature of the former method. However, it is possible to analyze the pipelines produced in each problem, and determine the relative feature importance assigned by the TPOT pipelines. In the following analysis, we will compare the feature selection and feature importance values produced by the TPOT pipelines relative to the results obtained by the H2O pipelines when using Interpretability = 10. In particular, we focus on the set of common features, those chosen in all three problems by the H2O pipelines; these are the following nine features from Table 3: x_1, x_6, x_7, x_8 ,x_{27}, x_{35} x_{41}, x_{54}, and x_{61}.

For the Pitting problem, there are two feature selection operators applied to the original data. After this filtering process, a total of 27 features were used to perform the classification. Relative to the common set of features found by H2O, seven of the nine features (77%) are also used by the TPOT pipeline in this problem, with the only exceptions being x_8 and x_{27} .

For the Crack problem, we can obtain a feature importance score from the Gradient Boosting classifier. It is important to remember that the Gradient Boosting classifier used a combination of synthetic features generated by the MLP and the original raw features. Based on this, we can rank the top 25 features used by the Gradient Boosting classifier, the same number of features used by the H2O pipeline. This ranking shows that all the top-ranked features are from the original feature set, without any of the synthetic features generated by the MLP. Moreover, among the top 25 features, the 9 common features from H2O are included (77%), with the only exceptions being x_{35} and x_{61}.

Finally, we can perform the same analysis for the BT problem, considering the top 15 features (based once again on the H2O results). In this case, we use the relative feature importance produced by both stacked models on the original feature set, namely the Gradient Boosting and linear SVC. In this case, the overlap with the common features produced by H2O is 8 out of 9 (88%), with the only exception x_{27}.

These results show that, overall, both AutoML pipelines converge to a very similar set of features for all problems, clearly indicating which features are more relevant in this domain.

6. Discussion

The previous section shows the proper performance of the original time-domain condition indicators for fault severity classification with the model obtained by the AutoML systems. The results are compared to alternative ones obtained by the authors when the best set of features and classification model is obtained by manually adjusting the classification parameters and taking the individual ranking provided by the feature ranking algorithm [80,81].

In [80], the same dataset of the pitting damage was analyzed for vibration signals and a subset of 24 time-domain condition indicators. Feature selection was conducted by using Chi-square-based ranking and a KNN classifier. Results show that 6 features can provide over 95% of accuracy and 12 features offer over 96%.

The method proposed in [81] was adopted to perform feature selection by using relief-based feature ranking and tested on a RF classifier, with the same dataset used in this work. Results are show in Table 6 and Figure 10.

Table 6. Accuracy by using RF-based classifier and feature ranking with RelieF.

Number of Features	Pitting	Broken Tooth	Crack
1	0.32	0.29	0.32
2	0.56	0.52	0.55
3	0.80	0.80	0.82
4	0.88	0.89	0.93
5	0.92	0.91	0.94
6	0.92	0.93	0.95
7	0.93	0.93	0.96
8	0.92	0.94	0.97
9	0.94	0.95	0.97
10	0.93	0.96	0.96
11	0.94	0.97	0.97
12	0.95	0.97	0.97
13	0.96	0.97	0.97
14	0.96	0.97	0.97
15	0.96	0.96	0.97
16	0.96	0.96	0.97
17	0.97	0.96	0.97
18	0.96	0.97	0.97
19	0.97	0.97	0.97
20	0.97	0.97	0.96

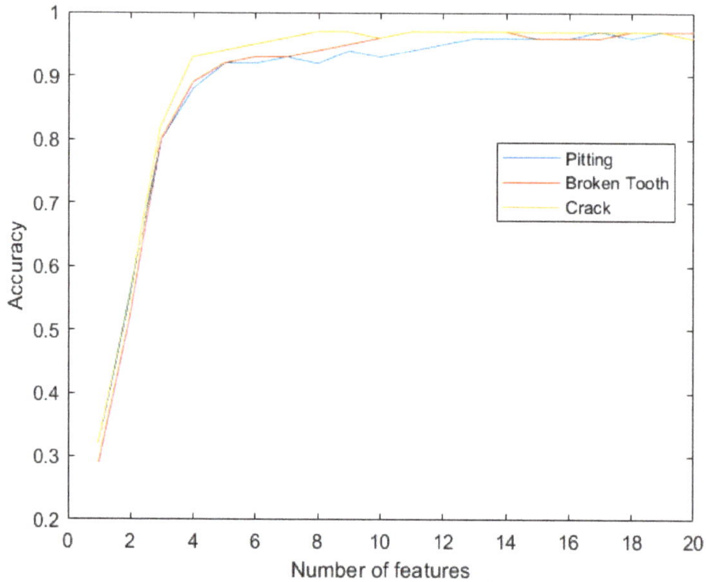

Figure 10. Accuracy trend by using RF based classifier and feature ranking with RelieF.

Results in Table 6 show that, for all the three failures modes, the accuracy remains above 97% with more than 13 features. By comparing to Figure 8, this trend is similar, and the accuracy is around 96% for pitting, 98% for crack, and 95% for Broken Tooth, when test examples are used. By comparing to Figure 9, when common features are used, the average accuracy over 98% is attained for pitting, 96% for crack and 97% for Broken tooth, by using over 13 features and test examples.

These results are similar for feature selection and classification using AutoML. The aggregate value of using AutoML is that the search is guided by an optimization problem, simplifying the underlying design process of the ML pipeline. However, at least on the tested problems, the overall performance of the resulting pipeline is not substantially different from those achieved by the standard pipeline design approach. On the other hand, given the simplified approach to ML development, this work was able to show that the same set of features can be used to solve three different fault diagnosis problems in gearboxes, which was previously unknown and not reported by other works in our knowledge.

7. Summary and Conclusions

This paper presents the application of two AutoML systems, H2O DAI and TPOT, for obtaining fault severity ML pipelines for spur gears under three different failures modes at different severity levels. The case study in this work is for fault severity assessment in gearboxes, treated as classification problems for three failure modes: pitting, crack and broken tooth. Fault severity assessment in gearboxes is not a trivial problem, as the gearboxes are mechanical systems with high non-linear and chaotic behaviors that usually work under changes in load and speed.

The use of AutoML to solve fault detection in gearboxes has not been previously reported in the literature, making this work the first to show the power of this approach to generate optimized ML models for this problem. Both AutoML systems are easy to use and provide both optimization modules and feature engineering modules (H2O explicitly, while TPOT mainly does this implicitly), which simplifies the design process of specialized ML pipelines.

The results and main conclusions in this paper are summarized in two ways. Regarding the general results of using AutoML in this domain:

- The setting of H2O DAI in the process of feature engineering is more explicit for the user. This is particularly useful when testing the creation of new features;
- Results of the evaluation and comparison between H2O DAI and TPOT show that both platforms select common features, regardless of the selected model by each platform. The size of the feature space used by each system varies, and neither of them is consistently more or less efficient in this regard;
- Classification accuracy when using all the features, without feature selection, remains very close for both systems, over 96%;
- The accuracy achieved by AutoML can be increased relative to a hand-tuned classification model, particularly by adjusting the feature selection technique.

Regarding the feature analysis of the generated AutoML pipelines, we can state the following:

- Time-domain statistical features are highly informative. This is verified by the fact that the feature engineering methods provided by the AutoML platforms do not substantially increase the classification accuracy of the ML pipelines. This particularity was identified because of the use of AutoML, and this discovery reduces the requirements of computing other complex features beyond the informative ones;
- Classification accuracy over 90% is obtained with 10 features, and over 95% with more than 13 features, for each failure mode, when problem-specific features are selected based on the relative feature importance. The use of AutoML permitted to set the proper number of features, and this directly improves the generalization capability of the ML model for fault diagnosis;
- Common features for all three failures modes can be selected based on average values of feature importance across all problems. These common features are highly informative as they achieve a classification accuracy over 96%. Moreover, the common set of features are ranked as highly informative for all problems by both AutoML systems. The analysis and use of the same set of features for all three failure modes has not been previously reported in the literature;
- The accuracy of the classifiers obtained by AutoML are highly competitive with the state-of-the-art in this domain, reaching 96% of accuracy and even 99% in some failure

modes. For comparison, accuracy by manual design of ML pipelines has been reported of up to 97% on the same datasets. This result verifies the power of the pipelines created from AutoML.

Future works can be focused on testing other AutoML platforms, like Neural Architecture Search (NAS) developed by Google DeepMind Team, to evaluate neural network architectures or a Bayesian approach, such as Auto-Sklearn [82], for the case study presented in this work.

Author Contributions: Conceptualization: M.C., L.T., R.V.S.; Methodology: J.C.M., D.C., D.E.H., H.A.C.Z.; Formal analysis and investigation: L.T., H.A.C.Z., J.C.M., D.E.H.; Writing—original draft preparation: L.T., R.V.S.; Writing—review and editing: M.C., D.E.H.; Funding acquisition: M.C., L.T., R.V.S.; Resources: M.C., L.T., R.V.S., D.C.; Supervision: M.C., L.T. All authors have read and agreed to the published version of the manuscript.

Funding: This work was supported by Universidad Politécnica Salesiana through the research group GIDTEC. Funding for this work was also provided by TecNM (Mexico) 2020 through the research project "Resolución de múltiples problemas de aprendizaje supervisado de manera simultánea con programación genética".

Conflicts of Interest: The authors declare that they have no conflict of interest.

References

1. Lei, Y.; Lin, J.; Zuo, M.J.; He, Z. Condition monitoring and fault diagnosis of planetary gearboxes: A review. *Measurement* **2014**, *48*, 292–305. [CrossRef]
2. Goyal, D.; Pabla, B.S.; Dhami, S.S. Condition monitoring parameters for fault diagnosis of fixed axis gearbox: A review. *Arch. Comput. Methods Eng.* **2017**, *24*, 543–556. [CrossRef]
3. Randall, R.B. *Vibration-Based Condition Monitoring: Industrial, Aerospace and Automotive Applications*; John Wiley & Sons: Hoboken, NJ, USA, 2011.
4. Li, F.; Li, R.; Tian, L.; Chen, L.; Liu, J. Data-driven time-frequency analysis method based on variational mode decomposition and its application to gear fault diagnosis in variable working conditions. *Mech. Syst. Signal Process.* **2019**, *116*, 462–479. [CrossRef]
5. Moradi, H.; Salarieh, H. Analysis of nonlinear oscillations in spur gear pairs with approximated modelling of backlash nonlinearity. *Mech. Mach. Theory* **2012**, *51*, 14–31. [CrossRef]
6. Litak, G.; Friswell, M.I. Dynamics of a gear system with faults in meshing stiffness. *Nonlinear Dyn.* **2005**, *41*, 415–421. [CrossRef]
7. Meng, Z.; Shi, G.; Wang, F. Vibration response and fault characteristics analysis of gear based on time-varying mesh stiffness. *Mech. Mach. Theory* **2020**, *148*, 103786. [CrossRef]
8. Jiang, Y.; Zhu, H.; Li, Z.; Peng, Z. The nonlinear dynamics response of cracked gear system in a coal cutter taking environmental multi-frequency excitation forces into consideration. *Nonlinear Dyn.* **2016**, *84*, 203–222. [CrossRef]
9. Cabrera, D.; Sancho, F.; Li, C.; Cerrada, M.; Sánchez, R.V.; Pacheco, F.; de Oliveira, J.V. Automatic feature extraction of time-series applied to fault severity assessment of helical gearbox in stationary and non-stationary speed operation. *Appl. Soft Comput.* **2017**, *58*, 53–64. [CrossRef]
10. Cerrada, M.; Li, C.; Sánchez, R.V.; Pacheco, F.; Cabrera, D.; Valente de Oliveira, J. A fuzzy transition based approach for fault severity prediction in helical gearboxes. *Fuzzy Sets Syst.* **2018**, *337*, 52–73. [CrossRef]
11. Park, S.; Kim, S.; Choi, J.H. Gear fault diagnosis using transmission error and ensemble empirical mode decomposition. *Mech. Syst. Signal Process.* **2018**, *108*, 262–275. [CrossRef]
12. Wang, D. K-nearest neighbors based methods for identification of different gear crack levels under different motor speeds and loads: Revisited. *Mech. Syst. Signal Process.* **2016**, *70–71*, 201–208. [CrossRef]
13. Lei, Y.; Zuo, M.J. Gear crack level identification based on weighted K nearest neighbor classification algorithm. *Mech. Syst. Signal Process.* **2009**, *23*, 1535–1547. [CrossRef]
14. Gharavian, M.; Almas Ganj, F.; Ohadi, A.; Heidari Bafroui, H. Comparison of FDA-based and PCA-based features in fault diagnosis of automobile gearboxes. *Neurocomputing* **2013**, *121*, 150–159. [CrossRef]
15. Lei, Y.; Zuo, M.J.; He, Z.; Zi, Y. A multidimensional hybrid intelligent method for gear fault diagnosis. *Expert Syst. Appl.* **2010**, *37*, 1419–1430. [CrossRef]
16. Hajnayeb, A.; Ghasemloonia, A.; Khadem, S.; Moradi, M. Application and comparison of an ANN-based feature selection method and the genetic algorithm in gearbox fault diagnosis. *Expert Syst. Appl.* **2011**, *38*, 10205–10209. [CrossRef]
17. Liao, Y.; Zhang, L.; Li, W. Regrouping particle swarm optimization based variable neural network for gearbox fault diagnosis. *J. Intell. Fuzzy Syst.* **2018**, *34*, 3671–3680. [CrossRef]
18. Ümütlü, R.C.; Hizarci, B.; Ozturk, H.; Kiral, Z. Classification of pitting fault levels in a worm gearbox using vibration visualization and ANN. *Sādhanā* **2020**, *45*, 22. [CrossRef]

19. Saravanan, N.; Siddabattuni, V.K.; Ramachandran, K. Fault diagnosis of spur bevel gear box using artificial neural network (ANN), and proximal support vector machine (PSVM). *Appl. Soft Comput.* **2010**, *10*, 344–360. [CrossRef]
20. Chen, F.; Tang, B.; Chen, R. A novel fault diagnosis model for gearbox based on wavelet support vector machine with immune genetic algorithm. *Measurement* **2013**, *46*, 220–232. [CrossRef]
21. Ray, P.; Mishra, D.P. Support vector machine based fault classification and location of a long transmission line. *Eng. Sci. Technol. Int. J.* **2016**, *19*, 1368–1380. [CrossRef]
22. Shen, Z.; Chen, X.; Zhang, X.; He, Z. A novel intelligent gear fault diagnosis model based on EMD and multi-class TSVM. *Measurement* **2012**, *45*, 30–40. [CrossRef]
23. Bordoloi, D.; Tiwari, R. Support vector machine based optimization of multi-fault classification of gears with evolutionary algorithms from time–frequency vibration data. *Measurement* **2014**, *55*, 1–14. [CrossRef]
24. Cerrada, M.; Zurita, G.; Cabrera, D.; Sánchez, R.V.; Artés, M.; Li, C. Fault diagnosis in spur gears based on genetic algorithm and random forest. *Mech. Syst. Signal Process.* **2016**, *70–71*, 87–103. [CrossRef]
25. Cabrera, D.; Sancho, F.; Sánchez, R.V.; Zurita, G.; Cerrada, M.; Li, C.; Vásquez, R.E. Fault diagnosis of spur gearbox based on random forest and wavelet packet decomposition. *Front. Mech. Eng.* **2015**, *10*, 277–286. [CrossRef]
26. Muralidharan, A.; Sugumaran, V.; Soman, K.; Amarnath, M. Fault diagnosis of helical gear box using variational mode decomposition and random forest algorithm. *Struct. Durab. Health Monit.* **2014**, *10*, 55.
27. Saravanan, N.; Ramachandran, K. Fault diagnosis of spur bevel gear box using discrete wavelet features and Decision Tree classification. *Expert Syst. Appl.* **2009**, *36*, 9564–9573. [CrossRef]
28. Sugumaran, V.; Jain, D.; Amarnath, M.; Kumar, H. Fault Diagnosis of Helical Gear Box using Decision Tree through Vibration Signals. *Int. J. Perform. Eng.* **2013**, *9*, 221.
29. Zhao, X.; Zuo, M.J.; Liu, Z. Diagnosis of pitting damage levels of planet gears based on ordinal ranking. In Proceedings of the 2011 IEEE Conference on Prognostics and Health Management, Montreal, QC, Canada, 20–23 June 2011; pp. 1–8.
30. Phinyomark, A.; Limsakul, C.; Phukpattaranont, P. A Novel Feature Extraction for Robust EMG Pattern Recognition. *J. Comput.* **2009**, *1*, 71–80.
31. Chowdhury, R.H.; Reaz, M.B.; Ali, M.A.B.M.; Bakar, A.A.; Chellappan, K.; Chang, T.G. Surface electromyography signal processing and classification techniques. *Sensors* **2013**, *13*, 12431–12466. [CrossRef]
32. Kim, S.; Choi, J.H. Convolutional neural network for gear fault diagnosis based on signal segmentation approach. *Struct. Health Monit.* **2019**, *18*, 1401–1415. [CrossRef]
33. Li, X.; Li, J.; Qu, Y.; He, D. Semi-supervised gear fault diagnosis using raw vibration signal based on deep learning. *Chin. J. Aeronaut.* **2020**, *33*, 418–426. [CrossRef]
34. Jing, L.; Zhao, M.; Li, P.; Xu, X. A convolutional neural network based feature learning and fault diagnosis method for the condition monitoring of gearbox. *Measurement* **2017**, *111*, 1–10. [CrossRef]
35. Singh, J.; Azamfar, M.; Ainapure, A.; Lee, J. Deep learning-based cross-domain adaptation for gearbox fault diagnosis under variable speed conditions. *Meas. Sci. Technol.* **2020**, *31*, 055601. [CrossRef]
36. Cabrera, D.; Sancho, F.; Cerrada, M.; Sánchez, R.V.; Li, C. Knowledge extraction from deep convolutional neural networks applied to cyclo-stationary time-series classification. *Inf. Sci.* **2020**, *524*, 1–14. [CrossRef]
37. Cabrera, D.; Sancho, F.; Long, J.; Sánchez, R.; Zhang, S.; Cerrada, M.; Li, C. Generative Adversarial Networks Selection Approach for Extremely Imbalanced Fault Diagnosis of Reciprocating Machinery. *IEEE Access* **2019**, *7*, 70643–70653. [CrossRef]
38. Monteiro, R.P.; Cerrada, M.; Cabrera, D.R.; Sánchez, R.V.; Bastos-Filho, C.J. Using a support vector machine based decision stage to improve the fault diagnosis on gearboxes. *Comput. Intell. Neurosci.* **2019**, *2019*, 1383752. [CrossRef] [PubMed]
39. Yao, Q.; Wang, M.; Chen, Y.; Dai, W.; Li, Y.F.; Tu, W.W.; Yang, Q.; Yu, Y. Taking Human out of Learning Applications: A Survey on Automated Machine Learning. *arXiv* **2018**, arXiv:1810.13306.
40. He, X.; Zhao, K.; Chu, X. AutoML: A Survey of the State-of-the-Art. *arXiv* **2019**, arXiv:1908.00709.
41. Petke, J.; Haraldsson, S.O.; Harman, M.; Langdon, W.B.; White, D.R.; Woodward, J.R. Genetic Improvement of Software: A Comprehensive Survey. *IEEE Trans. Evol. Comput.* **2018**, *22*, 415–432. [CrossRef]
42. Z-Flores, E.; Abatal, M.; Bassam, A.; Trujillo, L.; Juárez-Smith, P.; Hamzaoui, Y.E. Modeling the adsorption of phenols and nitrophenols by activated carbon using genetic programming. *J. Clean. Prod.* **2017**, *161*, 860–870. [CrossRef]
43. Pan, S.J.; Yang, Q. A Survey on Transfer Learning. *IEEE Trans. Knowl. Data Eng.* **2010**, *22*, 1345–1359. [CrossRef]
44. Weiss, K.; Khoshgoftaar, T.M.; Wang, D. A survey of transfer learning. *J. Big Data* **2016**, *3*, 9. [CrossRef]
45. Muñoz, L.; Trujillo, L.; Silva, S. Transfer learning in constructive induction with Genetic Programming. *Genet. Program. Evolvable Mach.* **2019**, *21*, 529–569. [CrossRef]
46. Liang, J.; Meyerson, E.; Hodjat, B.; Fink, D.; Mutch, K.; Miikkulainen, R. Evolutionary Neural AutoML for Deep Learning. In Proceedings of the Genetic and Evolutionary Computation Conference (GECCO '19), Prague, Czech Republic, 13–17 July 2019; ACM: New York, NY, USA, 2019; pp. 401–409.
47. Olson, R.S.; Bartley, N.; Urbanowicz, R.J.; Moore, J.H. Evaluation of a Tree-based Pipeline Optimization Tool for Automating Data Science. In Proceedings of the Genetic and Evolutionary Computation Conference 2016 (GECCO '16), Denver, CO, USA, 20–24 July 2016; ACM: New York, NY, USA, 2016; pp. 485–492.
48. Hall, P.; Kurka, M.; Bartz, A. Using H2O Driverless AI. Mountain View, CA, 2018. Available online: http://docs.h2o.ai (accessed on 30 November 2021).

49. Real, E.; Aggarwal, A.; Huang, Y.; Le, Q.V. Regularized evolution for image classifier architecture search. In Proceedings of the AAAI Conference on Artificial Intelligence, Honolulu, HI, USA, 27 January–1 February 2019; Volume 33, pp. 4780–4789.
50. Real, E.; Moore, S.; Selle, A.; Saxena, S.; Suematsu, Y.L.; Tan, J.; Le, Q.V.; Kurakin, A. Large-scale Evolution of Image Classifiers. In Proceedings of the 34th International Conference on Machine Learning-Olume 70 (ICML'17), Sydney, NSW, Australia, 6–11 August 2017; pp. 2902–2911.
51. Assunção, F.; Lourenço, N.; Machado, P.; Ribeiro, B. Evolving the Topology of Large Scale Deep Neural Networks. In *Genetic Programming*; Castelli, M., Ed.; Springer International Publishing: Cham, Switzerland, 2018; pp. 19–34.
52. Wong, C.; Houlsby, N.; Lu, Y.; Gesmundo, A. Transfer Learning with Neural AutoML. In Proceedings of the 32nd International Conference on Neural Information Processing Systems (NIPS'18), Montreal, QC, Canada, 3–8 December 2018; Curran Associates Inc.: Red Hook, NY, USA, 2018; pp. 8366–8375.
53. Stanley, K.O.; Miikkulainen, R. Evolving Neural Networks through Augmenting Topologies. *Evol. Comput.* **2002**, *10*, 99–127. [CrossRef]
54. Stanley, K.O.; D'Ambrosio, D.B.; Gauci, J. A Hypercube-Based Encoding for Evolving Large-Scale Neural Networks. *Artif. Life* **2009**, *15*, 185–212. [CrossRef] [PubMed]
55. Eiben, A.; Smith, J. *Introduction to Evolutionary Computing*; Springer: Berlin/Heidelberg, Germany, 2015. [CrossRef]
56. Pedregosa, F.; Varoquaux, G.; Gramfort, A.; Michel, V.; Thirion, B.; Grisel, O.; Blondel, M.; Prettenhofer, P.; Weiss, R.; Dubourg, V.; et al. Scikit-learn: Machine Learning in Python. *J. Mach. Learn. Res.* **2011**, *12*, 2825–2830.
57. Liang, X.; Zuo, M.J.; Feng, Z. Dynamic modeling of gearbox faults: A review. *Mech. Syst. Signal Process.* **2018**, *98*, 852–876. [CrossRef]
58. Jiang, H.; Liu, F. Mesh stiffness modelling and dynamic simulation of helical gears with tooth crack propagation. *Meccanica* **2020**, *55*, 1215–1236. [CrossRef]
59. Liang, X.H.; Liu, Z.L.; Pan, J.; Zuo, M.J. Spur gear tooth pitting propagation assessment using model-based analysis. *Chin. J. Mech. Eng.* **2017**, *30*, 1369–1382. [CrossRef]
60. Dadon, I.; Koren, N.; Klein, R.; Bortman, J. A step toward fault type and severity characterization in spur gears. *J. Mech. Des.* **2019**, *141*, 083301. [CrossRef]
61. Gecgel, O.; Ekwaro-Osire, S.; Dias, J.P.; Nispel, A.; Alemayehu, F.M.; Serwadda, A. Machine Learning in Crack Size Estimation of a Spur Gear Pair Using Simulated Vibration Data. In Proceedings of the 10th International Conference on Rotor Dynamics (IFToMM), Rio de Janeiro, Brazil, 23–27 September 2018; Cavalca, K.L., Weber, H.I., Eds.; Springer International Publishing: Cham, Switzerland, 2019; pp. 175–190.
62. Joshuva, A.; Sugumaran, V. Crack detection and localization on wind turbine blade using machine learning algorithms: A data mining approach. *Struct. Durab. Health Monit.* **2019**, *13*, 181. [CrossRef]
63. Pan, W.; He, H. An ordinal rough set model based on fuzzy covering for fault level identification. *J. Intell. Fuzzy Syst.* **2017**, *33*, 2979–2985. [CrossRef]
64. Wang, W.; Galati, F.A.; Szibbo, D. LSTM Residual Signal for Gear Tooth Crack Diagnosis. In *Advances in Asset Management and Condition Monitoring*; Ball, A., Gelman, L., Rao, B.K.N., Eds.; Springer International Publishing: Cham, Switzerland, 2020; pp. 1075–1090.
65. Qu, Y.; Zhang, Y.; He, M.; He, D.; Jiao, C.; Zhou, Z. Gear pitting fault diagnosis using disentangled features from unsupervised deep learning. *Proc. Inst. Mech. Eng. Part O J. Risk Reliab.* **2019**, *233*, 719–730. [CrossRef]
66. Elshawi, R.; Maher, M.; Sakr, S. Automated Machine Learning: State-of-The-Art and Open Challenges. *arXiv* **2019**, arXiv:1906.02287.
67. Patel, S.A.; Patel, S.P.; Adhyaru, Y.B.K.; Maheshwari, S.; Kumar, P.; Soni, M. Developing smart devices with automated Machine learning Approach: A review. In *Materials Today: Proceedings*; Elsevier: Amsterdam, The Netherlands, 2021.
68. Yi, H.; Bui, K.H.N. An automated hyperparameter search-based deep learning model for highway traffic prediction. *IEEE Trans. Intell. Transp. Syst.* **2020**, *22*, 5486–5495. [CrossRef]
69. Tokuyama, K.; Shimodaira, Y.; Kodama, Y.; Matsui, R.; Kusunose, Y.; Fukushima, S.; Nakai, H.; Tsuji, Y.; Toya, Y.; Matsuda, F.; et al. Soft-sensor development for monitoring the lysine fermentation process. *J. Biosci. Bioeng.* **2021**, *132*, 183–189. [CrossRef]
70. Zhu, J.; Shi, H.; Song, B.; Tao, Y.; Tan, S. Information concentrated variational auto-encoder for quality-related nonlinear process monitoring. *J. Process Control* **2020**, *94*, 12–25. [CrossRef]
71. Li, X.; Hu, Y.; Zheng, J.; Li, M. Neural Architecture Search For Fault Diagnosis. *arXiv* **2020**, arXiv:2002.07997.
72. Wang, R.; Jiang, H.; Li, X.; Liu, S. A reinforcement neural architecture search method for rolling bearing fault diagnosis. *Measurement* **2020**, *154*, 107417. [CrossRef]
73. Liu, T.; Kou, L.; Yang, L.; Fan, W.; Wu, C. A physical knowledge-based extreme learning machine approach to fault diagnosis of rolling element bearing from small datasets. In Proceedings of the 2020 ACM International Joint Conference on Pervasive and Ubiquitous Computing and Proceedings of the 2020 ACM International Symposium on Wearable Computers, Virtual Conference, 12–17 September 2020; pp. 553–559.
74. Listewnik, K.; Grzeczka, G.; Kłaczyński, M.; Cioch, W. An on-line diagnostics application for evaluation of machine vibration based on standard ISO 10816-1. *J. Vibroeng.* **2015**, *17*, 4248–4258.
75. Carvalho, D.V.; Pereira, E.M.; Cardoso, J.S. Machine Learning Interpretability: A Survey on Methods and Metrics. *Electronics* **2019**, *8*, 832. [CrossRef]

76. Juárez-Smith, P.; Trujillo, L.; García-Valdez, M.; Fernández de Vega, F.; Chávez, F. Local search in speciation-based bloat control for genetic programming. *Genet. Program. Evolvable Mach.* **2019**, *20*, 351–384. [CrossRef]
77. Hand, D.J.; Till, R.J. A Simple Generalisation of the Area Under the ROC Curve for Multiple Class Classification Problems. *Mach. Learn.* **2001**, *45*, 171–186. [CrossRef]
78. Chen, T.; Guestrin, C. XGBoost: A Scalable Tree Boosting System. In Proceedings of the 22nd ACM SIGKDD International Conference on Knowledge Discovery and Data Mining (KDD '16), San Francisco, CA, USA, 13–17 August 2016; Association for Computing Machinery: New York, NY, USA, 2016; pp. 785–794.
79. Sagi, O.; Rokach, L. Ensemble learning: A survey. *Wiley Interdiscip. Rev. Data Min. Knowl. Discov.* **2018**, *8*, e1249. [CrossRef]
80. Sánchez, R.V.; Lucero, P.; Vásquez, R.E.; Cerrada, M.; Cabrera, D. A comparative feature analysis for gear pitting level classification by using acoustic emission, vibration and current signals. *IFAC-PapersOnLine* **2018**, *51*, 346–352. [CrossRef]
81. Sánchez, R.V.; Lucero, P.; Vásquez, R.E.; Cerrada, M.; Macancela, J.C.; Cabrera, D. Feature ranking for multi-fault diagnosis of rotating machinery by using random forest and KNN. *J. Intell. Fuzzy Syst.* **2018**, *34*, 3463–3473. [CrossRef]
82. Feurer, M.; Eggensperger, K.; Falkner, S.; Lindauer, M.; Hutter, F. Auto-Sklearn 2.0: Hands-free AutoML via Meta-Learning. *arXiv* **2021**, arXiv:2007.04074.

Mathematical and Computational Applications

MDPI

Article

Attention Measurement of an Autism Spectrum Disorder User Using EEG Signals: A Case Study

José Jaime Esqueda-Elizondo [1,2], Reyes Juárez-Ramírez [2], Oscar Roberto López-Bonilla [1], Enrique Efrén García-Guerrero [1], Gilberto Manuel Galindo-Aldana [3], Laura Jiménez-Beristáin [2], Alejandra Serrano-Trujillo [2], Esteban Tlelo-Cuautle [4] and Everardo Inzunza-González [1,*]

1 Facultad de Ingeniería, Arquitectura y Diseño, Universidad Autónoma de Baja California, Carretera Transpeninsular Ensenada-Tijuana No. 3917, Ensenada C.P. 22860, Baja California, Mexico; jjesqueda@uabc.edu.mx (J.J.E.-E.); olopez@uabc.edu.mx (O.R.L.-B.); eegarcia@uabc.edu.mx (E.E.G.-G.)
2 Facultad de Ciencias Químicas e Ingeniería, Universidad Autónoma de Baja California, Calzada Universidad No. 14418, Parque Industrial Internacional, Tijuana C.P. 22390, Baja California, Mexico; reyesjua@uabc.edu.mx (R.J.-R.); ljimenezb@uabc.edu.mx (L.J.-B.); aserrano11@uabc.edu.mx (A.S.-T.)
3 Facultad de Ciencias de la Ingeniería y Tecnología, Universidad Autónoma de Baja California, Carretera Estatal No. 3, Gutiérrez, Mexicali C.P. 21720, Baja California, Mexico; gilberto.galindo.aldana@uabc.edu.mx
4 Departamento de Electrónica, Instituto Nacional de Astrofísica, Óptica y Electrónica, Luis Enrique Erro No. 1, Santa María Tonanzintla, Puebla C.P. 72840, San Andrés Cholula, Mexico; etlelo@inaoep.mx
* Correspondence: einzunza@uabc.edu.mx; Tel.: +52-646-175-0744

Citation: Esqueda-Elizondo, J.J.; Juárez-Ramírez, R.; López-Bonilla, O.R.; García-Guerrero, E.E.; Galindo-Aldana, G.M.; Jiménez-Beristáin, L.; Serrano-Trujillo, A.; Tlelo-Cuautle, E.; Inzunza-González, E. Attention Measurement of an Autism Spectrum Disorder User Using EEG Signals: A Case Study. *Math. Comput. Appl.* **2022**, *27*, 21. https://doi.org/10.3390/mca27020021

Academic Editors: Marcela Quiroz, Luis Gerardo de la Fraga, Adriana Lara, Leonardo Trujillo and Oliver Schütze

Received: 30 December 2021
Accepted: 24 February 2022
Published: 2 March 2022

Abstract: Autism Spectrum Disorder (ASD) is a neurodevelopmental life condition characterized by problems with social interaction, low verbal and non-verbal communication skills, and repetitive and restricted behavior. People with ASD usually have variable attention levels because they have hypersensitivity and large amounts of environmental information are a problem for them. Attention is a process that occurs at the cognitive level and allows us to orient ourselves towards relevant stimuli, ignoring those that are not, and act accordingly. This paper presents a methodology based on electroencephalographic (EEG) signals for attention measurement in a 13-year-old boy diagnosed with ASD. The EEG signals are acquired with an Epoc+ Brain–Computer Interface (BCI) via the Emotiv Pro platform while developing several learning activities and using Matlab 2019a for signal processing. For this article, we propose to use electrodes F3, F4, P7, and P8. Then, we calculate the band power spectrum density to detect the Theta Relative Power (TRP), Alpha Relative Power (ARP), Beta Relative Power (BRP), Theta–Beta Ratio (TBR), Theta–Alpha Ratio (TAR), and Theta/(Alpha+Beta), which are features related to attention detection and neurofeedback. We train and evaluate several machine learning (ML) models with these features. In this study, the multi-layer perceptron neural network model (MLP-NN) has the best performance, with an AUC of 0.9299, Cohen's Kappa coefficient of 0.8597, Matthews correlation coefficient of 0.8602, and Hamming loss of 0.0701. These findings make it possible to develop better learning scenarios according to the person's needs with ASD. Moreover, it makes it possible to obtain quantifiable information on their progress to reinforce the perception of the teacher or therapist.

Keywords: autism; attention; ASD; learning activities; EEG; BCI; features; artificial intelligence; machine learning

1. Introduction

Scientists have always been captivated by the brain, and cognitive processes are also the most intriguing for most people. A fundamental part of these cognitive processes is the attention process. To obtain knowledge, first, the attention process is needed. Attention is a cognitive process that enables selecting, focusing on, and sustained information processing [1]. The object of attention can either be an environmental stimulus actively processed by sensory systems or associative information and response alternatives generated by the ongoing cognitive activity. This allows us to orient ourselves towards relevant stimuli, ignoring those not, and act accordingly. Moreover, it is the basis of learning, and it is necessary to have it, in order to begin the learning process. There have been many measuring techniques, such as using the response times or the number of clicks given while using particular software, the eye contact time measured from videos, Magnetic Resonance Imaging (MRI) or functional Magnetic Resonance Imaging (fMRI) studies, among other techniques. Autism Spectrum Disorder (ASD) is a neurodevelopmental life condition characterized by problems with social interaction, low verbal and non-verbal communication skills, and repetitive and restricted behavior [2,3]. People with ASD usually have variable attention levels because they have hypersensitivity and large amounts of environmental information are a problem for them.

There are many methods for measuring attention reported in the literature, such as eye-tracking/gaze [4,5], fMRI [6,7], using a program [8], biofeedback [9], and electroencephalographic (EEG) signals [10–12], among others. The last one delivers great advantages over other neuroimaging techniques due to its high temporal resolution [13], neurodevelopmental diagnosis accuracy [14], cognitive-related bioelectrical data [15], low cost [16], and non-invasive application methods [17]. The authors [8] show how the attention of 49 children with ASD and a group of 51 typical children is measured using a mindfulness-based program (MBP); in other words, this is a computerized attention test. This MBP software measures the accuracy and reaction times, but they did not directly measure. Another way to measure attention is by analyzing the facial expressions or measuring the timing of eye contact from video recordings. The study [5] shows a measuring technique based on the analysis of video recordings of 1756 toddlers from 12 to 72 months with ASD while watching selected short videos on an iPhone or an iPad. Their facial expressions are video-recorded and analyzed as they watch the videos. Reference [18] presents a study about the concentration measurement of a children group while interacting with an NAO robot and their teacher. In this case, the eye contact time was measured by analyzing the video recordings of the sessions obtained with two cameras at the posterior.

The study [19] shows an approach to the joint analysis of EEG and eye-tracking for children's ASD evaluation. First, the synchronization measures, information entropy, and time-frequency features of the multi-channel EEG are derived. Then, a random forest is applied to the eye-tracking recordings of the same subjects to single out the most significant features. A convolutional graph network (GCN) model naturally fuses the two groups of features to differentiate the children with ASD from the typically developed (TD) subjects. Reference [20] uses EEG activity (raw EEG and alpha power) to provide a time-resolved index of attentional orienting towards salient stimuli that either matched or did not match target-defining properties. In all of the references presented above, the use of feature extraction techniques helps to obtain information from the signals acquired. These feature extraction techniques can help us to obtain useful or descriptive information while eliminating or reducing redundant or unnecessary information, noise, or artifacts. Once the feature extraction stage has finished, the classification can quantify the signals. This paper also shows the feature extraction and the classification algorithms most frequently used.

Nowadays, intelligent systems that incorporate artificial intelligence (AI) frequently rely on machine learning (ML) [21,22]. ML is a term that refers to a system's ability to learn from problem-specific training data in order to automate the process of developing analytical models and completing associated tasks [23,24]. Deep learning (DL) is a paradigm in machine learning that is based on the use of artificial neural networks [25,26]. Commonly, the use of ML algorithms is centered in the diagnosis or detection of ASD, as is presented in [20]. The authors in [27] used EEG and eye-tracking features to identify children with ASD. In [28], the authors used deep convolutional architectures to detect ASD. Other studies [29] reported statistical features for ASD classification. In reference [30], they used an ML and a DL process for diagnosing ASD from time-frequency spectrogram images of EEG. The authors in [31] reported that it is possible to evaluate mental stress using DL and

EEG records. There are also studies such as [32], where they used the free artifact signal of two electrodes to detect ASD. In [33], they used a hybrid light-weighted feature extractor from signal to spectrogram images.

Recent studies have a focus on the relationship between human and machine behavior, based on the premise that diverse social and psychological backgrounds correspond in practice with different modalities of human–computer interaction [34]. In general, EEG feature extraction techniques have offered strong clinical consistency since the beginning of their use for assessing and diagnosing different cognitive and neurological domains in ASD [35], learning difficulties [36], and attention [37]. It is widely accepted that AI techniques are helpful for automatic diagnosis and rehabilitation procedures in ASD cases. For example, in [38], a review of DL methods focusing on neuroimaging-based approaches is presented. Furthermore, the authors report a review of studies based on DL networks for diagnosing ASD and the challenges in automatized detection and ASD rehabilitation. Nowadays, there are some DL applications for brain disease diagnoses, such as the ones presented in [39] , which presents a review of automated multiple sclerosis (MS) detection methods based on MRI. They notice that the most used architectures for MS detection are convolutional neural networks (CNNs), autoencoders (AEs), generative adversarial networks (GANs), and CNN-RNN models. Schizophrenia (Sz) is another brain disease detected with DL methods using EEG signal processing [40]. The authors compare their results with the traditional AI methods, such as support vector machine (SVM), k-nearest neighbors, decision tree, naïve Bayes, random forest, extremely randomized trees, and bagging. The DL models used are long short-term memories (LSTMs), one-dimensional convolutional networks (1D-CNNs), and 1D-CNN-LSTMs. Convolutional neural networks and LSTMs perform best, cross-validated with a k-fold of 5. Moreover, epileptic seizures are detectable by using EEG signal processing; for example, in [41], the authors present a novel diagnostic procedure that uses fuzzy theory and DL techniques. They propose an adaptive neuro-fuzzy inference system (ANFIS) with a breeding swarm optimization (BS) method. These ANFIS-BS methods present accuracy of 99.74 % in a two-class classification task. Appendix A summarizes in Tables A1 and A2 the state of the art and shows a comparison with the proposed method, considering the dataset, data source, preprocessing, methods/algorithm, main findings, and applications.

The research questions that motivate this paper are: (1) What brain regions activate on average when attention increases? At what levels? Depending on the type of activity to be developed? (2) Can the level of the attention span of a person with Autism Spectrum Disorder be quantified as a feature using time-frequency analysis methods? (3) Is there a relationship between the increment in the power of electroencephalographic signals and attention span in a child with Autism Spectrum Disorder?

In this paper, the hypothesis is that measuring and quantifying the brain's electrical activity (power spectrum density) makes it possible to assess the level of attention when performing various cognitive activities and interacting with different software or systems. Therefore, this article aims to detect when an ASD user has high attention levels while developing learning activities based on the EEG signals acquired by an Epoc+ Brain–Computer Interface (BCI). The novelty of this paper is the use of ML algorithms to classify the "Attention" and "No Attention" states of an ASD user. This research presents a new methodology based on EEG signals and ML algorithms for classifying the attention of a 13-year-old boy with ASD. This research formulates a method for processing electroencephalographic signals to determine attention lapses in people with ASD, tested by performing various learning activities and interacting with computer programs.

The rest of this paper is organized as follows. Section 2 presents the materials and the proposed methodology. Section 3 shows the findings of this paper. Section 4 presents the discussion. Finally, Section 5 summarizes our conclusions.

2. Materials and Methods

The approval of this research by the Ethics Committee and Research for Pre-Graduates and Post-Graduates of the Facultad de Ingeniería y Negocios Guadalupe Victoria de la Universidad Autónoma de Baja California was obtained on 8 October 2020, with the POSG/020-1-04 register. The EEG signals were acquired with an Epoc+ Brain–Computer Interface (BCI) [42,43] via the Emotiv Pro platform while the ASD user developed several learning activities, and data were processed with Matlab 2019a and Emotiv Pro software using the Student Version.

Figure 1 depicts the electrode location (left) and the Emotiv Epoc+ headset (right). According to the coherence analysis in attention [44,45], the selected electrodes were F3, F4, P7, and P8.

The proposed methodology and the simulations were performed on a personal computer with the following specifications: Intel(R) Core i5-8250U CPU @ 1.60 GHz, 1800 Mhz, 4 Cores, 8 Logical Processors, and 8 GB in RAM.

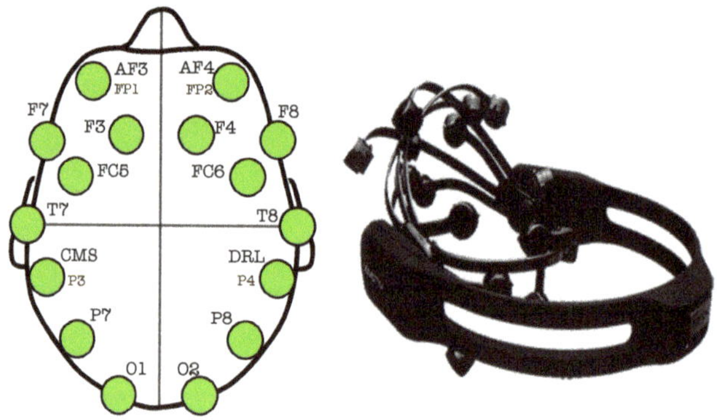

Figure 1. Electrode location (**left side**) of the Epoc+ headset (**right side**) of the Emotiv Inc., taken from Emotiv website https://emotiv.gitbook.io/epoc-user-manual/, accessed on 29 December 2021.

The signal was sampled at 2048 Hz, filtered with a dual-notch filter at 50 Hz and 60 Hz and a low-pass filter at 64 Hz, and then downsampled to 128 Hz for transmission. It was necessary to multiply the signal by 0.51 μ to convert it to a voltage.

The proposed data acquisition process is as follows:

Step 1. Place the headset with the electrodes hydrated on the test subject.
Step 2. Start the video recording and the EEG data acquisition.
Step 3. Give the worksheet to the test subject and the instructions.
Step 4. Let the test subject start the activity, and give him additional instructions if necessary, as in a regular school session.
Step 5. When the activity is over, stop video recording and data acquisition.

Figure 2 shows the EEG acquisition process and how the boy worked with the activity sheets using the Epoc+ headset.

Figure 2. Data acquisition process with the Emotiv Epoc+. The EEG recordings start once the localization of the headset is correct, and the signal quality, and the electrode contacts are verified and in green level. Pictures are from http://imagentv.uabc.mx/videos/electro-encefalograf%C3%ADas-y-autismo-uabc-no-se-detiene-imago, accessed on 29 December 2021.

2.1. Activity Sheets

Figures 3 and 4 depict examples of other activity sheets provided by the child's teachers, according to his knowledge and abilities. Figure 3 shows an activity sheet about reading, following instructions, and drawing. Figure 4 is a counting animal activity sheet. The school for children with ASD Eduke (https://www.facebook.com/EDUKE-123602824381330, accessed on 29 December 2021), located in Tijuana, Baja California, México, provided all the activity sheets used in this research.

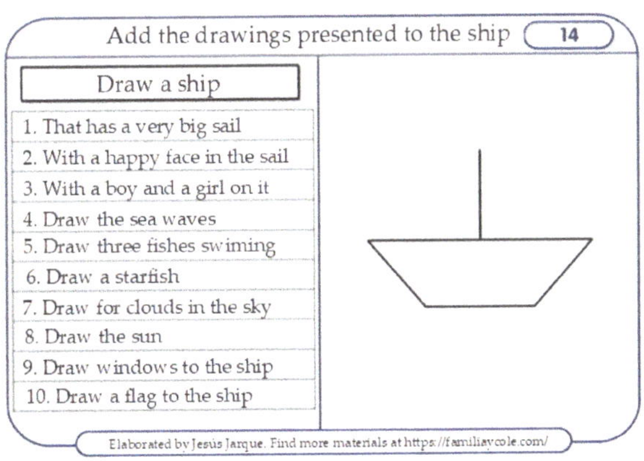

Figure 3. Example of reading, following instructions, and drawing activity sheet. This activity requires the child to read and follow instructions. The activity sheets are from https://familiaycole.com/, accessed on 29 December 2021.

Figure 4. Example of counting animals activity sheet. This activity requires the child to identify, classify, count the animals, and write the number in the white square. The activity sheets are from https://www.actividadesdeinfantilyprimaria.com/, accessed on 29 December 2021.

2.2. *Signal Processing Procedure*

Figure 5 depicts the block diagram of the procedure used for signal processing. The first step is preprocessing the EEG signal, and then the power spectrum density of signals is calculated and separated into bands. Next, we obtain the features presented in Table 1 and validate them. With these features, we train the machine learning algorithms. In the next section, we give more information about these steps.

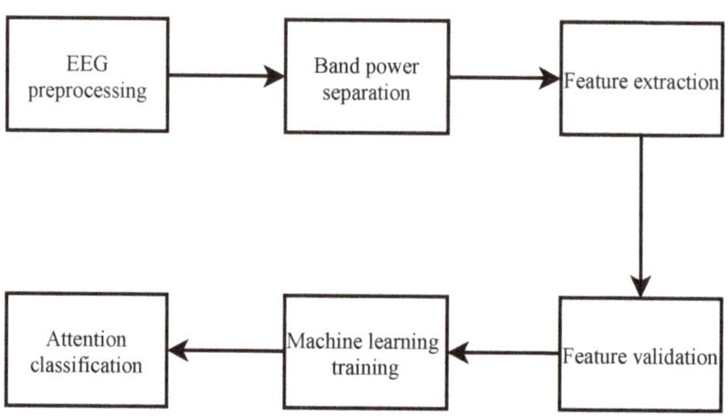

Figure 5. Block diagram of the proposed method. The first stage is signal preprocessing, after the band power separation, and then the feature extraction stage. Next is the feature validation process, the machine learning training stage, and finally, the attention quantification result.

2.2.1. Preprocessing of EEG Signal

The Emotiv software gives the recordings in a .csv file with integer numbers. It is necessary to convert the EEG signal acquired by the Epoc+ to its voltage equivalent by multiplying it by the factor 0.51×10^{-6}.

2.2.2. Band Power Separation

In EEG signal processing, it is common to separate the power spectrum density into the following bands: Delta (1–4 Hz), Theta (4–8 Hz), Alpha (8–12 Hz), Beta (12–30 Hz), and Ram (or Gamma) (30–50 Hz), depicted in Figure 6. These band powers [46] are the basis for calculating relative powers and ratios in the feature extraction stage. The Emotiv software gives the power of each band, except for the Delta band, and it gives the Beta band separated into Low Beta and High Beta powers [47]. For this research, we add both Beta band powers.

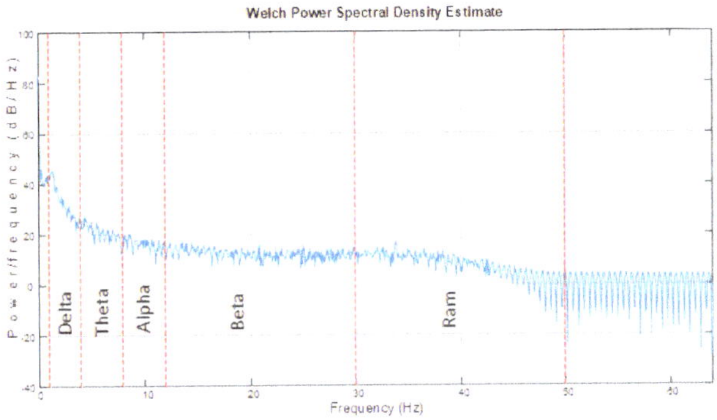

Figure 6. Band power separation example, Welch power spectral density estimate (illustrative figure).

The Emotiv software uses two-second windows to calculate the power spectrum density in absolute values, with units $\mu V^2/Hz$, and then separates it into bands. The two-second window involves 256 samples [47,48].

Figure 7 shows an example of band power separation. For this paper, we use the electrodes F3, P7, F4, and P8 because they show high coherence in attention tasks [44,45].

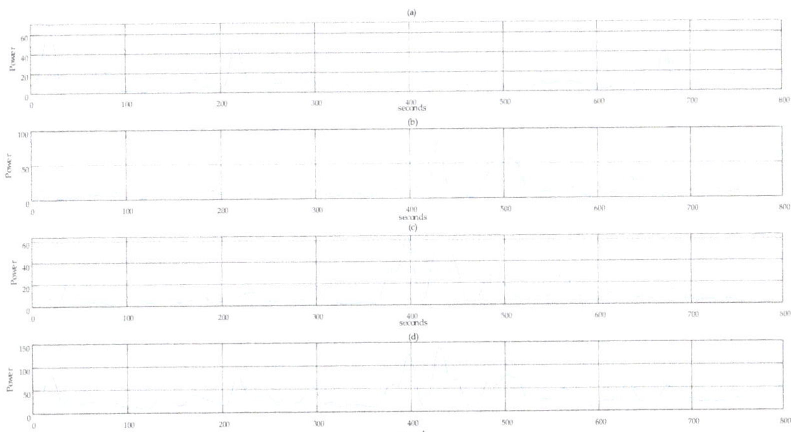

Figure 7. Band power separation example from F4 electrode. (**a**) Theta band power, (**b**) Alpha band power, (**c**) Beta band power, (**d**) Total band power.

2.2.3. Feature Extraction

To detect the Theta–Beta Ratio (TBR) and the Theta–Alpha Ratio (TAR), it is necessary first to calculate the band power spectrum density (PSD) of the EEG signal in two-second windows and for each channel or electrode. It is common to use the TBR features in attention detection and neurofeedback and the Theta Relative Power Beta and Theta/(Alpha + Beta), known as TBAR [48].

Table 1 presents the features calculated and their equations [48]. The next step is to use these features to train several machine learning models and evaluate their performance.

Table 1. Feature equations for attention detection.

Feature	Equation
Theta Relative Power	$\text{TRP} = \frac{\theta}{T}$
Alpha Relative Power	$\text{ARP} = \frac{\alpha}{T}$
Beta Relative Power	$\text{BRP} = \frac{\beta}{T}$
Theta–Beta Ratio	$\text{TBR} = \frac{\theta}{\beta}$
Theta–Alpha Ratio	$\text{TAR} = \frac{\theta}{\alpha}$
$\frac{\text{Theta}}{\text{Alpha + Beta}}$	$\text{TBAR}= \frac{\theta}{\beta+\alpha}$

$T = \theta + \alpha + \beta$ is the total power [48].

Figure 8 depicts the Theta, Alpha, and Beta relative powers (R.P.) obtained for the F4 electrode using the equations presented in Table 1. These R.P. values change with the time and function of the activity performance. Figure 9 shows the Theta–Beta Ratio, Theta–Alpha Ratio, and Theta/(Alpha–Beta Ratio) for the same F4 electrode.

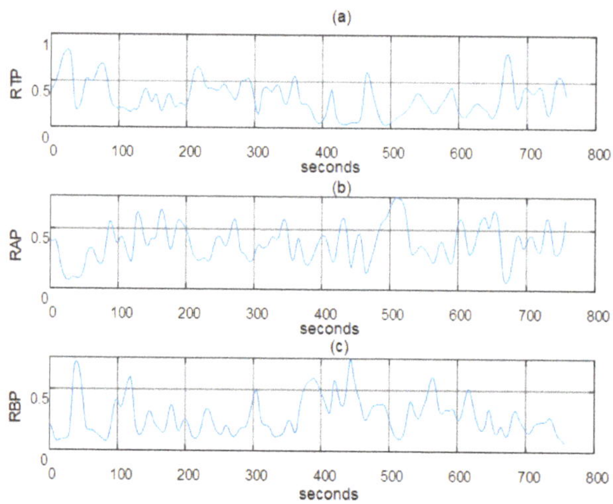

Figure 8. Example of relative powers obtained from F4 electrode. (**a**) F4 Theta relative power, (**b**) F4 Alpha relative power, and (**c**) F4 Beta relative power.

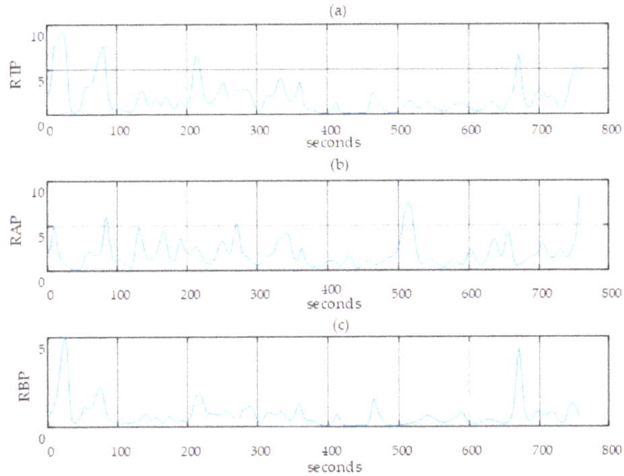

Figure 9. Example of ratios obtained from F4 electrode. (**a**) Theta–Beta Ratio, (**b**) Theta–Alpha Ratio, (**c**) Theta/(Alpha–Beta Ratio).

2.2.4. Dataset Preparation

The dataset consists of 24 features, 6 features for each electrode, with four electrodes (F3, F4, P7, and P8) and two classes: "Attention" and "No Attention". The dataset has 33,936 samples; it has 16,968 samples for each class to conserve balance. Figure A1 from Appendix B shows a fragment of the created dataset with 24 features acquired through the processing of EEG signals when the user is performing didactic activities and paying attention and when he is not paying attention to his learning process.

The Supplementary Materials dataset included six different Attention activities (counting, forming words, completing words, looking for differences between two figures, reading text, and answering simple questions from the reading), taken in 6 different moments. There are also No Attention samples recorded in non-learning activities such as watching cartoons, echolalia, doing nothing, and just sitting awake, trying to be as relaxed as possible.

2.2.5. Machine Learning Algorithm Training

In this paper, we chose eight ML algorithms to evaluate the classification of attention through the EEG signals of an ASD user. The chosen ML algorithms were naive Bayes (N.B.), stochastic gradient descent (SGD), decision trees (D.T.), support vector machine (SVM)-RBF, k-nearest neighbors (KNN), multi-layer perceptron neural network (MLP-NN), random forest (R.F.), and extra trees (E.T.). These ML models are part of the Scikit Learn library [49]. Figure 10 shows the flowchart to perform the training test of the ML algorithms. First, it is necessary to import the libraries or toolboxes required, such as Scikit Learn, Pandas, and Seaborn. Then, the features dataset is loaded; subsequently, separating the input data (features) from the output data is necessary. Next, we randomly divide the dataset, 80% for training and 20% for tests. Then, the data are scaled between 0 and 1 to obtain optimized results. Then, the machine learning model is trained. Then, we perform the scoring of the ML model, i.e., using the confusion matrix and performance metrics to evaluate the ML models.

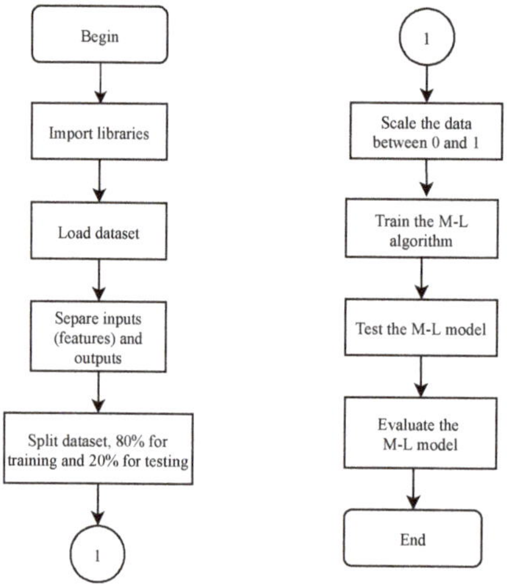

Figure 10. Flowchart for training and testing of ML algorithms.

3. Results

To evaluate the ML models, we rely on the metrics of the Scikit Learn library [49]. The metrics used to evaluate the scoring of the ML models are the confusion matrix (true positives, true negatives, false positives, false negatives), accuracy, F1 score, precision, sensitivity/recall, and specificity.

Table 2 shows the scoring parameters obtained for the ML models tested in this paper. The first four parameters correspond to the results of the confusion matrix. Naive Bayes with an accuracy of 0.7628, SGD with 0.8619, decision tree with 0.8697, SVM-RBF with 0.8940, KNN with 0.8968, MLP-NN with 0.9298, random forest with 0.9291, and finally extra trees with an accuracy of 0.9270. Therefore, the extra trees model has the best accuracy score.

Regarding the F1 score parameter, it is observable that naive Bayes, SGD, decision trees, and SVM-RBF obtained a score lower than 0.90. Meanwhile, the KNN, MLP-NN, random forest, and extra trees models obtained a score greater than 0.90, with extra trees achieving the highest score. Regarding the specificity/precision, we observed that the naive Bayes model was the lowest, while the extra trees and MLP-NN models were the highest, with 0.8896 and 0.9155, respectively. Regarding the sensitivity/recall score, all the models obtained a result greater than 0.90, except decision trees with 0.8720, and the extra trees model achieved the best result with 0.9738.

Table 3 shows the performance metrics obtained for each ML model. The metrics used to evaluate the performance of the ML models were the Area Under the Curve (AUC), the Cohen's Kappa coefficient, Hamming loss, and the Matthews correlation coefficient. Regarding the AUC metric, we notice that the naive Bayes, stochastic gradient descent, and decision trees models are the lowest, with 0.7642, 0.8624, and 0.8697, while the support vector machine (SVM)-RBF, KNN, extra trees, MLP-NN, and random forest (R.F.) models are the ones that obtained the best AUC, with 0.8944, 0.8972, 0.9274, 0.9299, and 0.9294, respectively, with the MLP-NN model obtaining a better AUC. This measure compares labelings by different human annotators, not a classifier versus ground truth, regarding Cohen's Kappa coefficient. The Kappa score is a number between −1 and 1. Scores above 0.8 indicate good agreement; zero or lower means no agreement (practically random labels). We observe that the naive Bayes, stochastic gradient descent, decision trees, support vector machine (SVM)-RBF, and KNN models obtained a Kappa coefficient less than 0.80 but greater than zero. However, the extra trees, MLP-NN, and random forest (R.F.) models obtained Kappa coefficients of 0.8542, 0.8597, and 0.8583, respectively, which are more significant than 0.80. Therefore, it means that these ML models have good agreement. We notice that the model MLP-NN is the one that obtained the highest Cohen's Kappa coefficient.

Table 2. Scoring parameters of the ML algorithms evaluated in this study.

	Machine-Learning Algorithm							
Scoring Parameters	**Naive Bayes**	**SGD**	**Decision Trees**	**(SVM)-RBF**	**KNN**	**MLP-NN**	**Random Forest (RF)**	**Extra Trees**
True positive	1984	2720	2967	2874	2892	3126	3039	3013
True negative	3194	3131	2937	3195	3196	3186	3268	3280
False positive	1436	700	453	546	528	294	381	407
False negative	174	237	431	173	172	182	100	88
Accuracy	0.7628	0.8619	0.8697	0.8940	0.8968	0.9298	0.9291	0.9270
F1 Score	0.7986	0.8698	0.8691	0.8988	0.9012	0.9304	0.9314	0.9278
Specificity/Precision	0.6898	0.8172	0.8663	0.8540	0.8582	0.9155	0.8955	0.8896
Sensitivity/Recall	0.9483	0.9296	0.8720	0.9486	0.9489	0.9459	0.9703	0.9738

Table 3. Performance metrics of the eight ML algorithms evaluated in this study.

	Performance Metrics			
Machine Learning Algorithm	**AUC**	**Cohen's Kappa Coefficient**	**Hamming Loss**	**Matthews Correlation Coefficient**
Naive Bayes	0.7642	0.5269	0.2371	0.5674
Stochastic Gradient Descent	0.8624	0.7241	0.1380	0.7310
Decision Trees	0.8697	0.7395	0.1302	0.7395
Support Vector Machine (SVM)-RBF	0.8944	0.7883	0.1059	0.7931
KNN	0.8972	0.7939	0.1031	0.7983
Extra Trees	0.9274	0.8542	0.0729	0.8580
MLP-NN	0.9299	0.8597	0.0701	0.8602
Random Forest (RF)	0.9294	0.8583	0.0708	0.8613

Regarding the Hamming loss, this Hamming loss should be zero; that is, the closer it is to zero, the model tends to be perfect or ideal. In this case, the extra trees, MLP-NN, and random forest (R.F.) models have the lowest Hamming loss. The MLP-NN model has the lowest Hamming loss, with 0.0701. We use in machine learning the Matthews correlation coefficient (MCC) or phi coefficient as a measure of the quality of binary (two-class) classifications, introduced by biochemist Brian W. Matthews [50]. In this case, the three best models are extra trees, MLP-NN, and random forest (R.F.), with 0.8580, 0.8602, and 0.8613, respectively, with random forest being the best (R.F.).

Figure 11 depicts the ROC curve of the top five ML models trained for attention classification using EEG data. The ROC curve shows the trade-off between sensitivity (TPR) and specificity (1-FPR). Classifiers that give curves closer to the top-left corner indicate better performance. The closer the curve comes to the 45-degree diagonal of the ROC space, the less accurate the test is. The SVM-RBF and KNN models are closer to the 45-degree diagonal, resulting in less accuracy. On the other hand, the random forest, extra trees, and MLP-NN models are closest to the upper left. Therefore, they are the ones with the best performance.

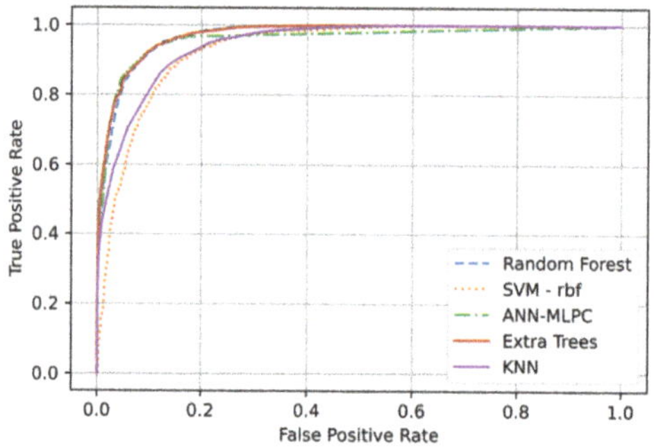

Figure 11. The receiver operating characteristic curve (ROC) of the top five ML models trained for attention classification using EEG data.

Figure 12 depicts the training time of the eight ML models tested in this study. The N.B., SGD, KNN, and D.T. models have the shortest training time. However, according to the results shown in Tables 2 and 3, they have the lowest performance metrics. In contrast, the SVM-RBF, R.F., and MLP-NN models have a longer training time of 17.01, 21.14, and 73.10 s, with the MLP-NN model having a longer training time. However, the model also has better performance metrics, as shown in Tables 2 and 3. Therefore, the classifier designer must conduct a cost–benefit analysis in terms of accuracy and processing time. In most cases, programmers prefer better accuracy, sacrificing training time since this process (training) is only done once and only uses the trained model. For this reason, in this study, it would be more convenient to choose the MLP-NN model.

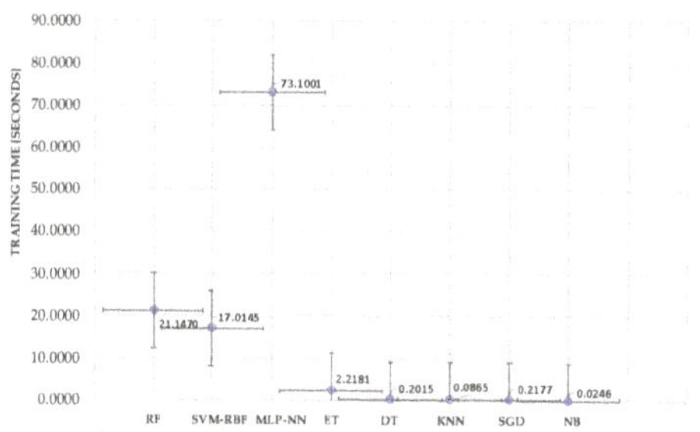

Figure 12. Training time of the eight ML models evaluated in this study.

4. Discussion

In this research, we observed that the power spectrum density (PSD) is helpful for attention detection, as proposed in the hypothesis. The features based on band PSD, such as Relative Theta Power (RTP), Relative Alpha Power (RAP), Relative Beta Power (RBP), Theta–Beta Ratio (TBR), Theta–Alpha Ratio (TAR), and the TBAR are good features for attention classification. With these features, the multi-layer perceptron neural network model (MLP-NN) achieved the best performance, with an AUC of 0.9299, Cohen's Kappa coefficient of 0.8597, Matthews correlation coefficient of 0.8602, and Hamming loss of 0.0701. Nevertheless, MLP-NN requires a longer training time of up to 73.1 s. However, the results presented in Tables 2 and 3 and Figures 11 and 12 show that the random forest and extra trees models have good performance metrics and a training time of 21.14 and 2.21, respectively. Therefore, the classifier designer must perform a cost–benefit analysis in terms of accuracy and processing time. In most cases, designers prefer better accuracy, sacrificing training time since this process (training) is only performed once, and then only the trained model is used. For this reason, in this study, it would be more convenient to choose the MLP-NN model.

Furthermore, feature extraction improves the acquisition of relevant information for accuracy for diagnosis and has been widely applied to different neuropsychological and neurophysiological fields [51]; the type of waveforms and definition of the morphology of EEG patterns increases the amount of available information for clinical decision making from brain dysfunction [52] to cognitive impairment [53]. Particular interest has been historically directed to the frontal areas in attention measurement, as they correspond to the brain regions responsible for activity direction and orientation. Classification of features may help to describe cortical connectivity, particularly for attentional deficits associated with frontal theta in children [36]. Other research refers to frontal bilateral theta waves in resting EEG in children with learning difficulties and an association with bilateral synchronous frontal theta waves [37], which closely relates to techniques for brain activity description in this study.

Limitations of the Study

One of the limitations of this research is that a BCI is required. The ASD user should not have much hair. The BCI must be pleasant and tolerated by them. Moreover, the electrodes must be kept hydrated with saline solution. It also depends on the battery life of the BCI. The emotional state of the ASD user is essential because good measurements will not be obtained if altered. Activities should be done in a scenario with learning conditions without distractions, such as a classroom.

5. Conclusions

In this paper, a methodology for the classification of attention by EEG signals of an ASD user was presented. The EEG data acquisition was performed while the ASD user performed some didactic learning activities. In addition, our dataset was created for the post-processing of the information and training of the ML algorithms. To create the dataset, it was necessary to perform preprocessing, filtering, and feature extraction. The proposed features can be used to train and evaluate several ML models to classify attention using EEG signals.

On the other hand, with these findings, therapists, teachers, and psychologists can develop better learning scenarios according to the cognitive needs of ASD users. In addition, diagnosis accuracy can be improved by acquiring individual EEG features, which provide relevant information for differential clinical neurodevelopmental symptomatology classification. Furthermore, with the proposed methodology, one can obtain quantifiable information about the performance of ML models when an ASD user performs didactic/learning activities, the above with the purpose of reinforcing the perception of the teacher or therapist.

The future work will involve implementing the proposed method on a real-time embedded system—for example, a stand-alone version using an edge device, novel deep learning methods, and internet of things (IoT). It is possible to explore the feasibility of a mobile-based platform that links with a BCI, instead of a computer. Furthermore, future replication of this methodology is needed to approach a broad spectrum of attention processes and standard estimation.

Supplementary Materials: The following supporting information can be downloaded at: https://www.mdpi.com/article/10.3390/mca27020021/s1.

Author Contributions: Conceptualization, R.J.-R. and E.I.-G.; Data curation, L.J.-B. and A.S.-T.; Formal analysis, E.E.G.-G., L.J.-B. and A.S.-T.; Funding acquisition, J.J.E.-E. and O.R.L.-B.; Investigation, J.J.E.-E. and E.T.-C.; Methodology, J.J.E.-E., G.M.G.-A. and E.I.-G.; Project administration, O.R.L.-B.; Resources, J.J.E.-E. and O.R.L.-B.; Software, J.J.E.-E.; Supervision, R.J.-R. and E.I.-G.; Validation, G.M.G.-A.; Visualization, A.S.-T.; Writing—original draft, J.J.E.-E.; Writing—review and editing, E.E.G.-G., E.T.-C. and E.I.-G. All authors have read and agreed to the published version of the manuscript.

Funding: This research had funds provided by the Universidad Autónoma de Baja California (UABC) through the grants number 679 and 300/2610.

Institutional Review Board Statement: The study was conducted according to the guidelines of the Declaration of Helsinki, and approved by the Ethics Committee and Research for Pre-Graduates and Post-Graduates of the Facultad de Ingeniería y Negocios Guadalupe Victoria of the Universidad Autónoma de Baja California; it was approved on 8 October 2020, with the POSG/020-1-04 code.

Informed Consent Statement: Written informed consent has been obtained from the patient to publish this paper.

Data Availability Statement: We share the dataset as Supplementary Material.

Acknowledgments: We want to thank Escudero Garrido Elena Patricia (Interdisciplinary Professional Unit of Biotechnology of the National Polytechnic Institute, UPIBI-IPN), Gutiérrez Montiel Ixchel (Interdisciplinary Professional Unit of Biotechnology of the National Polytechnic Institute, UPIBI-IPN), Macarena de Haro Iliana Elizabeth (Interdisciplinary Professional Unit of Biotechnology of the National Polytechnic Institute, UPIBI-IPN), Martínez Amaya Laura Fernanda (Manizalez Autonomous University, Colombia), Martínez Oliva Gonzalo Guillermo (Interdisciplinary Pro-fessional Unit of Biotechnology of the National Polytechnic Institute, UPIBI-IPN), López Rivas Andrea (Autonomous University of Baja California, UABC) and Becerril Valenzuela Brando (Autonomous University of Baja California, UABC) and Martinez Verdin Annette Sofia (Autonomous University of Baja California, UABC), for their participation in this project via the Summer of Scientific and Technological Research of the Pacific or Research Activities of UABC. We want to thank the Chemical Sciences and Engineering Faculty of the Autonomous University of Baja California (UABC) for supporting the project with grant number 300/2610. The authors would like to thank INAOE for accepting researcher Everardo Inzunza-González to carry out his sabbatical stay. Thanks are given to PRODEP (Professional Development Program for Professors) for supporting the academic groups to increase their degree of consolidation. We also want to thank the Tijuana Special Education School Eduke and the teachers Jessica Avelar, Letycia Gutiérrez, and Atziri Torres, for supplying the activity sheets and guiding us in working with the ASD person.

Conflicts of Interest: The authors declare no conflict of interest. The funders had no role in the design of the study; in the collection, analyses, or interpretation of data; in the writing of the manuscript, or in the decision to publish the results.

Abbreviations

The following abbreviations are used in this manuscript:

AI	Artificial Intelligence
ANFIS	Adaptive Neuro-Fuzzy Inference System
ASD	Autism Spectrum Disorder
BS	Breeding Swarm
CNN	Convolutional Neural Networks
DL	Deep Learning
fMRI	Functional Magnetic Resonance Imaging
GAN	Generative Adversarial Networks
LSTM	Long Short-Term Memories
MBP	Mindfulness-Based Program
ML	Machine Learning
MRI	Magnetic Resonance Imaging
PSD	Power Spectrum Density
RAP	Relative Alpha Power
RBP	Relative Beta Power
RNN	Recurrent Neural Network
RTP	Theta Relative Power
TAR	Theta–Alpha Ratio
TBR	Theta–Beta Ratio
TD	Typically Developed

Appendix A. Comparison of the Proposed Method with the State of the Art

Table A1. State of the art, part 1.

Reference	Dataset	Data Source	Preprocessing	Method/Algorithm	Main Findings	Application
Ref. [20]	12 ASD users and 12 typical children	EEG	N/A	Trial-averaged phase-locking value (PLV) approach and cubic support vector machine (SVM)	95.8% Accuracy, 100% Sensitivity, and 92% Specificity	ASD Classification/Detection
Ref. [27]	97 children aged from 3 to 6	EEG, eye-tracking tests individually on own-race and other-race stranger faces stimuli.	Data were band-pass filtered between 0.5 and 45 Hz. To improve computing speed, EEG data were then down-sampled to 250 Hz. Power line noise in EEG was removed by a notch filter centered at 50 Hz. Artifacts in EEG were removed using an ICA approach (EEGLab).	SVM, minimum-redundancy-maximum-relevance (MRMR).	Classification Accuracy from combining two types of data reached a maximum of 85.44%, AUC 0.93, when 32 features were selected.	ASD Classification/Detection
Ref. [28]	10 typically developing children (6 Male and 6 Female) and 10 autistic children (6 Male and 4 Female).	Natus Nihon Ohden MEB9000 version 05–81.	22 channels, sampling frequency of 500 Hz and filtered with a low pass filter and a high pass filter at a frequency range of [0.53, 70] Hz. After filtering out the signal at a frequency range of 0.53 to 70 Hz, the ocular artifacts in the EEG signal were removed by thresholding. The threshold was set based on the average value of the amplitude of the eye blink signal. The eye blink signal was observed for 10 seconds with the eye open and eye close event.	ResNet50	Average Accuracy of 81%	ASD Classification/Detection
Ref. [29]	N/A	EEG	Wavelet Transform, reduction of dimensionality, removal of irrelevant data.	K-Nearest Neighbour (KNN), A Correlation-based Feature Selection (CFS), Minimum Redundancy Maximum Relevance (MRMR) and the Information Gain (IG).	N/A	ASD Classification/Detection
Ref. [30]	Dataset from King Abdulaziz University (KAU) Hospital, Saudi Arabia. It is a public available dataset found in Sixteen subjects with twelve from ASD group (3 girls and 9 boys, age 6–20 years old) and four subjects from control group (all boys, 9–13 years old).	Spectrogram of EEG	Artifacts were removed from raw EEG data with re-referencing, filtering and normalization. Common average referencing (CAR) is used for re-referencing. IIR filter is used to low pass filter the signal at 40 Hz cut off frequency and finally the filtered signals from each electrode is normalized to the interval $[-1, 1]$. signals are segmented into 3.5 second window frames for each subject to the dataset. Using Short-Time Fourier Transform (STFT) for each of the above segments, the spectrogram plot is generated in the last step and saved as image.	NB, LDA, RF, kNN, LR and SVM. Ten-fold cross validation. Three different CNN models.	The proposed DL based model achieves higher accuracy (99.15%) compared to the ML based model (95.25%) on an ASD EEG dataset and also outperforms existing methods.	ASD Classification/Detection
Ref. [31]	5 Neurotypical and 8 ASD.	EEG	High-pass filtered at 1 Hz to remove slow trends and subsequently low-pass filtered at 50 Hz to remove line noise. The routine clinical bandwidth for EEG is from 0.5 to 50 Hz.	ML classifiers, namely support vector machine (SVM) and deep learning methods.	Multiclass two-layer LSTM RNN deep learning classifier is capable of identifying mental stress from ongoing EEG with an overall accuracy of 93.27%.	ASD Mental Stress
Ref. [32]	Study 1, 15 teenagers. Study 2, 20 subjects diagnosed with ASD and 20 subjects diagnosed with other neuropsychiatric disorders.	Artifact-free EEG data.	Raw EEG time-series were analyzed with a features extraction algorithm to extract 794 quantitative features (TSFRESH Python package).	TWIST, Sine-net ANN and Back Propagation ANN.	Sine-net ANN reached the best predictive capability in distinguishing autistic cases from typicals in study 1, Accuracy of 100%. Back Propagation ANN reached the best predictive capability in distinguishing autistic cases from subjects affected by other neuropsychiatric disorders in study 2 with an overall accuracy of 94.95%.	ASD Classification/Detection

Table A2. State of the art, part 2.

Reference	Dataset	Data Source	Preprocessing	Method/Algorithm	Main Findings	Application
Ref. [33]	122 subjects.	EEG to image converted	Spectrogram image generation model is presented using a combination of 1D_LBP and STFT.	Decision Tree (DT) , Discriminant Analysis (DA), Logistic Regression (LR), SVM, K-Nearest Neighbor (kNN).	SVM classifier reached 96.44% Accuracy	ASD Classification/Detection
Ref. [38]	The disorder group comprises of 8 boys (10–16 years), and the normal group consists of 10 boys (9–16 years). Neuroimaging: ABIDE-I and ABIDE-II, which encompasses sMRI, rs-fMRI, and phenotypic data. ABIDE-I: a total of 1112 datasets, 539 individuals with ASD and 573 healthy individuals (ages 64–7). ABIDEII: 1114 datasets from 521 individuals with ASD and 593 healthy individuals (ages 5–64).	Neuroimaging, fMRI, MRI, EEG	Several preprocesing applied to all signals.	Supervised learning, unsupervised learning, and reinforcement learning (RL).	Presents challenges and performances of DL techniques	ASD Classification/ Rehabilitation
Ref. [39]	MICCAI 2008, MICCAI 2016, ISBI 2015, and eHealth Lab.	MRI	Low level and high level preprocessing methods in MRI.	Most popular DL architectures for MS detection: convolutional neural networks (CNNs), Autoencoders (AEs), generative adversarial networks (GANs), and CNN-RNN models.	The inaccessibility of huge sMRI datasets belonging to a diverse population and lack of access to fMRI modalities are among the most important dataset-related challenges which are discussed in detail. Moreover, DL-related challenges include researchers' lack of access to powerful hardware resources for MS diagnosis research.	Multiple Sclerosis Detection
Ref. [40]	Dataset of the Institute of Psychiatry and Neurology in Warsaw.	EEG	EEG signals were divided into 25 s time frames and then normalized by z-score or norm L2.	EEG signals Classification: support vector machine, k-nearest neighbors, Decision Tree, Naive Bayes, Random Forest, extremely randomized trees, and bagging. DL models: long short-term memories (LSTMs), one-dimensional convolutional networks (1D-CNNs), and 1D-CNN-LSTM.	CNN-LSTM model accuracy of 99.25%.	Schizophrenia/Diagnosis
Ref. [41]	Bonn University dataset with six classification combinations and the Freiburg dataset.	EEG	Tunable-Q wavelet transform (TQWT) for EEG signal decomposition. Feature extraction, 13 different fuzzy entropies calculated from TQWT. Six layers Autoencoder (AE) for dimensionality reduction.	Classification: Adaptive neuro-fuzzy inference system (ANFIS), and also its variants with grasshopper optimization algorithm (ANFIS-GOA), particle swarm optimization (ANFIS-PSO), and breeding swarm optimization (ANFIS-BS).	ANFIS-BS two classes classification Accuracy: 99.74%; an Accuracy of 99.46% in ternary classification on the Bonn dataset, and 99.28% on the Freiburg dataset.	Detection of epileptic seizures
Proposed method	1 Subject, 33936 samples	EEG and BCI	Scaling, 2 seconds Band Power Separation. Features: TRP, ARP, BRP, TBR, TAR, and Theta/(alpha+beta).	Naive Bayes, Stochastic Gradient Descent, Decision trees, SVM, KNN, MLP-NN, RF, Extra trees.	(MLP-NN) with Accuracy 92.98%, Sensitivity 94.59%, F1 Score 93.04%, Specificity 91.55%, AUC of 0.9299, Cohen's Kappa coefficient of 0.8597, Matthews correlation coefficient of 0.8602, and Hamming loss of 0.0701.	ASD Attention Classification

Appendix B. Fragment of the Dataset Created for This Study

	A	B	C	D	E	F	G	H	I	J	K	L	M	N	O	P	Q	R	S	T	U	V	W	X	Y
1	Rtp_F3	Rap_F3	Rbp_F3	Tbr_F3	Tar_F3	Tbar_F3	Rtp_P7	Rap_P7	Rbp_P7	Tbr_P7	Tar_P7	Tbar_P7	Rtp_F4	Rap_F4	Rbp_F4	Tbr_F4	Tar_F4	Tbar_F4	Rtp_P8	Rap_P8	Rbp_P8	Tbr_P8	Tar_P8	Tbar_P8	Classes
2	0.0102257	0.0080753	0.9816991	0.0104163	0.0082258	0.0103313	0.0182301	0.0173039	0.964466	0.0189017	0.0179414	0.0185686	0.4418397	0.3412736	0.2168867	2.0371914	1.5735111	0.7916	0.3602113	0.1162547	0.523534	0.6880381	0.2220576	0.5630161	0
3	0.0124157	0.0081978	0.9793865	0.012677	0.0083704	0.0125717	0.019533	0.0170307	0.9634362	0.0202743	0.0176771	0.0199222	0.4339794	0.35182	0.2142006	2.0260418	1.6424793	0.76672	0.357329	0.1170774	0.5255935	0.6798581	0.2227528	0.5560062	0
4	0.0137813	0.0079854	0.9782333	0.0140879	0.0081631	0.0139739	0.0228004	0.0164545	0.9607451	0.023732	0.0171268	0.0233324	0.4158206	0.3685092	0.2156702	1.9280392	1.7086701	0.7118029	0.3554891	0.1135679	0.530943	0.6695429	0.2138984	0.5515642	0
5	0.0137383	0.0075644	0.9786973	0.0140373	0.007729	0.0139297	0.0279555	0.0159336	0.9561109	0.0292388	0.016665	0.0287595	0.4025978	0.3809536	0.2164486	1.8600157	1.7600186	0.6739142	0.3551171	0.1080237	0.5368592	0.6614716	0.2012143	0.5506691	0
6	0.0127689	0.007206	0.9800251	0.0130292	0.0073528	0.0129341	0.0348482	0.0166259	0.9485259	0.0367393	0.0175282	0.0361064	0.4040353	0.3820412	0.2139235	1.8886906	1.7858775	0.6779518	0.343822	0.0985567	0.5576213	0.6165869	0.1767449	0.5239767	0
7	0.0114414	0.0068034	0.9817553	0.011654	0.0069298	0.0115738	0.0422853	0.0195124	0.9382023	0.0450706	0.0207976	0.0441523	0.431724	0.3619059	0.20637	2.0919899	1.7536748	0.7597084	0.3205963	0.0889285	0.5904752	0.5429463	0.150605	0.4718789	0
8	0.0102782	0.0061561	0.9835657	0.01045	0.006259	0.010385	0.0499361	0.0257875	0.9242764	0.0540272	0.0279003	0.0525608	0.4715165	0.3270261	0.2014574	2.3405267	1.6233016	0.8922065	0.28744	0.0850815	0.6274785	0.4580875	0.1355927	0.4033907	0
9	0.0096363	0.0053189	0.9850448	0.0097826	0.0053997	0.00973	0.0564655	0.0347005	0.908834	0.0621296	0.0381814	0.0598447	0.5021135	0.2932252	0.2046613	2.453388	1.4327344	1.0084899	0.257834	0.0903469	0.6518191	0.3955606	0.1386073	0.3474074	0
10	0.0097321	0.0045489	0.985719	0.0098731	0.0046148	0.0098278	0.060201	0.0437183	0.8960807	0.0671826	0.0487884	0.0640574	0.5034565	0.2790178	0.2175257	2.3144689	1.2826886	1.0139223	0.2474764	0.1038717	0.6486519	0.3815243	0.1601347	0.328862	0
11	0.0102577	0.0039654	0.9857768	0.0104057	0.0040226	0.010364	0.0593589	0.0494421	0.891199	0.0666057	0.0554782	0.0631047	0.4696732	0.2949334	0.2353934	1.9952689	1.2529384	0.8856296	0.2665594	0.1194774	0.6139631	0.4341619	0.1946004	0.363437	0
12	0.0108793	0.0035984	0.9855223	0.0110391	0.0036513	0.0109989	0.0546509	0.0507077	0.8946414	0.0610869	0.0566794	0.0578102	0.4092739	0.3387292	0.2519969	1.6241226	1.3441801	0.6928318	0.3125059	0.1304666	0.5570275	0.5610241	0.2342193	0.4545579	0
13	0.0114132	0.0034962	0.9850905	0.011586	0.0035492	0.011545	0.0476615	0.0480615	0.904277	0.0527068	0.0531491	0.0500468	0.3399286	0.3990302	0.2610412	1.3022029	1.5286102	0.5149876	0.3724082	0.1328731	0.4947188	0.7527674	0.268583	0.5933923	0
14	0.011693	0.0037608	0.9845462	0.0118766	0.0038199	0.0118314	0.0406044	0.0429516	0.916444	0.0443065	0.0468677	0.0423229	0.2864039	0.4581543	0.2554418	1.1212101	1.7935761	0.401353	0.4293552	0.1252955	0.4453492	0.9640866	0.2813422	0.7524037	0
15	0.0112239	0.0044795	0.9842966	0.011403	0.004551	0.0113513	0.0354243	0.0368054	0.9277703	0.0381821	0.0396708	0.0367252	0.2664664	0.4992771	0.2342566	1.1374979	2.1313259	0.363264	0.465237	0.1096699	0.425093	1.0944358	0.2579905	0.8699874	0
16	0.0094187	0.0053526	0.9852287	0.0095599	0.0054328	0.0095083	0.0330734	0.030314	0.9366126	0.0353117	0.0323655	0.0342047	0.300691	0.5009614	0.1983476	1.5159799	2.5256743	0.429983	0.461863	0.0911257	0.4470112	1.0332247	0.2038556	0.858263	0
17	0.0070177	0.0056709	0.9873114	0.0071079	0.0057438	0.0070673	0.034861	0.0249049	0.9402342	0.0370769	0.0264879	0.0361201	0.3601675	0.4763241	0.1635084	2.2027454	2.9131465	0.562909	0.4272422	0.0782682	0.4944896	0.8640063	0.1582808	0.7459386	0
18	0.005506	0.0053482	0.9891459	0.0055664	0.0054068	0.0055365	0.038034	0.0197272	0.9422388	0.0403656	0.0209365	0.0395378	0.4430177	0.4226684	0.1343139	3.2983748	3.1468695	0.7953891	0.3617783	0.072336	0.5658858	0.6393132	0.1278278	0.5668536	0
19	0.0049982	0.0047888	0.990213	0.0050476	0.0048361	0.0050233	0.0435218	0.0171621	0.9393161	0.0463335	0.0182709	0.0455021	0.5006572	0.3778448	0.1214979	4.1207059	3.1098871	1.0026324	0.3062104	0.0743903	0.6193993	0.4943667	0.1201007	0.4413592	0
20	0.0051411	0.0043519	0.990507	0.0051904	0.0043936	0.0051677	0.0460258	0.016129	0.9378453	0.0490761	0.0171979	0.0482463	0.5447706	0.3380112	0.1172182	4.6474925	2.8836074	1.1966947	0.2644763	0.0772228	0.6583008	0.401756	0.1173063	0.3595755	0
21	0.0058207	0.0041266	0.9900527	0.0058792	0.0041681	0.0058548	0.0495438	0.018041	0.9324152	0.0531349	0.0193487	0.0521263	0.557043	0.320445	0.122512	4.5468425	2.6156203	1.2575553	0.2509923	0.0817746	0.667233	0.3761689	0.1225578	0.3350998	0
22	0.0073671	0.0042264	0.9884065	0.0074535	0.004276	0.0074217	0.0510271	0.0207348	0.9282381	0.054972	0.0223378	0.0537709	0.564282	0.306124	0.1295939	4.3542324	2.3621791	1.2950626	0.2460561	0.0853953	0.6685487	0.3680451	0.1277323	0.3263586	0
23	0.0101763	0.0047204	0.9851033	0.0103302	0.0047918	0.0102809	0.054881	0.0245947	0.9205243	0.0596193	0.0267182	0.0580678	0.5690571	0.2915452	0.1393978	4.0822544	2.0914625	1.3204929	0.240471	0.0905182	0.6690108	0.3594427	0.1353016	0.3166055	0
24	0.0143229	0.0055827	0.9800944	0.0146138	0.0056961	0.014531	0.060537	0.0281295	0.9113335	0.0664268	0.0308663	0.0644378	0.5847976	0.2664661	0.1487363	3.9317736	1.7915336	1.4084637	0.2237892	0.097089	0.6791218	0.3295273	0.1429626	0.2883098	0
25	0.0187164	0.0065112	0.9747724	0.0192008	0.0066797	0.0190734	0.0698402	0.031085	0.8990748	0.07768	0.0345745	0.075084	0.6071146	0.2343	0.1585854	3.8283129	1.4774371	1.5452715	0.1986098	0.1071318	0.6942584	0.2860748	0.1543112	0.2478316	0
26	0.0211509	0.0069246	0.9719245	0.0217619	0.0071246	0.0216079	0.0821744	0.0329856	0.88484	0.0928692	0.0372786	0.0895316	0.6330781	0.1986249	0.168297	3.7616724	1.1802047	1.7253758	0.172614	0.1231824	0.7042036	0.2451195	0.1749244	0.2086257	0
27	0.0209215	0.0065909	0.9724876	0.0215134	0.0067774	0.0213686	0.0943953	0.0338323	0.8717724	0.1082798	0.0388087	0.1042346	0.6522901	0.1683934	0.1793165	3.6376471	0.9390847	1.8759609	0.1493887	0.1447674	0.705844	0.2116455	0.2050982	0.1756251	0
28	0.0199334	0.0059587	0.9741079	0.0204633	0.006117	0.0203388	0.1072521	0.0361022	0.8566458	0.1252	0.0421436	0.120137	0.6477593	0.1523461	0.1998946	3.2405039	0.7621322	1.8389675	0.1334291	0.1740861	0.6924848	0.1926816	0.2513934	0.1539736	0
29	0.0202276	0.005485	0.9742873	0.0207615	0.0056298	0.0206453	0.1101597	0.03885	0.8509903	0.1294488	0.0456527	0.1237971	0.6242423	0.1454705	0.2302872	2.7107118	0.6316916	1.6612893	0.1252188	0.206933	0.6678482	0.1874959	0.3098504	0.143143	0
30	0.0223704	0.0052405	0.9723891	0.0230056	0.0053893	0.0228823	0.1083944	0.0440203	0.8475853	0.1278861	0.0519362	0.1215721	0.5776638	0.1451681	0.2771681	2.0841638	0.5237547	1.3677817	0.1272812	0.2439911	0.6287277	0.2024424	0.3880712	0.1458444	0
31	0.0260355	0.0051935	0.968771	0.0268748	0.0053609	0.0267315	0.0985104	0.0478026	0.853687	0.1153941	0.0559954	0.1092751	0.542735	0.1392828	0.3179822	1.7068097	0.4380209	1.1869157	0.134648	0.2749728	0.5903792	0.2280705	0.4657563	0.1555992	0
32	0.031128	0.0055411	0.9633309	0.0323129	0.005752	0.0321281	0.0877064	0.0506965	0.8615971	0.1017952	0.0588402	0.0961384	0.5093662	0.1363757	0.3542581	1.4378394	0.3849615	1.0381801	0.1475369	0.3010823	0.5513808	0.2675771	0.5460515	0.1730713	0
33	0.0354484	0.0063008	0.9582508	0.0369928	0.0065753	0.0367511	0.0789929	0.0530242	0.8679829	0.0910074	0.061089	0.0857679	0.4969496	0.1333259	0.3697245	1.344108	0.3606089	0.9878724	0.1611231	0.3151137	0.5237632	0.3076258	0.601634	0.19207	0
34	0.0358521	0.0074341	0.9567138	0.0374742	0.0077704	0.0371853	0.0710942	0.0532618	0.875644	0.0811908	0.0608259	0.0765354	0.4774386	0.1361777	0.3863838	1.2356589	0.3524414	0.9136506	0.1770675	0.3219488	0.5009837	0.3534396	0.6426332	0.2151665	0
35	0.0298397	0.0084185	0.9617418	0.0310268	0.0087534	0.0307575	0.0685476	0.0543052	0.8771472	0.0781483	0.0619111	0.0735922	0.4555648	0.1413245	0.4031107	1.1301232	0.3505847	0.8367659	0.1928804	0.3212049	0.4859148	0.3969428	0.6610313	0.2389737	0
36	0.0210839	0.0090167	0.9698994	0.0217382	0.0092966	0.021538	0.0677271	0.0523913	0.8798816	0.076973	0.0595436	0.0726473	0.4346541	0.1465537	0.4187922	1.0378753	0.3499438	0.7688285	0.2020514	0.3086883	0.4892603	0.4129733	0.6309286	0.2532136	0
37	0.0135885	0.0092488	0.9771627	0.013906	0.0094649	0.0137757	0.0730188	0.0494948	0.8774864	0.0832136	0.0564052	0.0787705	0.4290578	0.1468533	0.4240889	1.0117166	0.3462795	0.7514908	0.1968479	0.2823358	0.5208163	0.3779604	0.5421025	0.2450942	0
38	0.0088801	0.0092361	0.9818838	0.0090439	0.0094065	0.0089596	0.0781066	0.0414949	0.8803985	0.0887173	0.047132	0.0847241	0.4606392	0.1368442	0.4025166	1.1443981	0.3399716	0.8540465	0.174063	0.2430136	0.5829234	0.2986035	0.4168878	0.2107461	0
39	0.0069396	0.0091711	0.9838893	0.0070532	0.0093213	0.006988	0.0780478	0.0309431	0.8910091	0.0875949	0.0347282	0.084655	0.5204878	0.1205016	0.3590107	1.4497836	0.335649	1.0854526	0.1460077	0.2065428	0.6474495	0.2255121	0.3190099	0.1709708	0
40	0.0071394	0.0090914	0.9837691	0.0072572	0.0092414	0.0071908	0.0680201	0.0221527	0.9098271	0.0747616	0.0243483	0.0729845	0.5896496	0.103153	0.3071975	1.919448	0.3357872	1.4369415	0.1242805	0.1812023	0.6945172	0.1789451	0.2609041	0.1419181	0
41	0.0091097	0.0090057	0.9818846	0.0092778	0.0091719	0.0091935	0.0542515	0.0183296	0.927419	0.0584972	0.0197641	0.0573635	0.6461257	0.0902613	0.263613	2.4510391	0.3424008	1.8258624	0.1125915	0.1668085	0.7206	0.1562469	0.2314855	0.1268768	0

Figure A1. Fragment of the dataset created for this study.

References

1. Howe, T.R.; Trotter, J.S.; Davis, A.S.; Schofield, J.W.; Allen, L.; Millians, M.; Bolt, N. *Attention Span*; Springer: New York, NY, USA, 2011. [CrossRef]
2. American Psychiatric Association. *Diagnostic and Statistical Manual of Mental Disorders: DSM-5*, 5th ed.; American Psychiatric Publishing: Washington, DC, USA, 2013; p. 947.
3. Goqvkqpcn, E.; Ogpvcn, E. Autism Spectrum Disorder. *Nat. Rev. Dis. Prim.* **2020**, *6*, 6. [CrossRef]
4. Ishizaki, Y.; Higuchi, T.; Yanagimoto, Y.; Kobayashi, H.; Noritake, A.; Nakamura, K.; Kaneko, K. Eye gaze differences in school scenes between preschool children and adolescents with high-functioning autism spectrum disorder and those with typical development. *BioPsychoSoc. Med.* **2021**, *15*, 2. [CrossRef] [PubMed]
5. Egger, H.L.; Dawson, G.; Hashemi, J.; Carpenter, K.L.H.; Espinosa, S.; Campbell, K.; Brotkin, S.; Schaich-Borg, J.; Qiu, Q.; Tepper, M.; et al. Automatic emotion and attention analysis of young children at home: A ResearchKit autism feasibility study. *NPJ Digit. Med.* **2018**, *1*, 20. [CrossRef] [PubMed]
6. Son, J.; Ai, L.; Lim, R.; Xu, T.; Colcombe, S.; Franco, A.R.; Cloud, J.; LaConte, S.; Lisinski, J.; Klein, A.; et al. Evaluating fMRI-Based Estimation of Eye Gaze During Naturalistic Viewing. *Cereb. Cortex* **2020**, *30*, 1171–1184. [CrossRef]
7. Lawrence, S.J.D.; Formisano, E.; Muckli, L.; De Lange, F.P. Laminar fMRI: Applications for cognitive neuroscience. *NeuroImage* **2019**, *197*, 785–791. [CrossRef]
8. Ridderinkhof, A.; De Bruin, E.I.; Driesschen, S.v.d.; Bögels, S.M. Attention in Children with Autism Spectrum Disorder and the Effects of a Mindfulness-Based Program. *J. Atten. Disord.* **2018**, *24*, 681–692. [CrossRef] [PubMed]
9. Ababkova, M.; Leontieva, V.; Trostinskaya, I.; Pokrovskaia, N. Biofeedback as a cognitive research technique for enhancing learning process. *IOP Conf. Ser. Mater. Sci. Eng.* **2020**, *940*, 012127. [CrossRef]
10. Lau-Zhu, A.; Lau, M.; McLoughlin, G. Mobile EEG in research on neurodevelopmental disorders: Opportunities and challenges. *Dev. Cogn. Neurosci.* **2019**, *36*, 100635. [CrossRef]
11. Mehmood, F.; Ayaz, Y.; Ali, S.; Amadeu, R.D.C.; Sadia, H. Dominance in Visual Space of ASD Children Using Multi-Robot Joint Attention Integrated Distributed Imitation System. *IEEE Access* **2019**, *7*, 168815–168827. [CrossRef]
12. Wang, H.; Song, Q.; Ma, T.; Cao, H.; Sun, Y. Study on Brain-Computer Interface Based on Mental Tasks. In Proceedings of the 5th Annual IEEE International Conference on Cyber Technology in Automation, Control and Intelligent Systems, Shenyang, China, 8–12 June 2015; pp. 841–845. [CrossRef]
13. Ismail, L.E.; Karwowski, W. Applications of EEG indices for the quantification of human cognitive performance: A systematic review and bibliometric analysis. *PLoS ONE* **2020**, *15*, e0242857. [CrossRef] [PubMed]
14. Niemarkt, H.J.; Jennekens, W.; Maartens, I.A.; Wassenberg, T.; van Aken, M.; Katgert, T.; Kramer, B.W.; Gavilanes, A.W.; Zimmermann, L.J.; Bambang Oetomo, S.; et al. Multi-channel amplitude-integrated EEG characteristics in preterm infants with a normal neurodevelopment at two years of corrected age. *Early Hum. Dev.* **2012**, *88*, 209–216. [CrossRef] [PubMed]
15. Micoulaud-Franchi, J.A.; Batail, J.M.; Fovet, T.; Philip, P.; Cermolacce, M.; Jaumard-Hakoun, A.; Vialatte, F. Towards a Pragmatic Approach to a Psychophysiological Unit of Analysis for Mental and Brain Disorders: An EEG-Copeia for Neurofeedback. *Appl. Psychophysiol. Biofeedback* **2019**, *44*, 151–172. [CrossRef] [PubMed]
16. Singh, M.I.; Singh, M. Development of low-cost event marker for EEG-based emotion recognition. *Trans. Inst. Meas. Control* **2017**, *39*, 642–652. [CrossRef]
17. Yang, L.; Wilke, C.; Brinkmann, B.; Worrell, G.A.; He, B. Dynamic imaging of ictal oscillations using non-invasive high-resolution EEG. *Neuroimage* **2011**, *56*, 1908–1917. [CrossRef]
18. Ismail, L.I.; Shamsudin, S.; Yussof, H.; Hanapiah, F.A.; Zahari, N.I. Estimation of concentration by eye contact measurement in Robot-based Intervention Program with autistic children. *Procedia Eng.* **2012**, *41*, 1548–1552. [CrossRef]
19. Zhang, S.; Chen, D.; Tang, Y.; Zhang, L. Children ASD Evaluation Through Joint Analysis of EEG and Eye-Tracking Recordings with Graph Convolution Network. *Front. Hum. Neurosci.* **2021**, *15*, 651349. [CrossRef] [PubMed]
20. Alotaibi, N.; Maharatna, K. Classification of Autism Spectrum Disorder from EEG-Based Functional Brain Connectivity Analysis. *Neural Comput.* **2021**, *33*, 1914–1941. [CrossRef] [PubMed]
21. Cerrada, M.; Trujillo, L.; Hernández, D.E.; Correa Zevallos, H.A.; Macancela, J.C.; Cabrera, D.; Vinicio Sánchez, R. AutoML for Feature Selection and Model Tuning Applied to Fault Severity Diagnosis in Spur Gearboxes. *Math. Comput. Appl.* **2022**, *27*, 6. [CrossRef]
22. Enríquez Zárate, J.; Gómez López, M.d.l.A.; Carmona Troyo, J.A.; Trujillo, L. Analysis and Detection of Erosion in Wind Turbine Blades. *Math. Comput. Appl.* **2022**, *27*, 5. [CrossRef]
23. Janiesch, C.; Zschech, P.; Heinrich, K. Machine learning and deep learning. *Electron. Mark.* **2021**, *31*, 685–695. [CrossRef]
24. Fong-Mata, M.; García-Guerrero, E.; Mejia-Medina, D.; López-Bonilla, O.; Villarreal-Gomez, L.; Zamora-Arellano, F.; López-Mancilla, D.; Inzunza-González, E. An Artificial Neural Network Approach and a Data Augmentation Algorithm to Systematize the Diagnosis of Deep-Vein Thrombosis by Using Wells' Criteria. *Electronics* **2020**, *9*, 1810. [CrossRef]
25. Choi, R.Y.; Coyner, A.S.; Kalpathy-Cramer, J.; Chiang, M.F.; Campbell, J.P. Introduction to Machine Learning, Neural Networks, and Deep Learning. *Transl. Vis. Sci. Technol.* **2020**, *9*, 14. [CrossRef]
26. Navarro-Espinoza, A.; López-Bonilla, O.R.; García-Guerrero, E.E.; Tlelo-Cuautle, E.; López-Mancilla, D.; Hernández-Mejía, C.; Inzunza-González, E. Traffic Flow Prediction for Smart Traffic Lights Using Machine Learning Algorithms. *Technologies* **2022**, *10*, 5. [CrossRef]

27. Kang, J.; Han, X.; Song, J.; Niu, Z.; Li, X. The identification of children with autism spectrum disorder by SVM approach on EEG and eye-tracking data. *Comput. Biol. Med.* **2020**, *120*, 103722. [CrossRef] [PubMed]
28. Radhakrishnan, M.; Ramamurthy, K.; Choudhury, K.K.; Won, D.; Manoharan, T.A. Performance analysis of deep learning models for detection of Autism Spectrum Disorder from EEG signals. *Trait. Signal* **2021**, *38*, 853–863. [CrossRef]
29. Thirumal, S.; Thangakumar, J. Investigation of Statistical Feature Selection Techniques for Autism Classification Using EEG Signals. *J. Adv. Res. Dyn. Control Syst.* **2020**, *12*, 1254–1263. [CrossRef]
30. Tawhid, M.N.A.; Siuly, S.; Wang, H.; Whittaker, F.; Wang, K.; Zhang, Y. A spectrogram image based intelligent technique for automatic detection of autism spectrum disorder from EEG. *PLoS ONE* **2021**, *16*, e0253094. [CrossRef]
31. Sundaresan, A.; Penchina, B.; Cheong, S.; Grace, V.; Valero-Cabré, A.; Martel, A. Evaluating deep learning EEG-based mental stress classification in adolescents with autism for breathing entrainment BCI. *Brain Inform.* **2021**, *8*. [CrossRef]
32. Grossi, E.; Valbusa, G.; Buscema, M. Detection of an Autism EEG Signature From Only Two EEG Channels Through Features Extraction and Advanced Machine Learning Analysis. *Clin. EEG Neurosci.* **2021**, *52*, 330–337. [CrossRef] [PubMed]
33. Baygin, M.; Dogan, S.; Tuncer, T.; Datta Barua, P.; Faust, O.; Arunkumar, N.; Abdulhay, E.W.; Emma Palmer, E.; Rajendra Acharya, U. Automated ASD detection using hybrid deep lightweight features extracted from EEG signals. *Comput. Biol. Med.* **2021**, *134*, 104548. [CrossRef] [PubMed]
34. Hagendorff, T. Linking Human And Machine Behavior: A New Approach to Evaluate Training Data Quality for Beneficial Machine Learning. *Minds Mach.* **2021**, *31*, 563–593. [CrossRef] [PubMed]
35. Swatzyna, R.J.; Boutros, N.N.; Genovese, A.C.; MacInerney, E.K.; Roark, A.J.; Kozlowski, G.P. Electroencephalogram (EEG) for children with autism spectrum disorder: Evidential considerations for routine screening. *Eur. Child Adolesc. Psychiatry* **2019**, *28*, 615–624. [CrossRef] [PubMed]
36. Kurgansky, A.V.; Machinskaya, R. Bilateral frontal theta-waves in EEG of 7–8-year-old children with learning difficulties: Qualitative and quantitative analysis. *Hum. Physiol.* **2012**, *38*, 255–263. [CrossRef]
37. Machinskaya, R.; Semenova, O.A.; Absatova, K.A.; Sugrobova, G.A. Neurophysiological factors associated with cognitive deficits in children with ADHD symptoms: EEG and neuropsychological analysis. *Psychol. Neurosci.* **2014**, *7*, 461–473. [CrossRef]
38. Khodatars, M.; Shoeibi, A.; Sadeghi, D.; Ghaasemi, N.; Jafari, M.; Moridian, P.; Khadem, A.; Alizadehsani, R.; Zare, A.; Kong, Y.; et al. Deep learning for neuroimaging-based diagnosis and rehabilitation of Autism Spectrum Disorder: A review. *Comput. Biol. Med.* **2021**, *139*, 104949. [CrossRef]
39. Shoeibi, A.; Khodatars, M.; Jafari, M.; Moridian, P.; Rezaei, M.; Alizadehsani, R.; Khozeimeh, F.; Gorriz, J.M.; Heras, J.; Panahiazar, M.; et al. Applications of deep learning techniques for automated multiple sclerosis detection using magnetic resonance imaging: A review. *Comput. Biol. Med.* **2021**, *136*, 104697. [CrossRef] [PubMed]
40. Shoeibi, A.; Sadeghi, D.; Moridian, P.; Ghassemi, N.; Heras, J.; Alizadehsani, R.; Khadem, A.; Kong, Y.; Nahavandi, S.; Zhang, Y.D.; et al. Automatic Diagnosis of Schizophrenia in EEG Signals Using CNN-LSTM Models. *Front. Neuroinform.* **2021**, *15*. [CrossRef]
41. Shoeibi, A.; Ghassemi, N.; Khodatars, M.; Moridian, P.; Alizadehsani, R.; Zare, A.; Khosravi, A.; Subasi, A.; Rajendra Acharya, U.; Gorriz, J.M. Detection of epileptic seizures on EEG signals using ANFIS classifier, autoencoders and fuzzy entropies. *Biomed. Signal Process. Control* **2022**, *73*, 103417. [CrossRef]
42. Khng, K.H.; Mane, R. Beyond BCI—Validating a wireless, consumer-grade EEG headset against a medical-grade system for evaluating EEG effects of a test anxiety intervention in school. *Adv. Eng. Inform.* **2020**, *45*, 101106. [CrossRef]
43. Fouad, I. A robust and reliable online P300 based BCI system using Emotiv EPOC Headset. *J. Med. Eng. Technol.* **2021**, *45*, 94–114. [CrossRef]
44. Dubrovinskaya, N.; Machinska, R.; Kulakovsky, Y. Brain Organization of Selective Tasks Preceding Attention: Ontogenetic Aspects. In *Complex Brain Functions: Conceptual Advances in Russian Neurocience*; CRC Press: Boca Raton, FL, USA, 2000; Volume 1, pp. 169–180. [CrossRef]
45. Dubrovinskaya, N.V.; Machinskaya, R.I. Reactivity of Teta and Alpha EEG Frequency Bands in Voluntary Attention in Junior Schoolchildren. *Hum. Physiol.* **2002**, *28*, 522–527. [CrossRef]
46. Emotiv, I. Data Sample Object. Cortex API. Available online: https://emotiv.gitbook.io/cortex-api/data-subscription/data-sample-object (accessed on 29 December 2021).
47. Emotiv, I. Frequency Bands Emotiv PRO v3.0. Available online: https://emotiv.gitbook.io/emotivpro-v3/ (accessed on 29 December 2021).
48. Fahimi, F.; Guan, C.; Wooi, B.G.; Kai Keng, A.; Choon, G.L.; Tih, S.L. Personalized features for attention detection in children with Attention Deficit Hyperactivity Disorder. In Proceedings of the Annual International Conference of the IEEE Engineering in Medicine and Biology Society, EMBS, Jeju, Korea, 11–15 July 2017; pp. 414–417. [CrossRef]
49. Pedregosa, F.; Varoquaux, G.; Gramfort, A.; Michel, V.; Thirion, B.; Grisel, O.; Blondel, M.; Prettenhofer, P.; Weiss, R.; Dubourg, V.; et al. Scikit-Learn: Machine Learning in Python. 2011. Available online: https://scikit-learn.org (accessed on 29 December 2021).
50. Matthews, B.W. Comparison of the predicted and observed secondary structure of T4 phage lysozyme. *Biochim. Biophys. Acta (BBA)—Protein Struct.* **1975**, *405*, 442–451. [CrossRef]
51. Ein Shoka, A.A.; Alkinani, M.H.; El-Sherbeny, A.S.; El-Sayed, A.; Dessouky, M.M. Automated seizure diagnosis system based on feature extraction and channel selection using EEG signals. *Brain Inform.* **2021**, *8*, 1. [CrossRef] [PubMed]

52. Misiunas, A.V.M.; Meskauskas, T.; Samaitiene, R. Machine Learning Based EEG Classification by Diagnosis: Approach to EEG Morphological Feature Extraction. *AIP Conf. Proc.* **2019**, *2164*, 080005. [CrossRef]
53. Boroujeni, Y.K.; Rastegari, A.A.; Khodadadi, H. Diagnosis of attention deficit hyperactivity disorder using non-linear analysis of the EEG signal. *IET Syst. Biol.* **2019**, *13*, 260–266. [CrossRef] [PubMed]

Mathematical and Computational Applications

MDPI

Article

Variable Decomposition for Large-Scale Constrained Optimization Problems Using a Grouping Genetic Algorithm

Guadalupe Carmona-Arroyo *, Marcela Quiroz-Castellanos * and Efrén Mezura-Montes

Artificial Intelligence Research Institute, Universidad Veracruzana, Campus Sur, Calle Paseo Lote II, Sección Segunda 112, Nuevo Xalapa, Veracruz 91097, Mexico; emezura@uv.mx
* Correspondence: gcarmonaarroyo@gmail.com (G.G.-A.); maquiroz@uv.mx (M.Q.-C.)

Abstract: Several real optimization problems are very difficult, and their optimal solutions cannot be found with a traditional method. Moreover, for some of these problems, the large number of decision variables is a major contributing factor to their complexity; they are known as Large-Scale Optimization Problems, and various strategies have been proposed to deal with them. One of the most popular tools is called Cooperative Co-Evolution, which works through a decomposition of the decision variables into smaller subproblems or variables subgroups, which are optimized separately and cooperate to finally create a complete solution of the original problem. This kind of decomposition can be handled as a combinatorial optimization problem where we want to group variables that interact with each other. In this work, we propose a Grouping Genetic Algorithm to optimize the variable decomposition by reducing their interaction. Although the Cooperative Co-Evolution approach is widely used to deal with unconstrained optimization problems, there are few works related to constrained problems. Therefore, our experiments were performed on a test benchmark of 18 constrained functions under 100, 500, and 1000 variables. The results obtained indicate that a Grouping Genetic Algorithm is an appropriate tool to optimize the variable decomposition for Large-Scale Constrained Optimization Problems, outperforming the decomposition obtained by a state-of-the-art genetic algorithm.

Keywords: Grouping Genetic Algorithm; variable decomposition; Large-Scale Constrained Optimization

Citation: Carmona-Arroyo, G.; Quiroz-Castellanos, M.; Mezura-Montes, E. Variable Decomposition for Large-Scale Constrained Optimization Problems Using a Grouping Genetic Algorithm. *Math. Comput. Appl.* **2022**, *27*, 23. https://doi.org/10.3390/mca27020023

Academic Editor: Claudia Schillings

Received: 26 January 2022
Accepted: 1 March 2022
Published: 3 March 2022

1. Introduction

A constrained numerical optimization problem is defined by finding the vector $\mathbf{x} \in \mathbb{R}^D$ that minimizes the objective function $Obj(\mathbf{x})$ subject to inequality $g_j(\mathbf{x})$ and equality $h_k(\mathbf{x})$ constraints [1]. This is described by Equation (1).

$$\begin{aligned} &\text{minimize} \quad Obj(\mathbf{x}) \\ &\text{subject to} \\ &\qquad g_j(\mathbf{x}) \le 0, \quad j = 1, \ldots, q \\ &\qquad h_k(\mathbf{x}) = 0, \quad k = 1, \ldots, r. \end{aligned} \tag{1}$$

where q and r represent the number of inequality and equality constraints, respectively, $\mathbf{x} = (x_1, \ldots, x_D)$, and the search space $\mathbb{S}$ is defined by the lower limits l and upper limits u ($l_i \le x_i \le u_i$), while the feasible region is defined as the subset of solutions that satisfy the constraints of the problem $\mathbb{F} \subset \mathbb{S}$.

To handle the constrained problem, the constraint violation sum cvs [2], is calculated by Equation (2).

$$cvs(\mathbf{x}) = \sum_j^q \max(0, g_j(\mathbf{x})) + \sum_k^r \max(0, |h_k(\mathbf{x})| - \epsilon) \tag{2}$$

where $|h_k(\mathbf{x})| - \epsilon$ is the transformation of equality constraints into inequality constraints $|h_k(\mathbf{x})| - \epsilon \leq 0$ with $\epsilon = 1 \times 10^{-4}$.

According to the specialized literature [3,4], a Large-Scale Optimization Problem consists of 100 or more variables, while benchmark functions for sessions and competitions on the field include thousands of decision variables. Algorithms that solve Large-Scale Optimization Problems are usually affected by the curse of dimensionality; i.e., these problems are more complex to solve when the number of decision variables increases. One of the best-known approaches to deal with these problems is the one proposed by Potter and De Jong called Cooperative Co-Evolution (CC) [5], which is based on the divide-and-conquer strategy. This CC approach works in three stages: (1) first, the problem is decomposed into subcomponents of less dimension and complexity; then, (2) each subproblem is optimized separately; and finally, (3) the solutions of each subproblem cooperate to create the solution of the original problem.

Although many of the approaches to solve Large-Scale Optimization Problems have implemented the CC approach, the first problem that arises is to find the adequate decomposition of the subgroups since the interaction among the variables must be taken into account to divide the problem. In other words, if two or more variables interact with each other, they must remain in the same subcomponent, just as the variables that do not interact with others must be part of subcomponents with just one variable. The decomposition of the subgroups can be evaluated considering definitions of problem separability and partial separability, as explained in Section 3.2. If the interacting variables are not grouped into the same subgroup, CC tends to find a solution that is not the optimum of the original problem but a local optimum introduced by an incorrect problem decomposition [6].

Several strategies have been proposed in the literature to deal with the problem of creating an adequate decision variable decomposition, ranging from random approaches to strategies that study the interaction among variables to optimize this decomposition. When the original problem is decomposed into subproblems, we aim for the interaction between them to be at minimum. For this reason, we can work with the decomposition through optimization strategies, where the objective is to group variables that interact with each other in the same subcomponent.

One of the first works related to the optimization of the variable decomposition for Large-Scale Constrained Problems was proposed by Aguilar-Justo et al. [7], who presented a Genetic Algorithm (GA) to handle the interaction minimization in the subcomponents. This GA and its operators, such as crossover and mutation, work under an integer genetic encoding, which is one of the most popular ways of representing a solution as a chromosome in this type of algorithm.

In this work, we resort to a Grouping Genetic Algorithm (GGA) to solve the decomposition problem, since these algorithms have proven to be some of the best when it comes to combinatorial optimization problems where the optimization of elements in groups is involved [8]. This proposal aims to show the benefits of using a GGA and its group-based representation, for the creation of subcomponents, compared against a genetic algorithm. In addition, to the best of the authors' knowledge, our proposal is the first GGA approach to handle decomposition in Large-Scale Constrained Optimization Problems.

We chose similar main operators and parameters to the genetic algorithm proposed by Aguilar-Justo et al. [7] in order to evaluate the impact of the representation schemes for the decomposition problem and to make a fair comparison of the performance.

Both algorithms were evaluated on a set of 18 test functions proposed by Sayed et al. [3], which are problems with 1, 2, and 3 constraints with 100, 500, and 1000 variables, respectively. Experimental results show that the proposed GGA obtains a suitable variable decomposition when compared against the GA of Aguilar-Justo et al. [7] for variable decomposition in Large-Scale Constrained Optimization Problems, especially where the separation is more complicated, such as in non-separable problems.

The work continues as follows: in the next section, we show related work regarding Decomposition Methods and Grouping Genetic Algorithms. In Section 3, we show our proposed GGA and describe each of its components in detail. Section 4 contains the experiments and results of our algorithm compared to a genetic algorithm and a brief analysis of the performance of the GGA. Finally, in Section 5, we describe the conclusions and future work corresponding to our research.

2. Related Works

2.1. Decomposition Methods

According to Ma et al. [6], several variable grouping strategies have been proposed in the specialized literature for variable decomposition to deal with Large-Scale Unconstrained Optimization Problems. They classify them into the next classes: static variable grouping, random variable grouping, variable grouping based on interaction learning, variable grouping based on domain knowledge, overlap and hierarchical variable grouping, and finally, some hybrids of them. In the next paragraphs, we describe the works related to each class.

Static variable grouping methods do not rely on any intelligent procedure to create the variable decomposition. Instead, they preliminarily decompose the decision vector into a set of low-dimensional subcomponents and fix the variable grouping during the process of optimization. Among the works that perform static grouping of variables are the works of Potter and De Jong [5] and Liu et al. [9], which show good performance with fully separable problems. For non-separable problems, Van den Bergh and Engelbrecht [10] propose a static sequential decomposition. However, the decomposition process is dependent on an adequate number of subcomponents that must be adequate from the beginning of the strategy, and the static decomposition process has poor performance in many problems.

For that reason, several authors proposed *random variable grouping* strategies, such as the case of Yang et al. [11], who present random decomposition with a fixed number of subcomponents. Moreover, the same authors propose in [12] a random decomposition with a dynamic number of subcomponents. Omidvar et al. [13,14] improve these random strategies by integrating the probability of interaction between variables in the grouping technique.

In addition, if an algorithm can learn the structure of the problem and decompose it, the difficulty in solving the problem can be significantly reduced (*variable grouping based on interaction learning*). Many approaches have been proposed to detect variable interactions, and we can subdivide them into those based on perturbation, statistical models, distribution models, approximate models, and linkage adaptation. For example, in some cases, the interaction is captured by perturbing the decision variables and measuring the change in fitness caused by the perturbations, like in the work of Xu et al. [15]. Furthermore, Differential Grouping (DG) and Differential Grouping version 2 from Odmivar et al. [16,17], respectively, are based on perturbation as well, which are among the most popular decomposition algorithms and have been highly studied, which has led to various improvements, such as recursive decomposition [18–21]. Moreover, Delta Grouping [14] is classified as a decomposition method based on statistical models, where all variables and the objective functions are considered as random variables. Statistical analyses of variables or objective functions are performed first, and then the variables are grouped. In a distribution model, the set of promising solutions is first used to estimate the variable distributions and variable interactions, and then it is taken to generate new candidate solutions, based on the learned variable distributions and variable interactions: such is the case of Estimation of Distribution Algorithm (EDA), as is the case with the work of Sopov [22], where a genetic algorithm is combined with an EDA for collecting statistical data based on the past search experience to provide the problem decomposition by fixing genes in chromosomes, as well as other representatives of such methods [23,24]. As an example of an approximate model, the fitness evaluation of a Large-Scale Continuous Optimization Problem is converted to the evaluation of a simpler, partially separable problem in [25]. Linkage adaption methods use specially designed evolution operators, representations, and mechanisms to divide variables into groups [26].

When it comes to *overlap and hierarchy variable grouping*, which are usually interspersed in the decision vector, more elaborate strategies have to be used; Goh et al. [27,28] suggested assigning each variable to several subcomponents, each of which contains more than one variable, and the subcomponents compete with each other to represent shared variables. Furthermore, Strasser et al. [29] introduce some overlapping grouping strategies, including random overlapping grouping, neighbor overlapping grouping, centered overlapping grouping, and more.

Regarding the *variable grouping based on domain knowledge*, before CC is implemented to solve specific real-world problems, domain knowledge can be harnessed to reduce the complexity of the problems. The conflicting probability of two flights was used by Guan et al. [30] to learn the variable interaction in solving the flight conflicting avoidance problem, which is one of several examples of real optimization problems where domain knowledge can be a good tool.

All the works mentioned above focus on unconstrained problems. One of the first decomposition methods for solving Large-Scale Constrained Optimization Problems is an extension of the work of Sayed et al. [25]; this new version is known as the Variable Interaction Identification Technique for Constrained Problems (VIIC) [3]. This method can find the interaction between variables in problems of 100, 500, and 1000 dimensions with inequality constraints. Sayed et al. proposed to measure each decomposition of variables by minimizing the absolute difference between the full evaluation and the sum of each evaluated subgroup based on the definition of the separability and partial separability of a function. The approach was tested at a new benchmark and compared to Random Grouping (RG), and the results showed that VIIC outperformed RG. Later, Aguilar-Justo and Mezura-Montes [31] improved the performance of the VIIC to achieve adequate decomposition for a fixed number of subgroups. They transformed the constrained problem into an unconstrained problem and used a neighborhood heuristic to guide the search by their proposed decomposition and then optimize it (called VIICN). After that, they proposed an improvement to their work where the principles of VIIC and VIICN are used to build a genetic algorithm that performs a dynamic decomposition which they called DVIIC, without establishing a fixed number of subcomponents [7]. Recently, Vakhnin and Sopov [32] proposed a method based on CC that increases the size of groups of variables at the decomposition stage (called iCC), working with a fixed number of subcomponents.

Intending to improve the existing methods for the decomposition of variables, we propose a genetic algorithm with a genetic encoding based on groups, better known as the Grouping Genetic Algorithm, to optimize the variables decomposition. Experimental results demonstrate the benefits of using a group-based encoding scheme for this problem and its advantages over the genetic algorithm with an integer-based encoding scheme (DVIIC [7]).

2.2. Grouping Genetic Algorithms

As we have mentioned before, the variable decomposition problem is an optimization problem because we search for the best decomposition, in the sense of the variable interaction; that is, given a set $X = x_1, x_2, \ldots, x_D$ of D variables, we want to decompose said set into m disjoint groups, so that the variables within each group do not interact with the variables of the other groups. Therefore, we see our problem as a grouping problem.

According to the literature [8], grouping problems are a type of combinatorial optimization problem where a set X of D items is usually partitioned into a collection of m mutually disjoint subsets (groups) G_j, so that $X = \cup_{j=1}^{m} G_j$, and $G_j \cap G_k = \emptyset, j \neq k$. In this way, an algorithm designed to solve a grouping problem seeks the best possible distribution of the D items of the set X in m different groups ($1 \leq m \leq D$), such that each item is exactly in one group.

Kashan et al. [33] organized the grouping problems in three categories, using as criteria the number of groups, the type of groups, and the dependence on the order of the groups. First, using the number of groups as the criterion, grouping problems can be classified as either constants or variables. In this sense, if the number of groups required is known, the problem is constant. On the other hand, if the number of groups is unknown, the problem is variable. Second, these problems can be divided into identical and non-identical groups, considering the characteristics of their groups. In this classification, if the quality of a solution is modified by exchanging all the items of two groups, that problem belongs to the non-identical grouping class. Otherwise, the problem is part of the identical category. Finally, grouping problems are called order-dependent when the solution quality depends on the groups' order. Consequently, grouping problems without this dependency belong to the not order-dependent class. Thus, we can say that the decomposition problem is a variable, identical, and not order-dependent grouping problem.

Grouping problems are very difficult to solve. Most of them are NP-hard, which implies there is no algorithm capable of finding an optimal solution for every instance in polynomial time [34]. There are several NP-hard grouping problems, such as the Bin Packing Problem, Job Shop Scheduling Problem, etc. Ramos-Figueroa et al. [8] studied in their work the strategies that work with most problems like the ones mentioned before. They concluded that the best and the most popular strategies to solve such problems are the Grouping Genetic Algorithms or GGAs.

The GGA was designed in 1992 by Falkenauer [35] and is an extension to the traditional GA with the difference of using a group-based solutions representation scheme and variation operators working together with such solution encoding. Ramos-Figueroa et al. [36] remark that the encoding of a grouping problem solution into a chromosome is a key issue for obtaining good GGA performance; the authors also comment on the importance of integrating crossover and mutation operators adapted to work at the group level. They present a survey of different variation operators designed to work with GGAs that use different types of encoding, as well as their advantages to solve grouping problems.

The state-of-the-art [8] indicates that some of the best results when solving NP-hard grouping problems have been obtained by GGAs that combine grouping encoding schemes and special operators adapted to work with these genetic encodings. Moreover, GGAs have been highly studied for grouping problems that have similarities with the variable decomposition problem, which is also due to the exploration and exploitation of the search space that is given by the nature of the elements of evolutionary algorithms [37]. In this work, we propose, to the best of the authors' knowledge, the first GGA for the variable decomposition problem in Large-Scale Constrained Optimization Problems. A comparative study is conducted to evaluate the performance of our Grouping Genetic Algorithm versus the genetic algorithm DVIIC [7] on the decomposition of variables for Large-Scale Constrained Optimization Problems. To promote a fair comparison, we implement similar operators and equivalent parameter settings. The experiments were carried out using 18 test functions each with 100, 500, and 1000 variables. The obtained results allow us to validate the advantages of the group-based encoding over the integer-based encoding.

3. A Grouping Genetic Algorithm for the Variable Decomposition Problem

The variable decomposition problem can be classified as a grouping problem. We seek to optimize the separation into groups of the decision variables of the Large-Scale Problem; that is, to create the best partition of the decision variables into a collection of m mutually disjoint groups so that the variables belonging to each group have no interaction with the variables of another group.

To study the importance of the solution encoding in a genetic algorithm to solve the variable decomposition problem, we decided to develop a GGA with operators and parameters with similar features to the genetic algorithm DVIIC (proposed by Aguilar-Justo et al. [7]) so that the comparison is as fair as possible. The main difference between the two algorithms is the genetic encoding. The proposal of Aguilar-Justo et al. [7] includes an integer-based representation, where a chromosome has a fixed length that is equal to the number of variables, and each gene represents a variable and indicates the group where the variable is set. On the other hand, our GGA includes a group-based representation, where a chromosome can have a variable length, equal to the number of subcomponents, and each gene represents a subcomponent and indicates the variables that belong to this subset.

In Algorithm 1, we show the general steps of the GGA proposed in this work. The precise details are shown in the following subsections. The process begins by generating an initial population P of pop_size individuals created by the population initialization strategy (Line 1). After that, each of the individuals in the population is evaluated, and the best solution for the population is obtained (Line 2). Then, we iterate through a max_gen number of generations or until we find a value equal to zero in the decomposition evaluation. Within this cycle, the individuals to be crossed will be selected, and the offspring will be created through the grouping crossover operator (Lines 4–5). Similarly, the population is updated by the mutation of some individuals in the population (Lines 6–7). Finally, the population is evaluated again to update the population and the best global solution found so far (Lines 8–9).

Algorithm 1: Grouping Genetic Algorithm for variable decomposition algorithm.

```
1  Generate an initial population P
2  Evaluate population and save best_solution
3  while generation ≤ max_gen and grps_diff ≠ 0 do
4      Select n pairs of individuals for crossover
5      Apply grouping crossover operator
6      Select n individuals for mutation
7      Apply grouping mutation operator
8      Evaluate offspring and update best_solution
9      Apply replacement strategy of the population
10 Return best_solution
```

In the following subsections, we detail the components and operators of our GGA.

3.1. Genetic Encoding

One of the most important decisions to make while implementing a genetic algorithm is to decide the representation to use to represent the solutions. It has been observed that improper representation can lead to poor performance of the GA. Our GGA works with group-based representation, which is the main characteristic of the GGAs.

Each individual in the population is represented by the groups of variables. Figure 1 shows an example of an individual that represents a problem with 10 variables numbered from 0 to 9 randomly assigned to four subcomponents or groups. The groups of variables (genes) according to this individual are the following: $grps_1 = \{x_3, x_6, x_8\}$, $grps_2 = \{x_1, x_2, x_7\}$, $grps_3 = \{x_0, x_4\}$, and $grps_4 = \{x_5, x_9\}$. Note that the number V of variables in each subcomponent is variable and can be between 1 and D; in addition, the number of subcomponents m is between 1 and D.

3, 6, 8	1, 2, 7	0, 4	5, 9

Figure 1. Group-based chromosome, where each gene represents a subcomponent (set of variables).

3.2. Decomposition Evaluation

Each individual is evaluated to determine its fitness and to discover which one is the best within the population.

Sayed et al. [3] proposed a decomposition evaluation inspired by the definitions of problem separability [38] and partial separability [39]. The definition of problem separability states that a fully separable problem that has D variables can be written in the form of a linear combination of subproblems of the decision variables, where the evaluation of the complete problem, $F(\mathbf{x})$, is the same as the aggregation of the evaluation of the subproblems, $f(x_i)$, which means $F(\mathbf{x}) = \sum_{i=1}^{D} f(x_i)$. Additionally, a partially separable problem is defined as one which has D variables and that can be decomposed into m subproblems, where the summation of all subproblems equals the solution of the complete problem $F(x)$ such that $F(\mathbf{x}) = \sum_{k=1}^{m} f_k(x_v)$, $v = [1 + V \times (k-1), V \times k]$, where m is the number of subproblems and V is the number of variables in the k-th subproblem. Sayed et al. proposed to measure each decomposition of variables by minimizing the absolute difference between the full evaluation and the sum of each evaluated subgroup.

Algorithm 2 shows the decomposition evaluation procedure. First (in Line 1), we obtain $fit_{all_{c_1}}$ and $fit_{all_{c_2}}$ through the evaluations of Equations (3) and (4), where all the variables take the constant value c_1 and $c2$, respectively. After that, both evaluations are added and multiplied by the number of subgroups in the problem (m) to obtain $fit_{all_{c_1c_2}}$, and then we initialize $fit_{grps_{c_1c_2}}$ as 0 (Lines 2–3). Afterwards, we start a loop from $k = 1$ to the number of subgroups m in the individual (Lines 4–10), Within this loop, we create two arrangements of D variables in which each variable belonging to group k takes the value c_1, while the remaining variables take the value c_2 to evaluate this arrangement and obtain $fit_{grps_{k,c_1}}$. On the other hand, to obtain $fit_{grps_{k,c_2}}$, the variables in the k-th group take the value of c_2, and the remaining ones take the value of c_1 (Lines 5–8) according to Equations (5) and (6). After that, we calculate $fit_{grps_{k,c_1c_2}}$ as the sum of the previously calculated $fit_{grps_{k,c_1}}$ and $fit_{grps_{k,c_2}}$ (Line 9). Thus, to end the loop, we update $fit_{grps_{c_1c_2}}$ as the sum of $fit_{grps_{c_1c_2}}$ and $fit_{grps_{k,c_1c_2}}$. Finally, we obtain the evaluation of the decomposition by calculating the absolute difference $grps_{diff}$ shown in Line 11.

Algorithm 2: Decomposition evaluation.

Input: m, and $c_1 > 0$, $c_2 > 0$ random numbers
Output: $grps_{diff}$
1 Evaluate $fit_{all_{c_1}}$ and $fit_{all_{c_2}}$ according to Equations (3) and (4)
2 Calculate $fit_{all_{c_1c_2}} = m \times [fit_{all_{c_1}} + fit_{all_{c_2}}]$
3 $fit_{grps_{c_1c_2}} = 0$
4 **for** $k = 1$ *to* m **do**
5   Create arrangement $\mathbf{x}_{k,c_1}$ according to Equation (5)
6   Calculate $fit_{grps_{k,c_1}} = Obj(\mathbf{x}_{k,c_1}) + cvs(\mathbf{x}_{k,c_1})$
7   Create arrangement $\mathbf{x}_{k,c_2}$ according to Equation (6)
8   Calculate $fit_{grps_{k,c_2}} = Obj(\mathbf{x}_{k,c_2}) + cvs(\mathbf{x}_{k,c_2})$
9   Calculate $fit_{grps_{k,c_1c_2}} = fit_{grps_{k,c_1}} + fit_{grps_{k,c_2}}$
10   Update $fit_{grps_{c_1c_2}} = fit_{grps_{c_1c_2}} + fit_{grps_{k,c_1c_2}}$
11 Calculate $grps_{diff} = |fit_{all_{c_1c_2}} - fit_{grps_{c_1c_2}}|$

$$fit_{all_{c_1}} = Obj(\mathbf{x}) + cvs(\mathbf{x}),\ x_i = c_1,\ \forall i \in [1, D] \tag{3}$$

$$fit_{all_{c_2}} = Obj(\mathbf{x}) + cvs(\mathbf{x}),\ x_i = c_2,\ \forall i \in [1, D] \tag{4}$$

$$\mathbf{x}_{k,c_1} = \begin{cases} c_1 & \forall x_i \in grps_k \\ c_2 & \text{otherwise} \end{cases} \tag{5}$$

$$\mathbf{x}_{k,c_2} = \begin{cases} c_2 & \forall x_i \in grps_k \\ c_1 & \text{otherwise} \end{cases} \tag{6}$$

To clarify the decomposition evaluation procedure, we present an example below. Let $Obj(\mathbf{x}) + cvs(\mathbf{x}) = f(\mathbf{x}) = x_1x_2 + x_3x_4$ be the problem to decompose, and according to the arrangement decomposition given by $grps_1 = \{x_1\}$, $grps_2 = \{x_2, x_4\}$, and $grps_3 = \{x_3\}$, we have $m = 3$. Suppose $c_1 = 1$ and $c_2 = 2$. In the first step, we calculate $fit_{all_{c_1}}$ and $fit_{all_{c_2}}$. According to Equations (3) and (4),

$$fit_{all_{c_1}} = f(x_i = c_1) = 1 * 1 + 1 * 1 = 2$$

$$fit_{all_{c_2}} = f(x_i = c_2) = 2 * 2 + 2 * 2 = 8$$

Then, continuing with step 2,

$$fit_{all_{c_1 c_2}} = m \times [fit_{all_{c_1}} + fit_{all_{c_2}}] = 3 \times [2 + 8] = 30$$

In step 3, we initialize $fit_{grps_{c_1 c_2}} = 0$. Then, we start the for loop. At this point in the process, we have to create arrangement $\mathbf{x}_{k,c_1}$ according to Equation (5) in step 5 and evaluate it in step 6. Similarly, the process is performed for c_2 in steps 7 and 8. To calculate $fit_{grps_{k,c_1}}$ the variables of the k-th group will be evaluated in c_1, and the rest in c_2; for $k = 1$, the group is $grps_1 = \{x_1\}$, so $x_1 = 1$ and $x_2, x_3, x_4 = 2$. A similar calculation is performed with $fit_{grps_{k,c_2}}$, but evaluating the variables of the k-th group in c_2 and the rest in c_1. Therefore, following the steps into the loop,

For $k = 1$, $grps_1 = \{x_1\}$:

$$fit_{grps_{k,c_1}} = f_{grps_{k,c_1}} = 1 * 2 + 2 * 2 = 6$$

$$fit_{grps_{k,c_2}} = f_{grps_{k,c_2}} = 2 * 1 + 1 * 1 = 3$$

$$fit_{grps_{k,c_1 c_2}} = fit_{grps_{k,c_1}} + fit_{grps_{k,c_2}} = 6 + 3 = 9$$

$$fit_{grps_{c_1 c_2}} = fit_{grps_{c_1 c_2}} + fit_{grps_{k,c_1 c_2}} = 0 + 9 = 9$$

For $k = 2$, $grps_2 = \{x_2, x_4\}$:

$$fit_{grps_{k,c_1}} = f_{grps_{k,c_1}} = 2 * 1 + 2 * 1 = 4$$

$$fit_{grps_{k,c_2}} = f_{grps_{k,c_2}} = 1 * 2 + 1 * 2 = 4$$

$$fit_{grps_{k,c_1 c_2}} = fit_{grps_{k,c_1}} + fit_{grps_{k,c_2}} = 4 + 4 = 8$$

$$fit_{grps_{c_1 c_2}} = fit_{grps_{c_1 c_2}} + fit_{grps_{k,c_1 c_2}} = 9 + 8 = 17$$

For $k = 3$, $grps_3 = \{x_3\}$:

$$fit_{grps_{k,c_1}} = f_{grps_{k,c_1}} = 2 * 2 + 1 * 2 = 6$$

$$fit_{grps_{k,c_2}} = f_{grps_{k,c_2}} = 1 * 1 + 2 * 1 = 3$$

$$fit_{grps_{k,c_1 c_2}} = fit_{grps_{k,c_1}} + fit_{grps_{k,c_2}} = 6 + 3 = 9$$

$$fit_{grps_{c_1 c_2}} = fit_{grps_{c_1 c_2}} + fit_{grps_{k,c_1 c_2}} = 17 + 9 = 26$$

Finally,

$$grps_{diff} = |fit_{all_{c_1 c_2}} - fit_{grps_{c_1 c_2}}| = |30 - 26| = 4$$

The purpose of the problem decomposition is to create the best decomposition; that is, to create independent subcomponents, as well as to minimize the difference $grps_{diff}$. Therefore, we have adopted a similar evaluation to the one proposed by Aguilar-Justo et al. [7], which is defined next: to maximize the number of subproblems, the $grps_{diff}$ is updated as follows: (1) if the number of subgroups is one, then the $grps_{diff}$ takes an extreme greater value; (2) if the $grps_{diff}$ is zero, this means that the decomposition is perfect, and it is rewarded by subtracting the number of subgroups (m) of the individual; (3) in another case, the $grps_{diff}$ value does not change. Since the use of the previous evaluation function benefits a decomposition with a high number of subcomponents, in some cases, a complete decomposition (as many groups as numbers of variables) would be presented as optimal when in reality it is not the case. For this reason, a modification in the evaluation function is necessary (avoiding the benefit to a high number of groups). Therefore, the evaluation function has been modified in our algorithm. This change is summarized in Equation (7).

$$grps_{diff} = \begin{cases} infinite & \text{if } m = 1 \\ grps_{diff} & \text{otherwise} \end{cases} \tag{7}$$

3.3. Population Initialization

The initial population in most GGAs is generally generated by obtaining random partitions of the elements to group. In our GGA, to create a new chromosome, a random number between 1 and the dimension of the problem (D) is generated—i.e., $m \in [1, D]$—which represents the number of subcomponents m, and then each variable is randomly assigned to one of these groups. First, we ensure that each group contains at least one variable, and this is done by shuffling the variables and assigning the first m of them to each group. After that, the remaining variables are randomly assigned to one of the created groups. This is done because in the genetic algorithm DVIIC [7], a random number is chosen to determine the number of groups, and each variable is randomly assigned to a group (under the integer-based representation).

3.4. Grouping Crossover Operator

After choosing the individuals that are subject to the crossover operator, each pair of these individuals, called parents, will create two new individuals (offspring) through a mating strategy. There are several crossover operators for GGAs; however, for comparison purposes, we have chosen the two-point crossover operator that is analogous to the one used in the genetic algorithm DVIIC [7]. This operator works as follows: two crossing points (a and b) between 1 and the number of genes in the individual minus one ($m - 1$) are selected randomly to define the crossing section of both parents (P_1 and P_2). In this way, the first child (C_1) is generated with a copy of P_1, injecting and replacing the groups between the crossing points (a and b) of P_2. Next, the groups copied from P_1 with duplicated items are identified, removing the groups and releasing the remaining variables (missing variables), among which are also those elements that were lost when eliminating the groups from the crossing section and were not in the inserted groups. It is important to note that the injected groups remain intact. Finally, the missing variables are re-inserted into a random number of new groups (between 1 and the number of missing variables) to form the complete individual. The second child (C_2) is generated with the same process but changing the role of the parents.

In Figure 2, we can see an example of crossover for two individuals with 10 variables. The crossing points a and b are marked in step (1); then, in step (2), the section between a and b of parent P_1 is inserted and replaced in the other parent (P_2) and vice versa. In step (3), we have the free variables that result from the groups eliminated for having repeated variables, such as, in the first child, the free element 8 that was in the group with 3 and 6, which were repeated elements, and the elements 2 and 4 that were lost variables (elements). Finally, in step (4), we have the offspring with the free elements re-inserted.

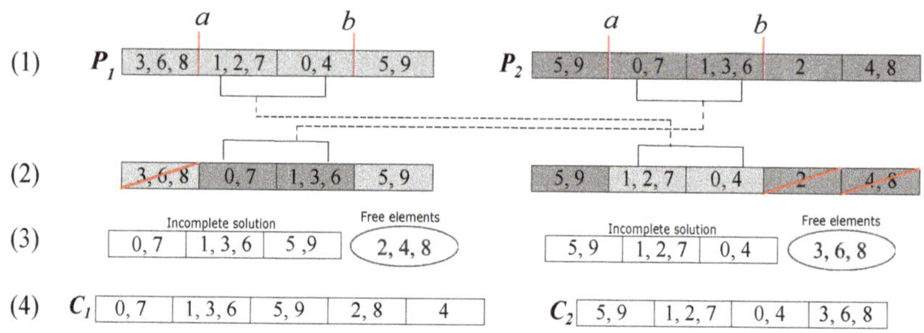

Figure 2. Two-point crossover operator.

3.5. Grouping Mutation Operator

The mutation operator used in the genetic algorithm DVIIC [7] is called uniform mutation, in which once an individual is selected to mutate, one of its genes is randomly selected and is changed from the group to which it belongs. Therefore, to take a similar operator, we have chosen to implement the group-oriented elimination mutation operator for GGAs. This operator works by eliminating a random group of the individual. Later, the deleted elements are re-inserted by adding a random number of groups between 1 and the number of free variables, with the variables randomly assigned to them (similar to how an individual is created). Figure 3 shows an example of the elimination operator. In step (1), the group marked in gray is the eliminated one; then, their elements pass to the free group of elements shown in step (2). Finally, the elements are re-inserted in step (3) with the aforementioned strategy.

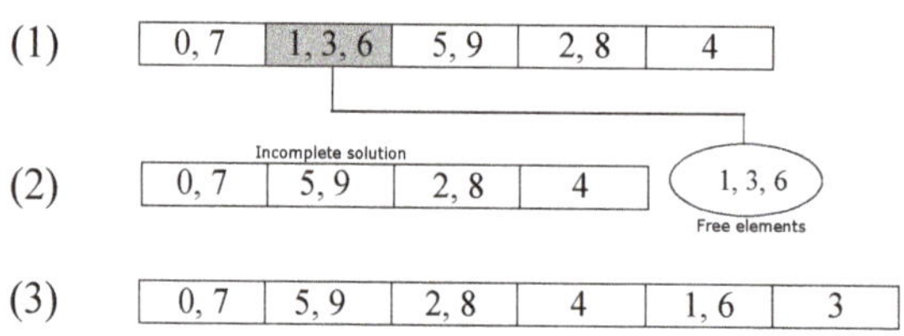

Figure 3. Elimination operator.

3.6. Selection and Replacement Strategies

In a Genetic Algorithm, we have to select the members of the population that will be candidates for crossover and mutation. A selection scheme decides which individuals are allowed to pass on their genes to the next generation, either through cloning, crossover, or mutation. Generally, selection schemes from the literature can be classified into three classes: proportional selection, tournament selection, and ranking selection. Usually, the selection is according to the relative fitness using the best or random individuals [40,41].

Several strategies have been proposed for the parent selection (individuals for crossover). In our GGA, we use a selection scheme similar to that included in the genetic algorithm DVIIC [7], and we carry out a shuffling of the population; for each pair of parents, a random number between 0 and 1 is created. This number determines if the pair of individuals is subject to crossover. That is, the crossover of both individuals is applied when the number is less than or equal to p_c.

In the same way, the selection of individuals to mutate has been studied, and there are various selection techniques for mutation. In this case, the selection method for mutation is similar to the selection method of the genetic algorithm DVIIC [7]. Given a mutation probability p_m, for each individual in the population, a random number between 0 and 1 is generated, and when this number is less or equal than p_m, the individual will be mutated.

In addition to the selection scheme, there must also be a criterion under which the population will be replaced in each generation. Generally, the replacement strategies can be split into three classes: age-based, fitness-based, and random-based (deleting the oldest, worst, or random individuals, respectively) [42]. Similar to the strategy of the genetic algorithm DVIIC [7], in our GGA, after crossover, the offspring replace the parents, and after the mutation, the mutated individuals replace the original ones. Elitism is adopted to always maintain the best solution of the population, replacing the worst individual of the new population.

4. Experiments and Results

In order to study the benefits of using a group-based against an integer encoding in a genetic algorithm, we compared our proposal with the decomposition strategy proposed by Aguilar-Justo et al. [31]. Therefore, we chose the same set of test functions the authors used. It is the first set for Large-Scale Constrained Optimization Problems and it was proposed by Sayed et al. in 2015 [3]. This test set has different separability complexity degrees, which are described in Table 1. It can be tested over three numbers of variables (100, 500, and 1000). These 18 functions were created by combining 6 objective functions with 1, 2, or 3 constraints. The 6 objective functions are based on 2 problems in the literature that have been used, for example, in the CEC 2008 benchmark problems [4], which are the Rosenbrock's function, which is multimodal and nonseparable, and the Sphere function, which is unimodal and separable. In addition, in Table 2, we can see the components of these 18 test functions; that is, the objective function and the constraints that make up each function. The details of the mathematical expression of each function can be consulted in the work of Sayed et al. [3].

Table 1. Characteristics of the objective functions and constraints.

	Description
Obj_1	Completely separable
Obj_2	Partially nonseparable
Obj_3	Partially nonseparable
Obj_4	Partially nonseparable and overlapping variables
Obj_5	Spliced nonseparable and overlapping variables
Obj_6	Spliced nonseparable and overlapping variables
g_1	Separable groups of 5 variables
g_2	Nonseparable groups of 3 variables
g_3	Spliced nonseparable pairs

We have compared the results of our GGA against the Dynamical Variable Interaction Identification Technique for Constrained Problems (DVIIC), in which Aguilar et al. [7] proposed a genetic algorithm for the decomposition of the 18 test functions. We computed 25 independent runs per each benchmark function, in 3 different numbers of variables (100, 500, and 1000). The parameters of our algorithm were set similarly as in the DVIIC work, to compare under equal conditions and perform the same number of function evaluations. These are as follows:

- Population size of 100 individuals;
- Crossover probability $p_c = 0.9$;
- Mutation probability $p_m = 0.1$;
- 10,000 function evaluations—i.e., 100 generations.

Such a configuration implies that the same number of evaluations is carried out by having 100 individuals in each generation for 100 generations, which is equal to 10,000 evaluations. These experiments were conducted on an Intel(R) Core(TM) i5 CPU with 2.50 GHz, Python 3.4, and Microsoft Windows 10.

In the following tables, we show the results of the execution of our proposed GGA and the genetic algorithm DVIIC. Both were executed for each of the 18 functions, 25 times in each dimension. Furthermore, each of the tables shows the results of the Wilcoxon Rank Sum test for each of the functions (column W). A checkmark (✓) means that there are significant differences in favor of the GGA; in addition, an equality symbol (=) represents there are no significant differences between both algorithms.

Table 2. Components of the 18 test functions.

Function	Objective	g_1	g_2	g_3
F_1		×		
F_2	Obj_1	×	×	
F_3		×	×	×
F_4		×		
F_5	Obj_2	×	×	
F_6		×	×	×
F_7		×		
F_8	Obj_3	×	×	
F_9		×	×	×
F_{10}		×		
F_{11}	Obj_4	×	×	
F_{12}		×	×	×
F_{13}		×		
F_{14}	Obj_5	×	×	
F_{15}		×	×	×
F_{16}		×		
F_{17}	Obj_6	×	×	
F_{18}		×	×	×

First, Table 3 contains the results according to the evaluation of the best individual for the 25 runs in the 18 functions under 100 variables. The best, median, and the standard deviation registered of the evaluation function ($grps_{diff}$) value are shown. We can observe that the GGA improves the decomposition evaluation function value in all of the cases compared to DVIIC. As we can see, unlike DVIIC, the GGA reaches the value of 0 in the best result in most cases. Furthermore, the median and standard deviation values obtained by the GGA are smaller in all cases. Such values equal to zero indicate that our algorithm found the best value for the evaluation function ($grps_{diff} = 0$) in the 25 runs for functions 1 to 12. Finally, the Wilcoxon Rank Sum Test reveals that there are significant differences in favor of the GGA in all cases.

Second, Table 4 shows the results of our algorithm to solve the same 18 functions, now with 500 variables. The GGA obtained the smallest values in most cases for the best, median, and standard deviation when compared against DVIIC. In a similar way as in Table 3, the standard deviation and median values equal to zero indicate that our algorithm found the best value for the evaluation function ($grps_{diff} = 0$) in the 25 runs for functions 1 to 12. Moreover, in the other test functions, the best, median, and standard deviation values are smaller when compared to DVIIC. On the other hand, the Wilcoxon Rank Sum test shows that there are no significant differences between the two algorithms in function 1 and shows significant differences in favor of the GGA in the remaining 17 functions. According to this test, in functions 2 to 18, the Wilcoxon Rank Sum test rejects the hypothesis that the DVIIC approach is as effective as the proposed GGA approach, and F_1 is trivial to solve by the two algorithms.

Table 3. Statistical results in dimension 100. Best results shown in boldface.

	GGA			DVIIC			
D = 100	Best	Median	Std	Best	Median	Std	W
F_1	**0.00×10^{0}**	**0.00×10^{0}**	**0.00×10^{0}**	**0.00×10^{0}**	2.18×10^{-11}	2.96×10^{-11}	✓
F_2	**0.00×10^{0}**	**0.00×10^{0}**	**0.00×10^{0}**	**0.00×10^{0}**	2.35×10^{4}	1.20×10^{4}	✓
F_3	**0.00×10^{0}**	**0.00×10^{0}**	**0.00×10^{0}**	1.38×10^{5}	1.38×10^{5}	6.77×10^{4}	✓
F_4	**0.00×10^{0}**	**0.00×10^{0}**	**0.00×10^{0}**	4.29×10^{4}	8.57×10^{4}	1.69×10^{4}	✓
F_5	**0.00×10^{0}**	**0.00×10^{0}**	**0.00×10^{0}**	1.27×10^{4}	1.78×10^{4}	2.48×10^{3}	✓
F_6	**0.00×10^{0}**	**0.00×10^{0}**	**0.00×10^{0}**	8.97×10^{5}	1.44×10^{6}	2.36×10^{5}	✓
F_7	**0.00×10^{0}**	**0.00×10^{0}**	**0.00×10^{0}**	2.51×10^{5}	1.01×10^{6}	2.62×10^{5}	✓
F_8	**0.00×10^{0}**	**0.00×10^{0}**	**0.00×10^{0}**	6.70×10^{5}	8.38×10^{5}	9.84×10^{4}	✓
F_9	**0.00×10^{0}**	**0.00×10^{0}**	**0.00×10^{0}**	2.13×10^{3}	3.20×10^{3}	3.58×10^{2}	✓
F_{10}	**0.00×10^{0}**	**0.00×10^{0}**	**0.00×10^{0}**	5.36×10^{5}	1.07×10^{6}	2.05×10^{5}	✓
F_{11}	**0.00×10^{0}**	**0.00×10^{0}**	**0.00×10^{0}**	4.12×10^{2}	5.84×10^{2}	9.04×10^{1}	✓
F_{12}	**0.00×10^{0}**	**0.00×10^{0}**	**0.00×10^{0}**	9.28×10^{3}	1.76×10^{4}	3.20×10^{3}	✓
F_{13}	**0.00×10^{0}**	**0.00×10^{0}**	**2.31×10^{4}**	6.31×10^{5}	1.09×10^{6}	1.08×10^{5}	✓
F_{14}	**0.00×10^{0}**	**0.00×10^{0}**	**5.35×10^{2}**	1.55×10^{7}	1.93×10^{7}	1.72×10^{6}	✓
F_{15}	**4.66×10^{-10}**	**4.66×10^{-10}**	**2.60×10^{2}**	5.42×10^{6}	8.72×10^{6}	9.33×10^{5}	✓
F_{16}	**1.79×10^{4}**	**5.38×10^{4}**	**1.72×10^{4}**	4.93×10^{5}	6.81×10^{5}	7.26×10^{4}	✓
F_{17}	**1.07×10^{5}**	**4.28×10^{5}**	9.52×10^{4}	4.83×10^{5}	6.99×10^{5}	**8.07×10^{4}**	✓
F_{18}	**1.63×10^{5}**	**4.88×10^{5}**	**1.40×10^{5}**	1.80×10^{6}	2.62×10^{6}	2.94×10^{5}	✓

Table 4. Statistical results in dimension 500. Best results shown in boldface.

	GGA			DVIIC			
D = 500	Best	Median	Std	Best	Median	Std	W
F_1	**0.00×10^{0}**	**0.00×10^{0}**	**0.00×10^{0}**	**0.00×10^{0}**	**0.00×10^{0}**	**0.00×10^{0}**	=
F_2	**0.00×10^{0}**	**0.00×10^{0}**	**0.00×10^{0}**	4.66×10^{-10}	4.43×10^{4}	1.16×10^{4}	✓
F_3	**0.00×10^{0}**	**0.00×10^{0}**	**0.00×10^{0}**	3.59×10^{4}	7.19×10^{4}	1.23×10^{4}	✓
F_4	**0.00×10^{0}**	**0.00×10^{0}**	**0.00×10^{0}**	1.17×10^{6}	1.24×10^{6}	2.98×10^{4}	✓
F_5	**0.00×10^{0}**	**0.00×10^{0}**	**0.00×10^{0}**	7.20×10^{6}	8.73×10^{6}	5.68×10^{5}	✓
F_6	**0.00×10^{0}**	**0.00×10^{0}**	**0.00×10^{0}**	3.34×10^{6}	4.12×10^{6}	2.09×10^{5}	✓
F_7	**0.00×10^{0}**	**0.00×10^{0}**	**0.00×10^{0}**	5.19×10^{4}	1.13×10^{5}	1.94×10^{4}	✓
F_8	**0.00×10^{0}**	**0.00×10^{0}**	**0.00×10^{0}**	8.79×10^{5}	9.75×10^{5}	4.19×10^{4}	✓
F_9	**0.00×10^{0}**	**0.00×10^{0}**	**0.00×10^{0}**	5.58×10^{4}	1.67×10^{5}	3.46×10^{4}	✓
F_{10}	**0.00×10^{0}**	**0.00×10^{0}**	**0.00×10^{0}**	9.69×10^{6}	1.19×10^{7}	5.03×10^{5}	✓
F_{11}	**0.00×10^{0}**	**0.00×10^{0}**	**0.00×10^{0}**	1.95×10^{6}	2.58×10^{6}	1.80×10^{5}	✓
F_{12}	**0.00×10^{0}**	**0.00×10^{0}**	**0.00×10^{0}**	3.50×10^{6}	3.84×10^{6}	1.10×10^{5}	✓
F_{13}	**0.00×10^{0}**	**0.00×10^{0}**	**2.56×10^{4}**	9.61×10^{5}	1.26×10^{6}	7.07×10^{4}	✓
F_{14}	**0.00×10^{0}**	**0.00×10^{0}**	**6.86×10^{4}**	9.10×10^{5}	1.35×10^{6}	1.08×10^{5}	✓
F_{15}	**0.00×10^{0}**	**5.96×10^{-8}**	**1.27×10^{5}**	7.97×10^{6}	9.88×10^{6}	4.48×10^{5}	✓
F_{16}	**8.13×10^{2}**	**3.25×10^{3}**	**7.04×10^{2}**	1.38×10^{7}	1.69×10^{7}	7.77×10^{5}	✓
F_{17}	**7.96×10^{4}**	**1.59×10^{5}**	**1.75×10^{4}**	2.26×10^{7}	2.43×10^{7}	1.48×10^{7}	✓
F_{18}	**3.83×10^{5}**	**7.66×10^{5}**	**1.46×10^{5}**	2.78×10^{7}	3.63×10^{7}	2.05×10^{6}	✓

Finally, Table 5 contains the results for the 18 functions with both algorithms implemented on 1000 variables. In this experiment, we observed that, in a similar way to the experiment with 500 variables, our GGA obtained the smallest best, median, and standard deviation values in most cases. On the other hand, the behavior of the median and standard deviation values allows us to see that we obtained the zero value in the 25 independent runs for the first 12 functions. Moreover, the Wilcoxon Rank Sum test results show that there are no significant differences between the two algorithms in function 1 and indicate significant differences in favor of the GGA in the remaining 17 functions. In a similar way as in the results with 500 variables, the Wilcoxon Rank Sum test determines that F_1 is a trivial case and rejects the hypothesis that the DVIIC approach is as effective as the proposed GGA approach for the other 17 remaining functions.

Table 5. Statistical results in dimension 1000. Best results shown in boldface.

		GGA			DVIIC		
D = 1000	Best	Median	Std	Best	Median	Std	W
F_1	**0.00 × 10^0**	**0.00 × 10^0**	**0.00 × 10^0**	**0.00 × 10^0**	**0.00 × 10^0**	**0.00 × 10^0**	=
F_2	**0.00 × 10^0**	**0.00 × 10^0**	**0.00 × 10^0**	4.18 × 10^3	1.25 × 10^4	5.39 × 10^3	✓
F_3	**0.00 × 10^0**	**0.00 × 10^0**	**0.00 × 10^0**	2.42 × 10^2	2.90 × 10^2	1.98 × 10^1	✓
F_4	**0.00 × 10^0**	**0.00 × 10^0**	**0.00 × 10^0**	3.93 × 10^6	4.86 × 10^6	2.33 × 10^5	✓
F_5	**0.00 × 10^0**	**0.00 × 10^0**	**0.00 × 10^0**	1.33 × 10^6	1.73 × 10^6	1.19 × 10^5	✓
F_6	**0.00 × 10^0**	**0.00 × 10^0**	**0.00 × 10^0**	1.05 × 10^6	1.36 × 10^6	7.85 × 10^4	✓
F_7	**0.00 × 10^0**	**0.00 × 10^0**	**0.00 × 10^0**	1.10 × 10^6	1.61 × 10^6	1.28 × 10^5	✓
F_8	**0.00 × 10^0**	**0.00 × 10^0**	**0.00 × 10^0**	4.16 × 10^6	4.91 × 10^6	1.82 × 10^5	✓
F_9	**0.00 × 10^0**	**0.00 × 10^0**	**0.00 × 10^0**	4.56 × 10^6	4.92 × 10^6	1.09 × 10^5	✓
F_{10}	**0.00 × 10^0**	**0.00 × 10^0**	**0.00 × 10^0**	2.16 × 10^6	2.47 × 10^6	8.18 × 10^4	✓
F_{11}	**0.00 × 10^0**	**0.00 × 10^0**	**0.00 × 10^0**	1.99 × 10^5	2.13 × 10^5	4.33 × 10^3	✓
F_{12}	**0.00 × 10^0**	**0.00 × 10^0**	**0.00 × 10^0**	2.93 × 10^5	3.18 × 10^5	8.43 × 10^3	✓
F_{13}	**0.00 × 10^0**	**0.00 × 10^0**	**3.51 × 10^4**	3.36 × 10^7	3.98 × 10^7	1.49 × 10^6	✓
F_{14}	**0.00 × 10^0**	**0.00 × 10^0**	**5.49 × 10^3**	4.60 × 10^7	1.76 × 10^8	6.74 × 10^7	✓
F_{15}	**0.00 × 10^0**	**0.00 × 10^0**	**2.00 × 10^5**	1.02 × 10^5	3.49 × 10^7	1.55 × 10^7	✓
F_{16}	**4.07 × 10^5**	**8.15 × 10^5**	**1.27 × 10^5**	7.71 × 10^7	8.72 × 10^7	2.37 × 10^6	✓
F_{17}	**4.32 × 10^4**	**8.63 × 10^4**	**2.52 × 10^4**	7.71 × 10^6	1.15 × 10^7	1.05 × 10^6	✓
F_{18}	**4.20 × 10^5**	**8.41 × 10^5**	**1.20 × 10^5**	3.03 × 10^7	4.39 × 10^7	3.07 × 10^6	✓

Given the previous tables, we observe that our algorithm presents better performance than DVIIC, obtaining better $grps_{diff}$ values in all cases (in comparison with the mentioned algorithm). An interesting behavior is observed in these experiments; it seems to be more difficult for our algorithm to find the minimum decomposition evaluation in the 18 test functions as the dimension decreases. Zero best, median, and standard deviation values of the 25 independent runs indicate a stable behavior of our algorithm in each execution of the first 12 functions of the benchmark (in the three experiments). However, these values increase with the complexity of the functions, and in the end, functions 16, 17, and 18 do not reach the minimum in any of the experiments.

Analyzing the Performance of the GGA

Due to the behavior observed in the previous experiments, a detailed study of the algorithm is necessary to improve it in future work. For this reason, we decided to make a brief study of the convergence of our algorithm.

In order to understand the on-line behavior of our algorithm, we carried out some plots of the GGA convergence for the most difficult functions of the benchmark. Figure 4 shows the convergence in functions F_{16}, F_{17}, and F_{18} through 100 generations for three dimension values.

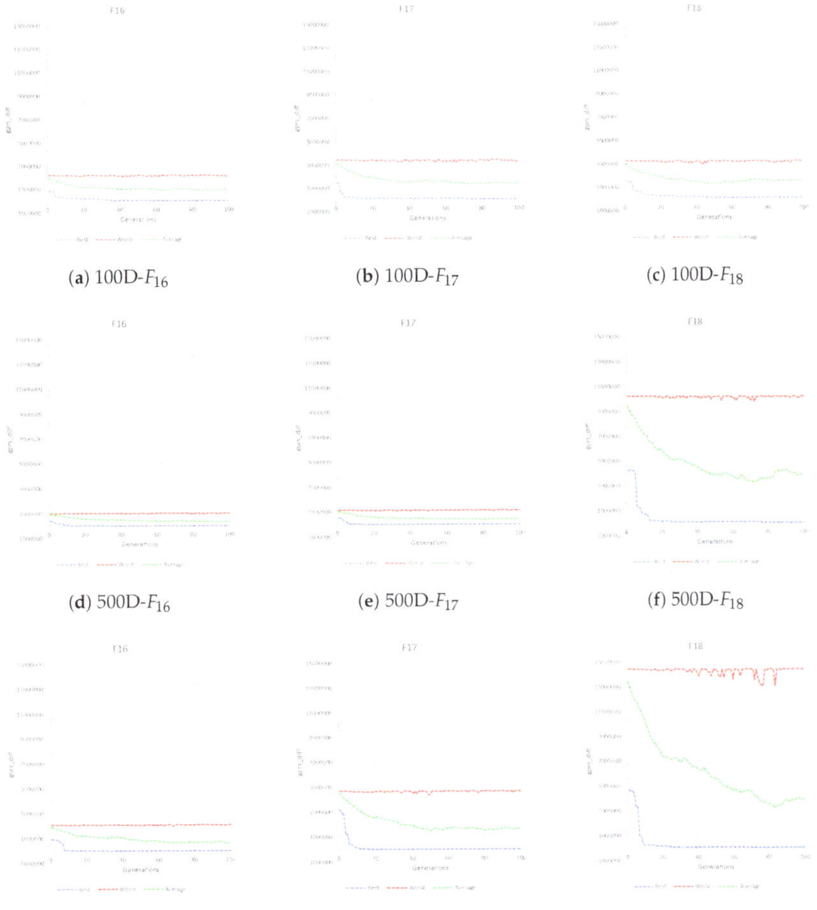

(**a**) 100D-F_{16} (**b**) 100D-F_{17} (**c**) 100D-F_{18}

(**d**) 500D-F_{16} (**e**) 500D-F_{17} (**f**) 500D-F_{18}

(**g**) 1000D-F_{16} (**h**) 1000D-F_{17} (**i**) 1000D-F_{18}

Figure 4. Convergence plots of 100 generations for functions F_{16}, F_{17}, and F_{18} with 100, 500, and 1000 variables.

Three convergence behaviors are shown in each of the graphs within Figure 4. First, we shown the convergence of the worst individual in the population—that is, the individual with the highest decomposition evaluation value (red color). After that, we show the behavior across the 100 generations of the average decomposition evaluation of the 100 individuals in the population (green color). Finally, we show the convergence across the 100 generations of the best individual in the population in terms of their decomposition evaluation value (blue color).

Figure 4a–c shows the convergence in the experiment with 100 variables from one of the 25 GGA runs. We can observe similar behavior in the three functions, with decomposition evaluation values below 4.0×10^7 in all three cases (best, worst, and average). It is important to note that the behavior of the best individual presents an early convergence in the three functions and how the decomposition evaluation of the worst individuals remains stable over the generations.

Regarding the convergence of the functions with 500 variables (Figure 4d–f), we can observe that function F_{18} shows the highest values of the decomposition evaluation for the three values (best, worst, and average), unlike the other functions. Similar to those functions with 100 variables, we see that this value does not converge to zero in any of the cases, and the best individual has a quick convergence. We can also induce, according to the graphs, that several individuals of the population do not converge in the neighborhood of the best solution.

In the case of the functions evaluated with 1000 variables (Figure 4g–i), we see a fast convergence of the best individual in the population. As in the previous graphs, the value of the decomposition evaluation in the worst individual has a continuous behavior, without converging to zero during the 100 generations, and the best value has a quick convergence.

The above discussed best, worst, and average values' convergence behaviors in the functions with spliced nonseparable and overlapping variables suggest that the included strategies in the GGA do not appear to lead to better solutions. We can see from the plots in Figure 4 that not the entire population converges to the neighborhood of the best solution, due to the low selective pressure of the selection and replacement strategies. We also observe that the GGA produces good solutions in the early stages but leads to the premature convergence of the best individual. This behavior can be related to the crossover and mutation operators that promoted the creation of new groups, which does not seem to be a suitable strategy for nonseparable functions. All these observations indicate that, although our algorithm performs well, it can still be further improved by analyzing its components.

5. Conclusions and Future Work

In this paper, we have proposed a Grouping Genetic Algorithm (GGA) to deal with the decomposition of variables in Large-Scale Constrained Optimization Problems to create subproblems of the original problem and thus reduce the dimension. To evaluate the impact of the representation scheme on the performance of a genetic algorithm, our GGA was designed in a similar way to a state-of-the-art genetic algorithm that works with the decomposition of variables and which includes an integer-based representation. The main difference between the two algorithms was the genetic encoding. The experiments were carried out in a benchmark of 18 functions with different complexity characteristics, and these functions were tested in 100, 500, and 1000 dimensions.

The obtained results confirm that the use of a group-based genetic encoding allows our GGA to obtain good and robust decompositions on test functions with different features and separability complexity degrees, outperforming in all the benchmark functions the results obtained by a genetic algorithm with an integer-based encoding.

We are aware that there are still test functions with spliced nonseparable and overlapping variables that show a high degree of difficulty; for these functions, the included strategies in the GGA do not appear to lead to better solutions. However, the GGA presented in this work does not include the state-of-the-art grouping genetic operators.

Future work will consist of studying the parameters of the GGA as well as the effect of each of the methods used in the crossover and mutation operators to identify the best strategies that work together with the grouping encoding scheme and the features of the functions. Furthermore, it is necessary to implement an efficient reproduction technique with a balance in selective pressure and population diversity to avoid the premature convergence of the best individuals and increase the algorithm's performance.

The introduction of a new decomposition method opens up an interesting range of possibilities for future research. Currently, we are working on including our GGA in the decomposition step of two Cooperative Co-Evolution methods that include different strategies for the optimization and cooperation of the subcomponents, with the respective feasibility and computational complexity analysis.

Finally, although the set of test functions analyzed in this work is varied concerning the characteristics of the functions, we would explore the proposal in other Large-Scale Constrained Optimization benchmarks.

Author Contributions: Conceptualization, G.C.-A., M.Q.-C. and E.M.-M.; methodology, M.Q.-C.; software, G.C.-A.; validation, M.Q.-C. and E.M.-M.; formal analysis, M.Q.-C.; investigation, G.C.-A., M.Q.-C. and E.M.-M.; resources, G.C.-A.; writing—original draft preparation, G.C.-A.; writing—review and editing, G.C.-A., M.Q.-C. and E.M.-M.; visualization, G.C.-A. and M.Q.-C.; supervision, M.Q.-C.; project administration, M.Q.-C. All authors have read and agreed to the published version of the manuscript.

Funding: This research received no external funding.

Conflicts of Interest: The authors declare no conflict of interest. The funders had no role in the design of the study; in the collection, analyses, or interpretation of data; in the writing of the manuscript, or in the decision to publish the results.

References

1. Deb, K. *Optimization for Engineering Design: Algorithms and Examples*; PHI Learning Pvt. Ltd.: New Delhi, India, 2012.
2. Smith, A.E.; Coit, D.W.; Baeck, T.; Fogel, D.; Michalewicz, Z. Penalty functions. In *Handbook of Evolutionary Computation*; CRC Press: Boca Raton, FL, USA, 1997.
3. Sayed, E.; Essam, D.; Sarker, R.; Elsayed, S. Decomposition-based evolutionary algorithm for large-scale constrained problems. *Inf. Sci.* **2015**, *316*, 457–486. [CrossRef]
4. Tang, K.; Yáo, X.; Suganthan, P.N.; MacNish, C.; Chen, Y.P.; Chen, C.M.; Yang, Z. *Benchmark Functions for the CEC'2008 Special Session and Competition on Large Scale Global Optimization*; Nature Inspired Computation and Applications Laboratory, USTC: Hefei, China, 2007; pp. 1–18.
5. Potter, M.A.; De Jong, K.A. A cooperative coevolutionary approach to function optimization. In *International Conference on Parallel Problem Solving from Nature*; Springer: Berlin/Heidelberg, Germany, 1994; pp. 249–257.
6. Ma, X.; Li, X.; Zhang, Q.; Tang, K.; Liang, Z.; Xie, W.; Zhu, Z. A survey on cooperative co-evolutionary algorithms. *IEEE Trans. Evol. Comput.* **2018**, *23*, 421–441. [CrossRef]
7. Aguilar-Justo, A.E.; Mezura-Montes, E.; Elsayed, S.M.; Sarker, R.A. Decomposition of large-scale constrained problems using a genetic-based search. In Proceedings of the 2016 IEEE International Autumn Meeting on Power, Electronics and Computing (ROPEC), Ixtapa, Mexico, 9–11 November 2016; IEEE: Piscataway, NJ, USA, 2016; pp. 1–6.
8. Ramos-Figueroa, O.; Quiroz-Castellanos, M.; Mezura-Montes, E.; Schütze, O. Metaheuristics to solve grouping problems: A review and a case study. *Swarm Evol. Comput.* **2020**, *53*, 100643. [CrossRef]
9. Liu, Y.; Yao, X.; Zhao, Q.; Higuchi, T. Scaling up fast evolutionary programming with cooperative coevolution. In Proceedings of the 2001 Congress on Evolutionary Computation (IEEE Cat. No. 01TH8546), San Francisco, CA, USA, 7–11 July 2001; IEEE: Piscataway, NJ, USA, 2001; Volume 2, pp. 1101–1108.
10. Van den Bergh, F.; Engelbrecht, A.P. A cooperative approach to particle swarm optimization. *IEEE Trans. Evol. Comput.* **2004**, *8*, 225–239. [CrossRef]
11. Yang, Z.; Tang, K.; Yao, X. Differential evolution for high-dimensional function optimization. In Proceedings of the 2007 IEEE Congress on Evolutionary Computation, Singapore, 25–28 September 2007; IEEE: Piscataway, NJ, USA, 2007; pp. 3523–3530.
12. Yang, Z.; Tang, K.; Yao, X. Large-scale evolutionary optimization using cooperative coevolution. *Inf. Sci.* **2008**, *178*, 2985–2999. [CrossRef]
13. Omidvar, M.N.; Li, X.; Yang, Z.; Yao, X. Cooperative co-evolution for large-scale optimization through more frequent random grouping. In Proceedings of the IEEE Congress on Evolutionary Computation, Barcelona, Spain, 18–23 July 2010; IEEE: Piscataway, NJ, USA, 2010; pp. 1–8.
14. Omidvar, M.N.; Li, X.; Yao, X. Cooperative co-evolution with delta grouping for large-scale non-separable function optimization. In Proceedings of the IEEE Congress on Evolutionary Computation, Barcelona, Spain, 18–23 July 2010; IEEE: Piscataway, NJ, USA, 2010; pp. 1–8.
15. Xu, B.; Zhang, Y.; Gong, D.; Guo, Y.; Rong, M. Environment sensitivity-based cooperative co-evolutionary algorithms for dynamic multi-objective optimization. *IEEE/ACM Trans. Comput. Biol. Bioinform.* **2017**, *15*, 1877–1890. [CrossRef] [PubMed]
16. Omidvar, M.N.; Li, X.; Mei, Y.; Yao, X. Cooperative co-evolution with differential grouping for large-scale optimization. *IEEE Trans. Evol. Comput.* **2013**, *18*, 378–393. [CrossRef]
17. Omidvar, M.N.; Yang, M.; Mei, Y.; Li, X.; Yao, X. DG2: A faster and more accurate differential grouping for large-scale black-box optimization. *IEEE Trans. Evol. Comput.* **2017**, *21*, 929–942. [CrossRef]
18. Sun, Y.; Kirley, M.; Halgamuge, S.K. A recursive decomposition method for large-scale continuous optimization. *IEEE Trans. Evol. Comput.* **2017**, *22*, 647–661. [CrossRef]

19. Sun, Y.; Omidvar, M.N.; Kirley, M.; Li, X. Adaptive threshold parameter estimation with recursive differential grouping for problem decomposition. In Proceedings of the Genetic and Evolutionary Computation Conference, Ser. GECCO '18, Kyoto, Japan, 15–19 July 2018; ACM: New York, NY, USA, 2018; pp. 889–896.
20. Sun, Y.; Li, X.; Ernst, A.; Omidvar, M.N. Decomposition for large-scale optimization problems with overlapping components. In Proceedings of the 2019 IEEE Congress on Evolutionary Computation (CEC), Wellington, New Zealand, 10–13 June 2019; pp. 326–333.
21. Yang, M.; Zhou, A.; Li, C.; Yao, X. An Efficient Recursive Differential Grouping for Large-Scale Continuous Problems. *IEEE Trans. Evol. Comput.* **2020**, *25*, 159–171. [CrossRef]
22. Sopov, E. *Large-Scale Global Optimization Using a Binary Genetic Algorithm with EDA-Based Decomposition*; Lecture Notes in Computer Science (Including Subseries Lecture Notes in Artificial Intelligence and Lecture Notes in Bioinformatics); Springer: Cham, Switzerland, 2016; Volume 9712; pp. 619–626.
23. Valdez, S.I.; Hernández, A.; Botello, S. A Boltzmann based estimation of distribution algorithm. *Inf. Sci.* **2013**, *236*, 126–137. [CrossRef]
24. Mühlenbein, H.; Paaß, G. From recombination of genes to the estimation of distributions I. Binary parameters. In *International Conference on Parallel Problem Solving from Nature*; Springer: Berlin/Heidelberg, Germany, 1996; pp. 178–187.
25. Sayed, E.; Essam, D.; Sarker, R. Dependency identification technique for large-scale optimization problems. In Proceedings of the 2012 IEEE Congress on Evolutionary Computation, Philadelphia, PA, USA, 7–11 July 2012; IEEE: Piscataway, NJ, USA, 2012; pp. 1–8.
26. Schaffer, J.D.; Morishima, A. An adaptive crossover distribution mechanism for genetic algorithms. In *Genetic Algorithms and Their Applications: Proceedings of the Second International Conference on Genetic Algorithms*; Lawrence Erlbaum Associates, Inc.: Hillsdale, NJ, USA, 1987; pp. 36–40.
27. Goh, C.K.; Tan, K.C. A competitive-cooperative coevolutionary paradigm for dynamic multiobjective optimization. *IEEE Trans. Evol. Comput.* **2008**, *13*, 103–127.
28. Goh, C.K.; Tan, K.C.; Liu, D.S.; Chiam, S.C. A competitive and cooperative co-evolutionary approach to multi-objective particle swarm optimization algorithm design. *Eur. J. Oper. Res.* **2010**, *202*, 42–54. [CrossRef]
29. Strasser, S.; Sheppard, J.; Fortier, N.; Goodman, R. Factored evolutionary algorithms. *IEEE Trans. Evol. Comput.* **2016**, *21*, 281–293. [CrossRef]
30. Guan, X.; Zhang, X.; Wei, J.; Hwang, I.; Zhu, Y.; Cai, K. A strategic conflict avoidance approach based on cooperative coevolutionary with the dynamic grouping strategy. *Int. J. Syst. Sci.* **2016**, *47*, 1995–2008. [CrossRef]
31. Aguilar-Justo, A.E.; Mezura-Montes, E. Towards an improvement of variable interaction identification for large-scale constrained problems. In Proceedings of the IEEE Congress on Evolutionary Computation, Vancouver, BC, Canada, 24–29 July 2016; IEEE Press: Piscataway, NJ, USA, 2016.
32. Vakhnin, A.; Sopov, E. Investigation of the iCC framework performance for solving constrained LSGO problems. *Algorithms* **2020**, *13*, 108. [CrossRef]
33. Kashan, A.H.; Akbari, A.A.; Ostadi, B. Grouping evolution strategies: An effective approach for grouping problems. *Appl. Math. Model.* **2015**, *39*, 2703–2720. [CrossRef]
34. Garey, M.R. *A Guide to the Theory of NP-Completeness. Computers and Intractability*; WH Freeman and Company: New York, NY, USA, 1979.
35. Falkenauer, E. A new representation and operators for genetic algorithms applied to grouping problems. *Evol. Comput.* **1994**, *2*, 123–144. [CrossRef]
36. Ramos-Figueroa, O.; Quiroz-Castellanos, M.; Mezura-Montes, E.; Kharel, R. Variation Operators for Grouping Genetic Algorithms: A Review. *Swarm Evol. Comput.* **2020**, *60*, 100796. [CrossRef]
37. Eiben, A.E.; Schippers, C.A. On evolutionary exploration and exploitation. *Fundam. Inform.* **1998**, *35*, 35–50. [CrossRef]
38. Mosk-Aoyama, D.; Shah, D. Fast distributed algorithms for computing separable functions. *IEEE Trans. Inf. Theory* **2008**, *54*, 2997–3007. [CrossRef]
39. Ray, T.; Yao, X. A cooperative coevolutionary algorithm with correlation-based adaptive variable partitioning. In Proceedings of the 2009 IEEE Congress on Evolutionary Computation, Trondheim, Norway, 18–21 May 2009; IEEE: Piscataway, NJ, USA, 2009; pp. 983–989.
40. Blickle, T.; Thiele, L. A comparison of selection schemes used in evolutionary algorithms. *Evol. Comput.* **1996**, *4*, 361–394. [CrossRef]
41. Zhang, B.T.; Kim, J.J. Comparison of selection methods for evolutionary optimization. *Evol. Optim.* **2000**, *2*, 55–70.
42. Smith, J. On replacement strategies in steady state evolutionary algorithms. *Evol. Comput.* **2007**, *15*, 29–59. [CrossRef]

Mathematical and Computational Applications

MDPI

Article

Evaluation of Machine Learning Algorithms for Early Diagnosis of Deep Venous Thrombosis

Eduardo Enrique Contreras-Luján [1], Enrique Efrén García-Guerrero [1], Oscar Roberto López-Bonilla [1], Esteban Tlelo-Cuautle [2], Didier López-Mancilla [3] and Everardo Inzunza-González [1,*]

1 Facultad de Ingeniería, Arquitectura y Diseño, Universidad Autónoma de Baja California, Carretera Transpeninsular Ensenada-Tijuana No. 3917, Ensenada C.P. 22860, Baja California, Mexico; eduardo.enrique.contreras.lujan@uabc.edu.mx (E.E.C.-L.); eegarcia@uabc.edu.mx (E.E.G.-G.); olopez@uabc.edu.mx (O.R.L.-B.)

2 Departamento de Electrónica, Instituto Nacional de Astrofísica, Óptica y Electrónica, Luis Enrique Erro No. 1, Santa María Tonanzintla, San Andrés Cholula, Puebla C.P. 72840, San Andrés Cholula, Mexico; etlelo@inaoep.mx

3 Centro Universitario de Los Lagos, Universidad de Guadalajara, Enrique Díaz de León, Col. Paseos de la Montaña, Lagos de Moreno C.P. 47460, Jalisco, Mexico; didier.lopez@academicos.udg.mx

* Correspondence: einzunza@uabc.edu.mx; Tel.: +52-646-175-0744

Abstract: Deep venous thrombosis (DVT) is a disease that must be diagnosed quickly, as it can trigger the death of patients. Nowadays, one can find different ways to determine it, including clinical scoring, D-dimer, ultrasonography, etc. Recently, scientists have focused efforts on using machine learning (ML) and neural networks for disease diagnosis, progressively increasing the accuracy and efficacy. Patients with suspected DVT have no apparent symptoms. Using pattern recognition techniques, aiding good timely diagnosis, as well as well-trained ML models help to make good decisions and validation. The aim of this paper is to propose several ML models for a more efficient and reliable DVT diagnosis through its implementation on an edge device for the development of instruments that are smart, portable, reliable, and cost-effective. The dataset was obtained from a state-of-the-art article. It is divided into 85% for training and cross-validation and 15% for testing. The input data in this study are the Wells criteria, the patient's age, and the patient's gender. The output data correspond to the patient's diagnosis. This study includes the evaluation of several classifiers such as Decision Trees (DT), Extra Trees (ET), K-Nearest Neighbor (KNN), Multi-Layer Perceptron Neural Network (MLP-NN), Random Forest (RF), and Support Vector Machine (SVM). Finally, the implementation of these ML models on a high-performance embedded system is proposed to develop an intelligent system for early DVT diagnosis. It is reliable, portable, open source, and low cost. The performance of different ML algorithms was evaluated, where KNN achieved the highest accuracy of 90.4% and specificity of 80.66% implemented on personal computer (PC) and Raspberry Pi 4 (RPi4). The accuracy of all trained models on PC and Raspberry Pi 4 is greater than 85%, while the area under the curve (AUC) values are between 0.81 and 0.86. In conclusion, as compared to traditional methods, the best ML classifiers are effective at predicting DVT in an early and efficient manner.

Keywords: DVT; early diagnosis; artificial intelligence; machine-learning; smart system; embedded system; edge computing; edge device

Citation: Contreras-Luján, E.E.; García-Guerrero, E.E.; López-Bonilla, O.R.; Tlelo-Cuautle, E.; López-Mancilla, D.; Inzunza-González, E. Evaluation of Machine Learning Algorithms for Early Diagnosis of Deep Venous Thrombosis. *Math. Comput. Appl.* **2022**, *27*, 24. https://doi.org/10.3390/mca27020024

Academic Editors: Marcela Quiroz, Luis Gerardo de la Fraga, Adriana Lara, Leonardo Trujillo and Oliver Schütze

Received: 29 December 2021
Accepted: 2 March 2022
Published: 4 March 2022

1. Introduction

Deep venous thrombosis (DVT) is a disorder in which blood clots form within the veins, obstructing the flow of blood through the circulatory system, and it affects people of all ages [1]. The cause of the disease is unknown; however, it is thought to be caused by a combination of variables, including genetic factors. Genetic factors are also thought to have a role in the diagnosis of the disorder. In the field of engineering, there are two major challenges: patients suspected of DVT have no visible symptoms, and failing to diagnose it could be fatal; without symptoms, the first test (D-dimer blood) is useless; and the use of ultrasound has high certainty but comes at a high cost and necessitates the use of many instruments [2,3]. DVT is a disease that must be recognized as soon as possible because the implications might be fatal for the patient. Several scientists have created various techniques and methods to diagnose the problem over the years, beginning in the 1970s with the development of ultrasonography [4], which marked a breakthrough in the timely diagnosis of clots in the lower limbs of the human body. Philip Wells, a renowned scientist, has stated on numerous occasions that technology, which has been revolutionized exponentially in recent years, will support the future in the early diagnosis of diseases. This, combined with new trends in the work of computer equipment, will enable great advances in science and human health.

Venous thromboembolism (VTE), the third most common vascular illness worldwide, is a complex condition impacted by various genetic and non-genetic risk factors [5]. The pathogenesis of VTE includes Virchow's triad, which provides for hypercoagulability, reduced blood flow or stasis, and damage to blood vessels due to disease or injury [6]; they are blood clots that can occur if the patient's blood flow changes or slows down somewhere in their body [7], putting the patient's life and health at risk. The annual incidence is 1 to 3 people per 1000 people. The prevalence of this condition varies with age. It can cause DVT or pulmonary embolism (PE) in some cases [1,8–10]; thrombosis can also develop in other veins such as the liver, cerebral sinus, retina, and mesenteric veins. Approximately one-third of VTE patients develop a PE, while two-thirds exclusively have DVT [11].

The Primary Care Unit (PCU) is the backbone of any health care system. The record of previous epidemics demonstrates the critical significance of PCU and necessitates PCU specialists' engagement in procedural decision making [12]. The PCU serves as the entry point to the Health System (HS), which is described as the primary level of health care. The "Health Unit Clinics" (Health Units that constitute Primary Health Care) are defined by their commitment to health promotion and protection, disease prevention, diagnosis, treatment, rehabilitation, harm reduction, and health maintenance on an individual and collective level, to provide comprehensive care that has a positive impact on the health status of communities [13]. They provide primary care services across the board, including the evaluation and diagnosis of acute illnesses and ongoing treatment for patients with chronic conditions [14].

Nowadays, numerous ways to determine the condition are available, such as statistical analysis and clinician scoring [15], D-dimer blood tests [16], infrared imaging [17], ultrasonography, and even the application of deep machine learning and Neural Networks (NN) [11,18]. Many countries, including the United States, Italy, the United Kingdom, Germany, and Canada, have pioneered artificial intelligence (AI) work in the diagnosis and prediction of DVT, with the percentage of accuracy and effectiveness steadily increasing over time as algorithms become more and more optimal and more data can be obtained from real cases. However, progress has been made in the development of NN in terms of debugging codes and developing new algorithms, but they have not been implemented outside of a computer. It should be noted that other types of prediction and analysis have very good effectiveness and accuracy, but the analysis is very expensive due to the difficulty of repeatability and reliability, since most diagnoses require two or more types of analysis.

On the other hand, it is well known that ultrasound is the standard test for the diagnosis of DVT and that it is one of the most accurate, and recently, they are also using ML techniques for the diagnosis of DVT [19]. However, the accuracy of the examinations improves with experience and the training that a sonographer gains in their working life, so the percentage is not always the same and is not very high at first [20,21]. Although there is research [22] that strives to combine Deep Learning (DL) and magnetic resonance imaging, with promising outcomes. Recently, it has been shown that the use of artificial neural network analysis can improve risk stratification of patients presenting with suspected DVT, the authors showed that an NN is able to diagnose DVT without the use of ultrasound, with a low false negative rate [23]. A new ML model was developed for the efficient, less intrusive, and reliable diagnosis of DVT. This is based on pattern recognition techniques that help with rapid diagnosis as well as well-trained machine learning models that help with decision making and validating whether or not someone is suffering from this ailment.

In recent years, the field of data science has been pioneered in the development of hardware and software for the application of Artificial Neural Networks (ANNs) in clinical analysis, which can be useful for the diagnosis of DVT and other diseases in general, for example, the use of ML models such as Decision Trees, Support Vector Machine (SVM), and Neural Networks [24–26]. Nowadays, there are alternative methods of DVT diagnosis, some of which use AI. For example, in [6], ML models for venous thromboembolism (VTE) risk assessment in China are compared to the Padua model, with the Random Forest (RF) model having a higher specificity and sensitivity than the Padua model. The authors in [27] reported an automatic diagnosis model by using effective ML to predict the important risk factors of VTE collecting patient data of the medical ward at King Chulalongkorn Memorial Hospital from Thailand. Other efforts are being dedicated for the prediction of VTE with ML techniques in young and middle-aged inpatients; for example, [28] develop VTE risk classifiers using models based on multi-kernel learning and random optimization [29]. However, a drawback is that these systems are expensive, big, heavy, and have moderate energy consumption.

On the other hand, edge computing can minimize the reaction time, increase the data processing capacity, ensure data security (since it is closer to end-users, it provides greater privacy) [30], be easy to design, and be cheap [31]. It has excellent application value and features such as high reliability, superior energy savings [32], low latency, and high real-time processing, increasing the overall data quality and utilization performance under the premise of efficient processing [33]. Accordingly, one can take advantage of edge devices such as Raspberry Pi 4 (RPi4), which are very useful for solving real-world problems across various fields of application [34–39]. In this paper, the well-known RPi4 is used as the edge-computing device to develop the ML models and to evaluate their performance in diagnosing DVT. The cost–benefit of a clinical pre-examination based on ML is noted in the research [7], reducing the expenses of medical units and labor acquired using the standard method. The authors of [40] describe the development of a device for the treatment of DVT that uses Bluetooth communication with a mobile app and sensors within the system to collect data for statistical analysis.

The ML algorithms have advanced in the early diagnosis of DVT and other applications [41–43], moving from binary Decision Trees developed by the team of [44] to more sophisticated algorithms that integrate image analysis by AI [18] and are also very complex in that they go into up to 68 variables to give a final verdict of this disease [45]. In some investigations with very big datasets, the predictors have an area under the receiver operating characteristic (AU-ROC) of 0.83 to 0.85 [46].

For the reasons stated above, the goal of this research is to propose several ML models that are trained by using a dataset of patients with the condition. It is collected from the state of the art [10] to have good judgment and clinical analysis to determine the diagnosis of DVT in a patient with the symptomatology of the condition, with the purpose of having a timely response and thus saving many lifes. In this research, the well-known Raspberry Pi 4 (RPi4) is employed as the edge-computing device to develop ML models and assess their performance in diagnosing DVT. This is to facilitate the development of smart, portable, reliable, and cost-effective instrumentation. All of this is possible thanks to pattern recognition algorithms that provide accurate diagnoses and well-trained ML algorithms that determine whether or not a patient has the condition. The assumption is that ML algorithms will outperform today's standard approaches as a means of early diagnosis for diagnostic aid in the health sector and primary care.

The paper is organized as follows. Section 2 presents the materials and methods used to develop the ML models. In Section 3, the scoring and performance metrics of every ML model are shown; furthermore, the usage scenario is described, and the performance comparison of PC and RPi4. Finally, in Section 4, the conclusions and future work are summarized.

2. Materials and Methods

2.1. Machine Learning Algorithm Training

In this paper, we propose six ML algorithms to evaluate the occurrence of DVT in a patient: Decision Trees (DT), K-Nearest Neighbors (KNN), Support Vector Machine (SVM), Random Forest (RF), Multi-Layer Perceptron Neural Network (MLP-NN), and Extra Trees (ET). All of these ML models may be found in the Scikit Learn library [47,48]. The Scikit Learn library is built on top of NumPy, and it can be used for any kind of project. It has a lot of tools that can be used for both the pre- and post-processing of data. The flow chart for performing the ML algorithm training and testing is shown in Figure 1. First, it imports the appropriate libraries or toolboxes, such as Scikit Learn, Pandas, and Seaborn. Next, the features dataset is loaded, and the input data (features) and output data must be separated. Then, the dataset is randomly divided, with 85% used for training and cross-validation, and the remaining 15% used for testing. Next, the data are scaled between 0 and 1 to produce optimum results. Then, the ML algorithms is trained. Next, the ML model is scored, i.e., the confusion matrix and performance metrics are used to evaluate the ML models.

In this paper, a PC and RPi4 are used to train the ML models for DVT diagnosis, with the goal of testing their performance on both hardware and confirming that the RPi4's scoring parameters and performance metrics are equally as trustworthy as those on a PC. Table 1 compares the RPi4's primary technical specifications to those of a PC. While the hardware of the PC (laptop) obviously outperforms that of the RPi4, it is vital to prove experimentally that the results produced with the RPi4 are competitive to those obtained with a PC. Additionally, it is observed that the RPi4 is significantly less expensive than a PC, which would significantly lower manufacturing costs in a process of large-scale production of intelligent devices, for example, in the manufacture of hundreds or millions of smart instruments.

Table 1. RPi4 versus PC technical specifications comparison.

Hardware	Raspberry Pi 4	PC (Laptop)
Central Processing Unit (CPU)	Broadcom BCM2711 Quad Core 1.5 GHz	AMD Ryzen 7 4800H 2.9 GHz
Graphics Processing Unit (GPU)	Video Core VI 500 MHz	Nvidia GeForce GTX 1660ti
Random Access Memory (RAM)	4 GB DDR4	8 GB DDR4
Networking	WiFi, Ethernet, Bluetooth	WiFi, Ethernet, Bluetooth
Storage	32 GB SD-Card	512 GB SSD
Operating system	Raspbian	Windows 10
Operating Voltage	5 V	19.5 V
Energy consumption	3–7 Wh	200 Wh
Weight	46 g	2.37 Kg
Cost (USD)	$55.00	$1599.00

The dataset for this study was compiled from the following sources [10]. Since these data had been used previously, and only 59 real cases from a public hospital had been obtained, a data augmentation algorithm was devised. They are used to construct a dataset of 10,000 synthetic examples, which will be used for later training, validation, and testing as well as validation of the proposed ML models. The following process was used, as shown in Figure 1, and it will be detailed in depth in each stage below.

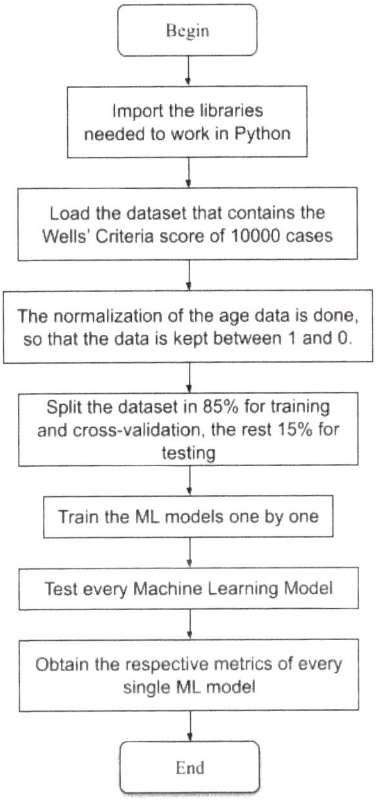

Figure 1. Proposed methodology for early diagnosis of DVT.

Data augmentation is a technique that is frequently used in machine learning to enhance the size of the dataset utilized in the learning process [10] . It entails producing new instances from the original data set while maintaining the data's pattern. It is mostly used in medical contexts to augment the image collection for image-based diagnosis; see, for example [49–52]. In this paper, Algorithm (1) reported in [10] was used. It performs the data augmentation to generate each case that will comprise the set of synthetic data for training and validation of the proposed ML models. Therefore, the first task to be performed is to calculate the percentage of positive and negative cases that are present for each type of risk probability of the occurrence of DVT in addition to the percentage for which each of the factors of the Wells Score was observed in the real cases to which we had access. To calculate the percentage of suspected cases of DVT in each type of risk proposed by Wells, historical data was taken, where it is mentioned that of all the cases observed, 19% were diagnosed as DVT, while the remaining 81% had a different diagnosis. Furthermore, it is mentioned that in the cases detected as Low Risk, only 5% of the cases were diagnosed as positive for DVT, while 17% were diagnosed in Medium Risk, and 53% were diagnosed as High Risk.

The Wells Criteria, as shown in Table 2, are used to train the ML algorithms for the prediction of DVT, in which the trained models are expected to perform well in order to reach a high accuracy in the prediction of this condition.

Table 2. Wells criteria for predicting deep vein thrombosis (DVT), taken from [10,53–55].

Clinical Feature	Score
Lower extremity paralysis, paresis, or recent cast immobilization	1
Deep vein thrombosis has previously been observed	1
Active cancer (patients who are undergoing cancer therapy or who have been diagnosed with cancer within the last 6 months)	1
Pitting edema limited to the affected leg	1
Recently hospitalized for three or more days, or recently had major surgery Anesthesia is required for a total of 12 weeks	1
Tenderness in a specific area of the deep venous system's distribution	1
Skinny veins on each sides (non-varicose)	1
The whole leg is swollen	1
Calf enlarged by at least 3 cm in comparison to the asymptomatic (ten centimeters below the tibial tuberosity)	1
Deep vein thrombosis is the most probable diagnosis; however, other possibilities exist.	−2
When a patient has symptoms in both legs, the leg with the most severe symptoms is utilized.	

2.2. Pre-Processing Data

Two criteria are taken into account that are not covered by the Wells rubric. The first is age, which is measured in numbers ranging from 1 to 9, each of which corresponds to one of the age groups listed in Table 3 [10]. The second factor is gender, which is assigned a value of 0 to males and 1 to females. They are being offered as a way to help patients with suspected DVT better stratify their risk, just as it is managed in [23], so that the data collected may be pre-processed and ML can detect the illness without difficulty. Since the input data are binary, that is, they are regarded as 1 (for positive comorbidity) or 0 (for negative comorbidity), working with them in a computer system is simple, as most media handle binary values, with the age range being the main distinction, as weighted in Table 3.

Table 3. Age factor pre-processing, taken from [10].

Age (Years Old)	Life Stage	Numerical Value
85–120	Advanced old age	9
70–84	Intermediate old age	8
60–69	Initial old age	7
50–59	Mature adults	6
40–49	Average adults	5
21–39	Young adults	4
13–20	Youth	3
6–12	Middle childhood	2
0–5	Childhood	1

The dataset is in Comma Separated Values (CSV) format in American Standard Code for Information Interchange (ASCII), so that it can be processed more easily in the Python environment, as well as in management so that it can be saved and extracted quickly. The data from the Excel file is fed into the software on the computer, using the Jupyter Notebook platform with Python, with each header referring to the DVT comorbidity in each of the columns. After that, using the Seaborn pairplot command, plots are generated between all of the data so that each of the values may be discriminated. It is required to normalize the values of 1–9 to values between 0 and 1 for good NN training.

Equation (1) is used to normalize the data in the age column so that this factor is not the most determinant or the one with the most weight within the ML model used. In this way, all the values of each clinical characteristic will be kept between 0 and 1 except for the age, which will be a floating value, and the others being integers.

$$Norm_{Age} = \frac{Age - Min_{Age}}{Max_{Age} - Min_{Age}} \tag{1}$$

A data description is created to note specific properties of each column of information as well as the primary statistics of the values in the constructed dataset. We noted that the data is in a huge imbalance as a result of this, as it contains 7562 negative genuine cases and 2438 positive real cases.

The *train test split* function divides the dataset into 85% for training and 15% for validation, where the 11 input factors are considered clinical characteristics of the Wells criteria (*cancer, immobilization, surgery, pain, leg swelling, ankle swelling, edema, superficial veins, and previously diagnosed DVT*) and the factors of age and gender, respectively, and the output will result in the DVT diagnosis.

The K-fold cross-validation (with K = 5) is used to evaluate the performance of the ML models and perform a comparative analysis to select the model that best fits the DVT classification problem [56].

An *early-stopping* function has been constructed so that if there is no change of 0.01 in accuracy after 5 epochs, the model's training is truncated and ended, so that the training does not take too long and the percentage of accuracy of the ML model employed during training does not vary much.

2.3. Hyperparameters of the ML Models

The use of an NN with table properties was first suggested during the creation of neural networks. It has an input layer with 11 predictors (cancer, immobility, surgery, pain, leg swelling, ankle swelling, edema, superficial veins, and previously diagnosed DVT). There is no magic formula for selecting the optimum number of hidden layers and neurons. However, some thumb rules are available for calculating the number of hidden layers and neurons. A rough approximation can be obtained by the geometric pyramid rule proposed by [57,58]. In this case, four hidden layers (32–64–32–16) were found for the best performance metrics, and an output layer with the DVT diagnosis is proposed, as shown in Table 4. Since the computer is binary, it is suggested that the number of neurons per layer be multiples of 2^N for optimal processing time, where N is an integer, and the number of neurons in the first hidden layer should be greater than the number of inputs, being multiples of 2^N, ascending in each hidden layer until a maximum of 2^N is reached and then descending with multiples of 2^N until the last hidden layer has a number of neurons slightly greater than the number of input neurons.

Table 4. Proposed sequential model (NN) for DVT diagnosis.

Input Layer with 11 Predictors	Hidden Layer	Output Layer
Gender, age, cancer, surgery, immobilization, tenderness, leg swollen, calf swollen, edema, superficial veins, previous DVT	32–64–32–16	DVT diagnosis

An input layer of 11 neurons, four hidden layers of 32, 64, 32, and 16 neurons, and an output layer representing the diagnostic result make up this ANN model. The input layer's activation function is a *relu* function, while the hidden layers' activation function is *tanh*, the learning rates of the classifier are defined as constant equal to 0.001, the max iteration number is 400, and with *Adam* as the optimizer.

The same ANN model is also trained on a Raspberry Pi 4 due to hardware limitations, and the hyperparameters of the classifiers are the same as the PC, only the results are slightly different and are discussed on Section 3. Furthermore, it is suggested that ML models be used in DVT diagnosis to compare each of the models, including SVM, KNN, DT, ET, and RF classifiers. The hyperparameters dealt with by the SVM classifier are as follows: a random state of 42, a *C parameter* of 1.0, a "linear" classifier kernel, a *degree* of 3, a *gamma* "scale", and a *random state* of 3. The hyperparameters for the KNN classifier are as follows: the number of neighbors is set to 50, the *weights* are set to "distance", and the *algorithm* is set to "ball tree".

The Decision Tree classifier *criteria* used is "entropy", the *splitter* is "random", the *minimum sample divisors* is 2, the *minimum leaf samples* are given by 1, the *maximum features* are given by "auto", the *max features* are 80, and there is a *random state* of 42.

The Extra Trees classifier is employed with a *random state* of 42 and many *estimators* of 200. The *criterion* utilized is "gini", the *minimum sample divisors* is 2, the *minimum leaf samples* is 1, the *maximum of features* is "auto", and the *max features* are given by 80.

Finally, the RF classifier has several *estimators* in 480, with "gini" as the *criterion*, 2 as the *minimum sample divisors*, 1 as the *minimum number of leaf samples*, "auto" as the *maximum number of features*, true *Bootstrap*, and 42 as the *random state*. All these hyperparameters are shown in the Table 5 for every simulation in this paper.

Table 5. Hyperparameters of ML models.

Hyperparameter	SVM	KNN	Decision Tree	Extra Trees	Random Forest
C	1.0	N/A	N/A	N/A	N/A
Kernel	"linear"	N/A	N/A	N/A	N/A
Degree	3	N/A	N/A	N/A	N/A
Gamma	"scale"	N/A	N/A	N/A	N/A
Random State	42	42	42	42	42
N Neighbors	N/A	50	N/A	N/A	N/A
Weights	N/A	"Distance"	N/A	N/A	N/A
Algorithm	SVM	"Ball tree"	DT	ET	RF
Criterion	N/A	N/A	"Entropy"	"gini"	"gini"
Splitter	N/A	N/A	"Random"	N/A	N/A
Min Samples Split	N/A	N/A	2	2	2
Min Samples Leaf	N/A	N/A	1	1	1
Max Features	N/A	N/A	"auto"	"auto"	"auto"
Max Depth	N/A	N/A	80	80	N/A
N Estimators	N/A	N/A	N/A	200	480
Bootstrap	N/A	N/A	N/A	N/A	True

To improve the performance of each classifier, the hyperparameters must be optimized. This can be accomplished using the GridSearchCV tool, which implements the standard estimator API. By "fitting" it to a dataset, all possible combinations of parameter values are evaluated, and only the best combination is kept. The chosen parameters maximize the score of the missing data unless an explicit score is given, in which case it is utilized instead of the default parameters for scoring [47,48]. The following classifiers have modified hyperparameters: the Random Forest and Extra Trees classifiers (Number of Estimators, Criterion, Max Features), the KNN classifier (Number of Neighbors, Weights, and Algorithm), the Decision Tree classifier (Criterion, Splitter, Max Features, and Max Depth), and the SVM classifier (C, Kernel, Degree (when using *rbf*, *poly*, and *sigmoid*)). They were distinct in each of them regarding the different possibilities dealt with. The optimization procedure is based on the "GridSearch" (GS) algorithm, which methodically calculates all possible combinations of hyperparameters. The main disadvantage is that it needs a significant amount of time and calculation [59].

The hardware used for the development of these experiments has the following specifications: AMD Ryzen 7 2.9 GHz CPU, 8 GB 3200 MHz DDR4 RAM, NVIDIA GeForce GTX 1660 Ti 6000 MB GPU, a Windows 10 operating system, and Python software with Anaconda 3. Similarly, a Raspberry Pi 4 SBC with the following specs: CPU 1.5 GHz Broadcom BXM2711, RAM 4GB 3200 MHz DDR4, Raspbian OS, and Python software with Thonny IDE was used to study the behavior of the ML in various contexts and system architectures. The following libraries utilized in this paper: Seaborn, which aids in statistical data visualization within Python; Pandas, which is a library to perform data analysis; NumPy, the platform's data manipulation library; and Matplotlib, which is Python's animated and interactive static visualization library. Scikit-Learn was used to create the ML models as well as custom neural networks.

3. Results

A two-class classification confusion matrix is developed to track the progress of the ML model trained, allowing the metrics to be validated and the process to be more dependable within the rubric by separating it into negative and positive DVT classifications, respectively.

Each of them is kept with the true diagnosis and the diagnosis predicted by the ML algorithm; the first is True Negative (TN), in which both the true diagnosis and the ML prediction are negative, and the second is False Negative (FN), in which the ML diagnosis is negative but the true diagnosis is positive for DVT. Another level of the confusion matrix is the False Positive (FP) criterion, which occurs when the algorithm diagnoses DVT as positive when the actual diagnosis is negative, and finally, the True Positive (TP) criterion, which occurs when the actual diagnosis is positive for the condition and the ML model prediction is positive, resulting in a True Positive (TP), as shown in Table 6.

Table 6. Confusion matrix of two-class classification, taken from [10,60].

	Predicted Diagnostic	
True Diagnostic	**Negative DVT**	**Positive DVT**
Negative DVT	True Negative	False Positive
Positive DVT	False Negative	True Positive

For each ML model trained, the values of Accuracy, F1 Score, Precision, Recall, Specificity, and the area under the curve (AUC) are acquired and printed using sklearn metrics, the acquisition of Accuracy, Precision, Recall, and the ROC curve was accomplished in the case of the Multi-Layer Perceptron NN (MLP-NN), and the Accuracy (2), F1 score (3), Specificity (4), and Recall (5) values are calculated using the following equations taken from [24,61].

$$Accuracy = \frac{True\ Positive + True\ Negative}{True\ Positive + False\ Positive + True\ Negative + False\ Positive} \quad (2)$$

$$F1 - Score = 2 \cdot \frac{Precision \cdot Recall}{Precision + Recall} \quad (3)$$

$$Specificity/Precision = \frac{True\ Negative}{False\ Positive + True\ Negative} \quad (4)$$

$$Sensitivity/Recall = \frac{True\ Positive}{False\ Negative + True\ Positive} \quad (5)$$

These actions are carried out for both PC and Raspberry Pi 4 metrics acquisition. Table 7 shows each of the scoring parameters gathered by each ML model (SVM, KNN, Decision Tree, Extra Trees, Random Forest), the ANN Multi-Layer Perceptron (MLP) model was also registered in the same way; each one includes the metrics (Accuracy, F1 Score, Precision/Specificity, and Recall/Sensitivity) as well as the values that each of the boxes of the two-class confusion matrix (True Positive, True Negative, False Positive and False Negative) gave, allowing you to see how these findings are obtained.

Table 7. Scoring parameters of the ML algorithms evaluated in this study using 15% of the separated data for testing.

			Machine Learning Algorithm on PC			
Scoring Parameters	**SVM**	**Decision Trees**	**Extra Trees**	**RF**	**MLP-NN**	**KNN**
True Positive	265	262	268	274	274	292
True Negative	1017	1061	1060	1064	1067	1064
False Positive	117	73	74	70	67	70
False Negative	101	104	98	92	92	74
Accuracy	0.8546	0.8820	0.8853	0.8920	0.8940	0.9040
F1 Score	0.7085	0.7475	0.7570	0.7718	0.7751	0.8021
Specificity/Precision	0.6937	0.7820	0.7836	0.7965	0.8035	0.8066
Sensitivity/Recall	0.7240	0.7158	0.7322	0.7486	0.7486	0.7978
			Machine Learning Algorithm on RPi 4			
Scoring Parameters	**SVM**	**Decision Trees**	**Extra Trees**	**RF**	**MLP-NN**	**KNN**
True Positive	265	262	268	274	274	292
True Negative	1017	1061	1060	1064	1067	1064
False Positive	117	73	74	70	67	70
False Negative	101	104	98	92	92	74
Accuracy	0.8546	0.8820	0.8853	0.8920	0.8940	0.9040
F1 Score	0.7085	0.7475	0.7570	0.7718	0.7751	0.8021
Specificity/Precision	0.6937	0.7820	0.7836	0.7965	0.8035	0.8066
Sensitivity/Recall	0.7240	0.7158	0.7322	0.7486	0.7486	0.7978

In terms of prediction model validation, there are two basic approaches utilized as selection criteria for a prediction model: (i) The hold-out model and (ii) the K-fold cross validation model. Both have the feature of utilizing a subset of the dataset for training and keeping a portion for validation. The K-fold cross-validation is a method that is utilized as a selection criterion for a prediction/classification model [56]. Essentially, it entails using a subset of the dataset to construct the model and leaving another portion of the dataset to validate it. The K-fold cross-validation procedure runs K times and averages the classification results for each interaction. As shown in Figure 2, it entails partitioning the dataset into k segments and selecting a different section to test the model K times. In contrast, the remaining K-1 elements are used to train the ML model [10,56]. The average values computed in the loop are the performance metrics supplied by K-fold cross-validation. This method is computationally expensive, but it does not waste a lot of data (unlike setting an arbitrary validation set), which is a big plus in applications such as inverse inference when the number of samples is small [47,48].

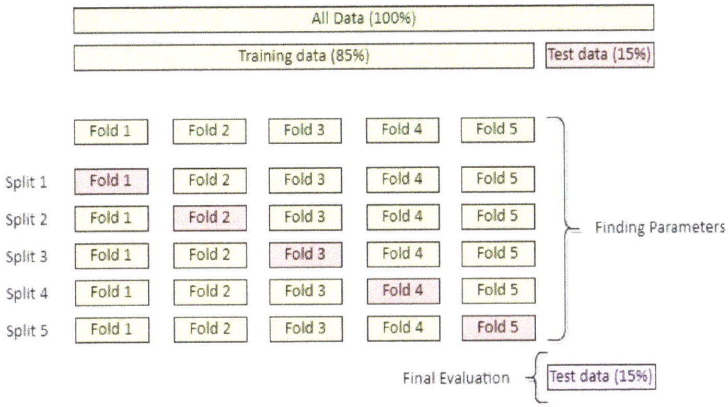

Figure 2. Scheme of the K-fold cross-validation for the proposed ML models, inspired from [47,48].

Table 8 shows the average results of K-fold cross-validation corresponding to each ML model. In this study, we use K = 5 folds, and this assisted in validating all the scoring parameters of each ML model. The Accuracy, F1 score, Precision, Recall, and ROC-AUC were the metrics that could be achieved through this cross-validation; the ML model that had the best overall performance was the KNN, with the best scoring parameters, followed by RF in second place. Later, we found the Extra Trees model, in fourth place is MLP-NN, in fifth place are the Decision Trees, and last but not least is the SVM classifier.

Table 8. Average results of K-fold cross-validation with K = 5 of the ML models trained on PC.

ML Model	K-Fold Cross-Validation with K = 5				
	Accuracy	F1 Score	Specificity/Precision	Sensitivity/Recall	ROC-AUC
SVM	0.8468	0.6906	0.6804	0.7012	0.8589
Decision Trees	0.8677	0.7154	0.7528	0.6819	0.7562
Extra Trees	0.8710	0.7248	0.7552	0.6969	0.8831
RF	0.8744	0.7347	0.7576	0.7133	0.8830
MLP-NN	0.8632	0.7107	0.7347	0.6887	0.7924
KNN	0.8823	0.7586	0.7586	0.7586	0.8906

Regarding the final evaluation (testing), Tables 7 and 9 show the the outcomes of performance metrics when the ML models were tested with 15% of the dataset that was randomly split from the original dataset; i.e., this 15% of the data was kept separate from the dataset used for training; therefore, they had influence neither during training nor in cross-validation. As indicated in Table 7, the best ML model for this issue using PC is KNN, which has the greatest metrics. MLP-NN is second, Random Forest (RF) is third, Extra Trees is fourth, Decision Trees is fifth, and lastly, we have the SVM model that is sixth. Within these rubrics, the accuracy values of the ML models are very good, ranging from 85.4% (SVM) to 90.4% (KNN), as well as the specificity of the model, which is somewhat deplorable in the case of the SVM, being the lowest with 69.37 %, but the others far exceed it, with the KNN classifier reaching 80.66%. It also possessed a superior comparative response to the angiologist physician metrics of 73.82% in accuracy and a specificity of 71.43% [10], proving to be a better technique to detect DVT.

Despite its restricted technology, the Raspberry Pi 4 achieves good results, attaining the same metrics as the PC with no differences, demonstrating the strength of the SystemOnChip (SoC), which is ideal for moving this type of diagnosis to smart devices. Despite the embedded system's limited computational capability, excellent metrics for a portable intelligent system are attained, outperforming the Wells technique in a typical approach.

Receiver Operating Characteristic (ROC) curves can be generated using the aforementioned data, which indicate how well the model can distinguish between two objects. They are key metrics for assessing an ML model's performance. Furthermore, they are employed in binary classification issues, i.e., problems with two distinct output classes. The connection between the model's True Positive Rate (TPR) and False Positive Rate (FPR) is depicted by the ROC curve.

Both the ROC curve on PC and the ROC curve on Raspberry Pi 4 have a similar response; the KNN model has a larger area under the curve, making it more visually appealing, which is complemented by the scoring metrics mentioned in Table 7, which is followed by the Random Forest (RF) classifier, as shown in Figure 3.

Similarly, it is possible to obtain PR curves (Precision–Recall curves), which are a useful measure for observing prediction success when classes are highly disequilibrated or unbalanced. In information retrieval, precision is a measure of the relevance of the results, whereas recall is a measure of the number of truly relevant results that are returned. The PR curve depicts the trade-off between Precision and Recall at various levels. A high area under the curve indicates both High Recall and High Precision, with High Precision corresponding to a low False Positive Rate and high Recall corresponding to a low False Negative Rate. High scores in both cases show that the classifier is delivering accurate (High Precision) results as well as the bulk of positive outcomes (High Recall).

Figure 4 shows that the PR curves of the Raspberry Pi 4 and PC are similar; the KNN classifier has a better PR curve over all classifiers in both cases, having a larger area under the curve covered within the graph, making it one of the best classifiers, followed by the Extra Trees classifier.

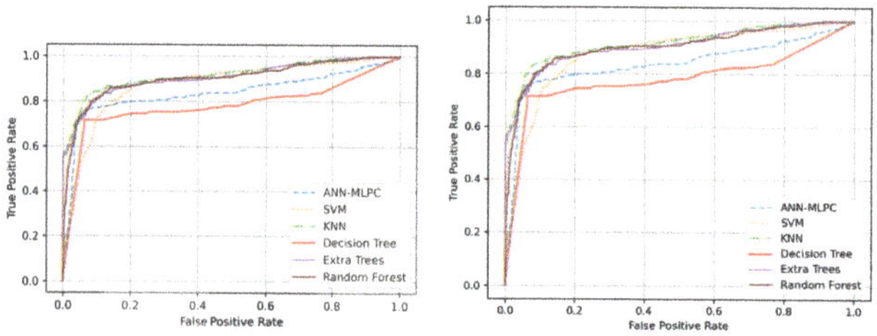

(a) (b)

Figure 3. ROC curves. (**a**) ROC curve on PC, and (**b**) ROC curve on Raspberry Pi 4.

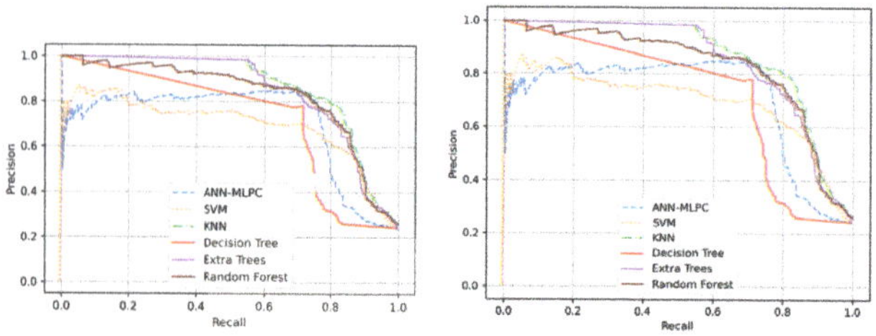

(a) (b)

Figure 4. PR curves. (**a**) PR curve on PC, and (**b**) PR curve on Raspberry Pi 4.

The next step is to obtain the performance metrics to evaluate the ML models; we rely on the metrics of the Scikit learn library. The performance metrics are the Area Under the Curve (AUC) using the trapezoidal method, the Cohen's Kappa coefficient, Hamming Loss, and Matthew's correlation coefficient, all of which are achieved on a PC and a Raspberry Pi 4 correspondingly, which are listed in Table 9.

According to Table 9, the KNN classifier is the best binary ML classifier for PC in terms of performance metrics, followed by Random Forest (RF) in second, the Extra Trees classifier in third, MLP-NN model in fourth, Decision Trees classifier in fifth, and last but not least, the SVM classifier. The system's measurements show an AUC ranging from 81.04% with the SVM model to 86.80% with the KNN classifier. The Hamming Loss is another metric that goes from SVM at 14.53% to the best with a lower percentage, the KNN model at 9.60%.

Table 9. Performance metrics of the six ML algorithms evaluated in this study using 15% of the separated data for testing.

			Performance Metrics on PC		
Machine Learning Algorithm	ROC-AUC	PR-AUC	Cohen's Kappa Coefficient	Hamming Loss	Matthew's Correlation Coeficient
SVM	0.8104	0.6960	0.6118	0.1453	0.6120
Decision Trees	0.8257	0.6432	0.6707	0.1180	0.6718
Extra Trees	0.8334	0.8461	0.6821	0.1146	0.6828
Random Forest (RF)	0.8434	0.8283	0.7011	0.1080	0.7017
MLP-NN	0.8447	0.7102	0.7058	0.1060	0.7066
KNN	0.8680	0.8619	0.7388	0.0960	0.7388
			Performance Metrics on RPi 4		
Machine Learning Algorithm	ROC-AUC	PR-AUC	Cohen's Kappa Coefficient	Hamming Loss	Matthew's Correlation Coeficient
SVM	0.8104	0.6954	0.6118	0.1453	0.6120
Decision Trees	0.8257	0.6432	0.6707	0.1180	0.6718
Extra Trees	0.8334	0.8461	0.6821	0.1146	0.6828
Random Forest (RF)	0.8434	0.8283	0.7011	0.1080	0.7017
MLP-NN	0.8447	0.7102	0.7058	0.1060	0.7066
KNN	0.8680	0.8619	0.7388	0.0960	0.7388

For its part, the Raspberry Pi 4 achieved good results, which are identical in theory to those acquired by the PC, when all of the parameters involved in each of the classifiers, during training, and determining the performance metrics and score were taken into account.

The average time of each of the proposed classifiers to be used within this problem was obtained both on PC and Raspberry Pi 4. Figure 5 shows the average time of ten training runs of each model with their respective characteristics, both on the computer and on the SoC, to analyze the cost–benefit of each of the proposed classifiers to be used within this problem. Due to factors of processing power operations, it turns out to be faster than the Raspberry Pi 4, with the Decision Trees model being the fastest, followed closely by the KNN, and further behind the Extra Trees, all with a training time of less than a second, then the SVM, before reaching 2 s, the RF classifier, and finally, the MLP-NN being the slowest of all with a time greater than 30 s. The training on the Raspberry Pi 4 takes two to five times longer due to the embedded system's processing limitations, with times of 0.02 s in Decision Trees, 0.1 s in KNN models, 2.81 s in Extra Trees, 3.12 s in SVM, 6.89 s in RF models, and finally 175.85 s in RF models.

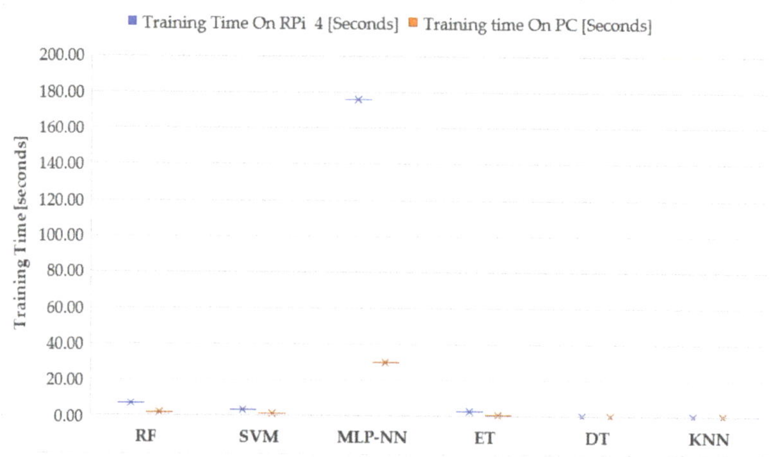

Figure 5. ML algorithms training time.

Usage Scenario

These ML models could be implemented on an embedded system such as the Raspberry Pi 4 (RPi4) to develop a DVT diagnostic smart system. This could be integrated with a color sensor (RGB), heart rate (BPM), and temperature (°C), as well as a user interface (GUI) that may include some questions according to the Wells criteria. Furthermore, the physician can acquire the raw Wells criteria for each patient to be diagnosed with this proposed system. The smart system will have a trained ML model into which the selected patient's data will be entered, and it will provide a diagnosis prediction for the patient's condition. The prospective apparatus proposed by this research can be shown in Figure 6. As discussed before and shown in Table 1, the RPi4 is less expensive than a PC, which would result in significant cost savings associated with the large-scale production of intelligent devices, such as the manufacture of hundreds or millions of smart instruments.

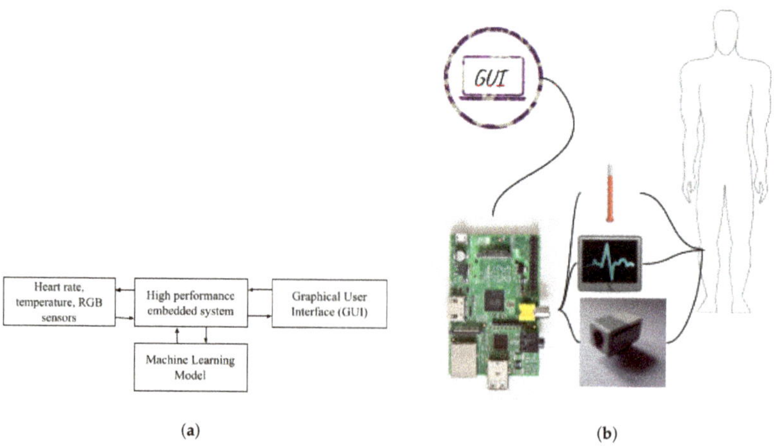

Figure 6. Suggested usage scenario. (**a**) Block diagram of proposed system and (**b**) proposed system outline.

4. Conclusions

Multiple ML classifiers were assessed for the prediction of DVT in the lower limbs according to Wells criteria. They were subjected to different score and performance metrics to assist with identifying the dependability of each one. The results of each of the created models were subjected to cross-validation. The experimental results show that the KNN model is the best in terms of performance and score metrics (higher accuracy (90.40%), higher specificity (80.66%), ROC-AUC (86.80%), and PR-AUC (86.16%)), but it is second in terms of execution time (0.01904 s) followed by the MLP-NN model, which is the slowest in terms of execution time (30.08 s), but gives us the second best accuracy (89.40%). The KNN classifier, on the other hand, among the models trained on the Raspberry Pi 4, has the same score and performance metrics as on the PC; the main difference is in the execution time, as it takes 0. 0951 s to train the model, making it the second in this category; however, in real terms, it is possible to wait a little longer for a portable result, and in second place is the MLP-NN classifier, with an execution time of 175.8485 s, making it the slowest. The accuracy of all trained models on PC and Raspberry Pi 4 is greater than 85%, while the AUC values are between 81 and 86%. In conclusion, as compared to traditional methods, the best ML classifiers were effective at predicting DVT diagnosis in a timely and efficient manner.

Author Contributions: Conceptualization, E.E.G.-G. and E.I.-G.; Data curation, D.L.-M.; Formal analysis, E.T.-C.; Funding acquisition, O.R.L.-B.; Investigation, E.E.C.-L. and E.I.-G.; Methodology, E.E.C.-L.; Project administration, O.R.L.-B.; Resources, O.R.L.-B.; Software, E.E.C.-L. and E.I.-G.; Supervision, E.E.G.-G. and E.I.-G.; Validation, E.E.G.-G., E.T.-C. and D.L.-M.; Visualization, E.E.G.-G. and O.R.L.-B.; Writing—original draft, E.E.C.-L.; Writing—review and editing, E.T.-C. and E.I.-G. All authors have read and agreed to the published version of the manuscript.

Funding: This research was funded by the Universidad Autónoma de Baja California (UABC) through the 22th internal call with grant number 679.

Data Availability Statement: The dataset, pair plot, and trained ML models are shared as supplementary material. The trained ML models can be loaded using the joblib library. Furthermore, supplemental material is available at: http://doi.org/10.17632/dpyngbvc47.1 (accessed on 14 February 2022), and https://data.mendeley.com/datasets/dpyngbvc47/1 (accessed on 14 February 2022).

Acknowledgments: The authors would like to thank INAOE and UDG for accepting researcher E. Inzunza-González to carry out his sabbatical stay. The authors are very grateful to Conacyt for the scholarship awarded to E.E.C.-L. Thanks are given to PRODEP for supporting the academic groups to increase their degree of consolidation. The authors appreciate the reviewers' remarks and recommendations to enhance the paper.

Conflicts of Interest: The authors declare no conflict of interest. The funders had no role in the design of the study; in the collection, analyses, or interpretation of data; in the writing of the manuscript, or in the decision to publish the results.

References

1. Yang, M.; Tan, T. Formation of Thrombosis and Its Potential Diagnosis and Treatment with Optoacoustic Technology. In Proceedings of the Third International Conference on Medical and Health Informatics 2019 (ICMHI 2019), Xiamen, China, 17–19 May 2019; pp. 1–5. [CrossRef]
2. Liu, K.; Chen, J.; Zhang, K.; Wang, S.; Li, X. A Diagnostic Prediction Model of Acute Symptomatic Portal Vein Thrombosis. *Ann. Vasc. Surg.* **2019**, *61*, 394–399. [CrossRef] [PubMed]
3. Kim, D.; Byyny, R.; Rice, C.; Faragher, J.; Nordenholz, K.; Haukoos, J.; Liao, M.; Kendall, J. Test Characteristics of Emergency Physician-Performed Limited Compression Ultrasound for Lower-Extremity Deep Vein Thrombosis. *J. Emerg. Med.* **2016**, *51*, 684–690. [CrossRef] [PubMed]
4. Moore, R.D.; Pryce, W.I.J.; Todd, J.M. The use of the ultrasonic Doppler test in the detection of deep vein thrombosis. *Phys. Med. Biol.* **1973**, *18*, 142–143. [CrossRef]
5. Penco, S.; Grossi, E.; Cheng, S.; Intraligi, M.; Maurelli, G.; Patrosso, M.; Marocchi, A.; Buscema, M. Assessment of Genetic Polymorphism Role in Venous Thrombosis Through Artificial Neural Networks. *Ann. Hum. Genet.* **2005**, *69*, 693–706. [CrossRef]
6. Wang, X.; Yang, Y.Q.; Liu, S.H.; Hong, X.Y.; Sun, X.F.; Shi, J.H. Comparing different venous thromboembolism risk assessment machine learning models in Chinese patients. *J. Eval. Clin. Pract.* **2019**, *26*, 26–34. [CrossRef]

7. Luo, L.; Kou, R.; Feng, Y.; Xiang, J.; Zhu, W. Cost-Effective Machine Learning Based Clinical Pre-Test Probability Strategy for DVT Diagnosis in Neurological Intensive Care Unit. *Clin. Appl. Thromb.* **2021**, *27*, 10760296211008650. [CrossRef]
8. Ferroni, P.; Zanzotto, F.M.; Scarpato, N.; Riondino, S.; Guadagni, F.; Roselli, M. Validation of a Machine Learning Approach for Venous Thromboembolism Risk Prediction in Oncology. *Dis. Markers* **2017**, *2017*, 8781379. [CrossRef]
9. Riondino, S.; Ferroni, P.; Zanzotto, F.M.; Roselli, M.; Guadagni, F. Predicting VTE in cancer patients: Candidate biomarkers and risk assessment models. *Cancers* **2019**, *11*, 95. [CrossRef]
10. Fong-Mata, M.; Garcia-Guerrero, E.; Mejia-Medina, D.; Lopez-Bonilla, O.; Villarreal-Gomez, L.; Zamora-Arellano, F.; Lopez-Mancilla, D.; Inzunza-Gonzalez, E. An Artificial Neural Network Approach and a Data Augmentation Algorithm to Systematize the Diagnosis of Deep-Vein Thrombosis by Using Wells' Criteria. *Electronics* **2020**, *9*, 1810. [CrossRef]
11. Segal, J.B.; Eng, J.; Tamariz, L.J.; Bass, E.B. Review of the Evidence on Diagnosis of Deep Venous Thrombosis and Pulmonary Embolism. *Ann. Fam. Med.* **2007**, *5*, 63. [CrossRef]
12. Smyrnakis, E.; Symintiridou, D.; Andreou, M.; Dandoulakis, M.; Theodoropoulos, E.; Kokkali, S.; Manolaki, C.; Papageorgiou, D.I.; Birtsou, C.; Paganas, A.; et al. Primary care professionals' experiences during the first wave of the COVID-19 pandemic in Greece: A qualitative study. *BMC Fam. Pract.* **2021**, *22*, 174. [CrossRef]
13. da Silva, L.G.R.; da Silva Pinto, A.W.; de Queiroz, W.E.; Coelho, C.C.; Blatt, C.R.; Oliveira, M.G.; de Lima Pimentel, A.C.; Elseviers, M.; Baldoni, A.O. Deprescribing clonazepam in primary care older patients: A feasibility study. *Int. J. Clin. Pharm.* **2022**. [CrossRef] [PubMed]
14. Győrffy, Z.; Békási, S.; Döbrössy, B.; Bognár, V.K.; Radó, N.; Morva, E.; Zsigri, S.; Tari, P.; Girasek, E. Exploratory attitude survey of homeless persons regarding telecare services in shelters providing mid- and long-term accommodation: The importance of trust. *PLoS ONE* **2022**, *17*, e0261145. [CrossRef] [PubMed]
15. Wells, P. Predictive analytics by deep machine learning: A call for next-gen tools to improve health care. *Res. Pract. Thromb. Haemost.* **2020**, *4*, 181–182. [CrossRef] [PubMed]
16. Yokomichi, A.; Rodrigues, V.; Moroz, A.; Bertanha, M.; Ribeiro, S.; Defune, E.; Moraes, M. Detection of Factor VIII and D-dimer biomarkers for venous thromboembolism diagnosis using electrochemistry immunosensor. *Talanta* **2020**, *219*, 121241. [CrossRef]
17. Kacmaz, S.; Ercelebi, E.; Zengin, S.; Cindoruk, S. The Use of Infrared Thermal Imaging in the Diagnosis of Deep Vein Thrombosis. *Infrared Phys. Technol.* **2017**, *86*, 120–129. [CrossRef]
18. Tanno, R.; Makropoulos, A.; Arslan, S.; Oktay, O.; Mischkewitz, S.; Noor, F.; Oppenheimer, J.; Mandegaran, R.; Kainz, B.; Heinrich, M. AutoDVT: Joint Real-Time Classification for Vein Compressibility Analysis in Deep Vein Thrombosis Ultrasound Diagnostics. In Proceedings of the 21st International Conference, Granada, Spain, 16–20 September 2018; Part II, pp. 905–912. [CrossRef]
19. Kainz, B.; Heinrich, M.P.; Makropoulos, A.; Oppenheimer, J.; Mandegaran, R.; Sankar, S.; Deane, C.; Mischkewitz, S.; Al-Noor, F.; Rawdin, A.C.; et al. Non-invasive diagnosis of deep vein thrombosis from ultrasound imaging with machine learning. *NPJ Digit. Med.* **2021**, *4*, 137. [CrossRef]
20. Lewiss, R.E.; Kaban, N.L.; Turandot, S. Point-of-Care Ultrasound for a Deep Venous Thrombosis. *Glob. Heart* **2013**, *8*, 329–333. [CrossRef]
21. Ly-Pen, D.; Penedo-Alonso, R.; Sánchez-Pérez, M. Comparison of the Accuracy of Emergency Department–Performed Point-of-Care Ultrasound in the Diagnosis of Lower-Extremity Deep Vein Thrombosis. *J. Emerg. Med.* **2018**, *55*, 716–717. [CrossRef]
22. Huang, C.; Tian, J.; Yuan, C.; Zeng, P.; He, X.; Chen, H.; Huang, Y.; Huang, B. Fully Automated Segmentation of Lower Extremity Deep Vein Thrombosis Using Convolutional Neural Network. *BioMed Res. Int.* **2019**, *2019*, 3401683. [CrossRef]
23. Willan, J.; Katz, H.; Keeling, D. The use of artificial neural network analysis can improve the risk-stratification of patients presenting with suspected deep vein thrombosis. *Br. J. Haematol.* **2019**, *185*, 289–296. [CrossRef] [PubMed]
24. Fong-Mata, M.B.; Inzunza-Gonzalez, E.; Garcia-Guerrero, E.; Mejia-Medina, D.; Morales Contreras, O.; Gomez-Roa, A. Trombosis venosa profunda en extremidades inferiores: Revisión de las técnicas de diagnóstico actuales y su simbiosis con el aprendizaje automático para un diagnóstico oportuno. *Rev. Cienc. Tecnol.* **2020**, *3*, 23–34. [CrossRef]
25. Litjens, G.; Kooi, T.; Ehteshami Bejnordi, B.; Adiyoso Setio, A.A.; Ciompi, F.; Ghafoorian, M.; Van Der Laak, J.A.W.M.; Van Ginneken, B.; Sánchez, C.I. A survey on deep learning in medical image analysis. *Med. Image Anal.* **2017**, *42*, 60–88. [CrossRef] [PubMed]
26. Agharezaei, L.; Agharezaei, Z.; Nemati, A.; Bahaadinbeigy, K.; Keynia, F.; Baneshi, M.; Iranpour, A.; Agharezaei, M. The Prediction of the Risk Level of Pulmonary Embolism and Deep Vein Thrombosis through Artificial Neural Network. *Acta Inform. Med.* **2016**, *24*, 354. [CrossRef] [PubMed]
27. Sukperm, A.; Rojnuckarin, P.; Akkawat, B.; Sa-Ing, V. Automatic Diagnosis of Venous Thromboembolism Risk based on Machine Learning. In Proceedings of the 2021 IEEE International IOT, Electronics and Mechatronics Conference (IEMTRONICS), Toronto, ON, Canada, 21–24 April 2021; pp. 1–4. [CrossRef]
28. Liu, H.; Yuan, H.; Wang, Y.; Huang, W.; Xue, H.; Zhang, X. Prediction of venous thromboembolism with machine learning techniques in young-middle-aged inpatients. *Sci. Rep.* **2021**, *11*, 12868. [CrossRef]
29. Ferroni, P.; Zanzotto, F.M.; Scarpato, N.; Riondino, S.; Nanni, U.; Roselli, M.; Guadagni, F. Risk Assessment for Venous Thromboembolism in Chemotherapy-Treated Ambulatory Cancer Patients: A Machine Learning Approach. *Med. Decis. Mak.* **2016**, *37*, 234–242. [CrossRef]
30. Egwuche, O.S.; Ganiyu, M.; Ibiyomi, M.A. A survey of mobile edge computing in developing countries: Challenges and prospects. *J. Phys. Conf. Ser.* **2021**, *2034*, 012004. [CrossRef]

31. Liu, L.; Chen, L.; Xu, S.; Xu, Y.; Shi, C. Design and implementation of intelligent monitoring terminal for distribution room based on edge computing. *Energy Rep.* **2021**, *7*, 1131–1138. [CrossRef]
32. Andrawes, A.; Nordin, R.; Albataineh, Z.; Alsharif, M.H. Sustainable delay minimization strategy for mobile edge computing offloading under different network scenarios. *Sustainability* **2021**, *13*, 12112. [CrossRef]
33. Teng, Y.; Cui, J.; Jiang, W. Research on application of edge computing in real-time environmental monitoring system. *J. Phys. Conf. Ser.* **2021**, *2010*, 012157. [CrossRef]
34. Alessio, K.; Tischer, B.; Voss, M.; Teixeira, I.; Brendler, B.; Duarte, F.; Helfer, G.; Costa, A.; Barin, J. Open source, low-cost device for thermometric titration with non-contact temperature measurement. *Talanta* **2020**, *216*, 120975. [CrossRef] [PubMed]
35. Nykvist, C.; Larsson, M.; Sodhro, A.; Gurtov, A. A lightweight portable intrusion detection communication system for auditing applications. *Int. J. Commun. Syst.* **2020**, *33*, e4327. [CrossRef]
36. Zamora-Arellano, F.; López-Bonilla, O.R.; García-Guerrero, E.E.; Olguín-Tiznado, J.E.; Inzunza-González, E.; López-Mancilla, D.; Tlelo-Cuautle, E. Development of a Portable, Reliable and Low-Cost Electrical Impedance Tomography System Using an Embedded System. *Electronics* **2021**, *10*, 15. [CrossRef]
37. Gautam, A.; Kumar, A.; Kinjalk, K.; Thangaraj, J.; Priye, V. A Low Cost FBG Based Online Weight Monitoring System. *IEEE Sens. J.* **2020**, *20*, 4207–4214. [CrossRef]
38. Nirmala, M.; Malarvizhi, K. Internet of things based solar powered truck. *Test Eng. Manag.* **2020**, *83*, 9358–9365.
39. Aguirre-Castro, O.; Inzunza-González, E.; García-Guerrero, E.; Tlelo-Cuautle, E.; López-Bonilla, O.; Olguín-Tiznado, J.; Cárdenas-Valdez, J. Design and Construction of an ROV for Underwater Exploration. *Sensors* **2019**, *19*, 5387. [CrossRef]
40. Dhatri, M.P.; Shivram, R. Development of a Functional Testing System for Test Automation and Statistical Analysis of the behavior of health care device used to treat Deep Vein Thrombosis. In Proceedings of the 2018 3rd IEEE International Conference on Recent Trends in Electronics, Information & Communication Technology (RTEICT), Bangalore, India, 18–19 May 2018; pp. 1715–1719. [CrossRef]
41. Cerrada, M.; Trujillo, L.; Hernández, D.E.; Correa Zevallos, H.A.; Macancela, J.C.; Cabrera, D.; Vinicio Sánchez, R. AutoML for Feature Selection and Model Tuning Applied to Fault Severity Diagnosis in Spur Gearboxes. *Math. Comput. Appl.* **2022**, *27*, 6. [CrossRef]
42. Enríquez Zárate, J.; Gómez López, M.d.l.A.; Carmona Troyo, J.A.; Trujillo, L. Analysis and Detection of Erosion in Wind Turbine Blades. *Math. Comput. Appl.* **2022**, *27*, 5. [CrossRef]
43. Esqueda-Elizondo, J.J.; Juarez-Ramirez, R.; Lopez-Bonilla, O.R.; Garcia-Guerrero, E.E.; Galindo-Aldana, G.M.; Jimenez-Beristain, L.; Serrano-Trujillo, A.; Tlelo-Cuautle, E.; Inzunza-Gonzalez, E. Attention Measurement of an Autism Spectrum Disorder User Using EEG Signals: A Case Study. *Math. Comput. Appl.* **2022**, *27*, 21. [CrossRef]
44. Nwosisi, C.; Sung-Hyuk, C.; Yoo, J.A.; Tappert, C.C.; Lipsitz, E. Constructing Binary Decision Trees for Predicting Deep Venous Thrombosis. In Proceedings of the 2010 2nd International Conference on Software Technology and Engineering, San Juan, PR, USA, 3–5 October 2010; Volume 1, pp. V1-121–V1-124. [CrossRef]
45. Nafee, T.; Gibson, C.M.; Travis, R.; Yee, M.K.; Kerneis, M.; Chi, G.; AlKhalfan, F.; Hernandez, A.F.; Hull, R.D.; Cohen, A.T.; et al. Machine learning to predict venous thrombosis in acutely ill medical patients. *Res. Pract. Thromb. Haemost.* **2020**, *4*, 230–237. [CrossRef]
46. Ryan, L.; Mataraso, S.; Siefkas, A.; Pellegrini, E.; Barnes, G.; Green-Saxena, A.; Hoffman, J.; Calvert, J.; Das, R. A Machine Learning Approach to Predict Deep Venous Thrombosis Among Hospitalized Patients. *Clin. Appl. Thromb.* **2021**, *27*, 1076029621991185. [CrossRef] [PubMed]
47. Pedregosa, F.; Varoquaux, G.; Gramfort, A.; Michel, V.; Thirion, B.; Grisel, O.; Blondel, M.; Prettenhofer, P.; Weiss, R.; Dubourg, V.; et al. Scikit-learn: Machine Learning in Python. *J. Mach. Learn. Res.* **2011**, *12*, 2825–2830.
48. Buitinck, L.; Louppe, G.; Blondel, M.; Pedregosa, F.; Mueller, A.; Grisel, O.; Niculae, V.; Prettenhofer, P.; Gramfort, A.; Grobler, J.; et al. API design for machine learning software: Experiences from the scikit-learn project. In Proceedings of the European Conference on Machine Learning and Principles and Practice of Knowledge Discovery in Databases (ECMLPKDD 2013), Prague, Czech Republic, 23–27 September 2013; pp. 108–122.
49. Liu, Z.; Cao, Y.; Li, Y.; Xiao, X.; Qiu, Q.; Yang, M.; Zhao, Y.; Cui, L. Automatic diagnosis of fungal keratitis using data augmentation and image fusion with deep convolutional neural network. *Comput. Methods Programs Biomed.* **2020**, *187*, 105019. [CrossRef] [PubMed]
50. Zuluaga-Gomez, J.; Al Masry, Z.; Benaggoune, K.; Meraghni, S.; Zerhouni, N. A CNN-based methodology for breast cancer diagnosis using thermal images. *Comput. Methods Biomech. Biomed. Eng. Imaging Vis.* **2020**, *9*, 131–145. [CrossRef]
51. de Souza, L.A.; Passos, L.A.; Mendel, R.; Ebigbo, A.; Probst, A.; Messmann, H.; Palm, C.; Papa, J.P. Assisting Barrett's esophagus identification using endoscopic data augmentation based on Generative Adversarial Networks. *Comput. Biol. Med.* **2020**, *126*, 104029. [CrossRef] [PubMed]
52. Gao, Z.; Li, J.; Guo, J.; Chen, Y.; Yi, Z.; Zhong, J. Diagnosis of Diabetic Retinopathy Using Deep Neural Networks. *IEEE Access* **2019**, *7*, 3360–3370. [CrossRef]
53. Wells, P.S.; Anderson, D.R.; Bormanis, J.; Guy, F.; Mitchell, M.; Gray, L.; Clement, C.; Robinson, K.S.; Lewandowski, B. Value of assessment of pretest probability of deep-vein thrombosis in clinical management. *Lancet* **1997**, *350*, 1795–1798. [CrossRef]
54. Wells, P.S.; Owen, C.; Doucette, S.; Fergusson, D.; Tran, H. Does this patient have deep vein thrombosis? *JAMA* **2006**, *295*, 199–207. [CrossRef]

55. Modi, S.; Deisler, R.; Gozel, K.; Reicks, P.; Irwin, E.; Brunsvold, M.; Banton, K.; Beilman, G.J. Wells criteria for DVT is a reliable clinical tool to assess the risk of deep venous thrombosis in trauma patients. *World J. Emerg. Surg.* **2016**, *11*, 24. [CrossRef]
56. Jung, Y. Multiple predicting K-fold cross-validation for model selection. *J. Nonparametr. Stat.* **2018**, *30*, 197–215. [CrossRef]
57. Pano-Azucena, A.; Tlelo-Cuautle, E.; Tan, S.D.; Ovilla-Martinez, B.; De la Fraga, L. FPGA-Based Implementation of a Multilayer Perceptron Suitable for Chaotic Time Series Prediction. *Technologies* **2018**, *6*, 90. [CrossRef]
58. Kaufmann, M. *Practical Neural Networks Recipes in C++*; Elsevier: Boston, MA, USA, 1993; p. 493. [CrossRef]
59. Vávra, J.; Hromada, M.; Lukáš, L.; Dworzecki, J. Adaptive anomaly detection system based on machine learning algorithms in an industrial control environment. *Int. J. Crit. Infrastruct. Prot.* **2021**, *34*, 100446. [CrossRef]
60. Sadrawi, M.; Sun, W.Z.; Ma, M.M.; Yeh, Y.T.; Abbod, M.; Shieh, J.S. Ensemble genetic fuzzy neuro model applied for the emergency medical service via unbalanced data evaluation. *Symmetry* **2018**, *10*, 71. [CrossRef]
61. Sokolova, M.; Japkowicz, N.; Szpakowicz, S. Beyond Accuracy, F-Score and ROC: A Family of Discriminant Measures for Performance Evaluation. In Proceedings of the 19th Australian Joint Conference on Artificial Intelligence, Hobart, Australia, 4–8 December 2006; pp. 1015–1021.

Mathematical and Computational Applications

MDPI

Article

Applications of ANFIS-Type Methods in Simulation of Systems in Marine Environments

Aakanksha Jain [1], Iman Bahreini Toussi [1,*], Abdolmajid Mohammadian [1], Hossein Bonakdari [1] and Majid Sartaj [2]

[1] Department of Civil Engineering, University of Ottawa, 161 Louis Pasteur Private, Ottawa, ON K1N 6N5, Canada; ajain016@uottawa.ca (A.J.); amohamma@uottawa.ca (A.M.); Hossein.Bonakdari@uottawa.ca (H.B.)
[2] Department of Environmental Engineering, University of Ottawa, 161 Louis Pasteur Private, Ottawa, ON K1N 6N5, Canada; msartaj@uottawa.ca
* Correspondence: ibahr094@uottawa.ca

Abstract: ANFIS-type algorithms have been used in various modeling and simulation problems. With the help of algorithms with more accuracy and adaptability, it is possible to obtain better real-life emulating models. A critical environmental problem is the discharge of saline industrial effluents in the form of buoyant jets into water bodies. Given the potentially harmful effects of the discharge effluents from desalination plants on the marine environment and the coastal ecosystem, minimizing such an effect is crucial. Hence, it is important to design the outfall system properly to reduce these impacts. To the best of the authors' knowledge, a study that formulates the effluent discharge to find an optimum numerical model under the conditions considered here using AI methods has not been completed before. In this study, submerged discharges, specifically, negatively buoyant jets are modeled. The objective of this study is to compare various artificial intelligence algorithms along with multivariate regression models to find the best fit model emulating effluent discharge and determine the model with less computational time. This is achieved by training and testing the Adaptive Neuro-Fuzzy Inference System (ANFIS), ANFIS-Genetic Algorithm (GA), ANFIS-Particle Swarm Optimization (PSO) and ANFIS-Firefly Algorithm (FFA) models with input parameters, which are obtained by using the realizable k-ε turbulence model, and simulated parameters, which are obtained after modeling the turbulent jet using the OpenFOAM simulation platform. A comparison of the realizable k-ε turbulence model outputs and AI algorithms' outputs is conducted in this study. Statistical parameters such as least error, coefficient of determination (R^2), Mean Absolute Error (MAE), and Average Absolute Deviation (AED) are measured to evaluate the performance of the models. In this work, it is found that ANFIS-PSO performs better compared to the other four models and the multivariate regression model. It is shown that this model provides better R^2, MAE, and AED, however, the non-hybrid ANFIS model provides reasonably acceptable results with lower computational costs. The results of the study demonstrate an error of 6.908% as the best-case scenario in the AI models.

Keywords: OpenFOAM; CFD; ANFIS; ANFIS (GA); ANFIS (PSO); ANFIS (FFA)

Citation: Jain, A.; Bahreini Toussi, I.; Mohammadian, A.; Bonakdari, H.; Sartaj, M. Applications of ANFIS-Type Methods in Simulation of Systems in Marine Environments. *Math. Comput. Appl.* **2022**, *27*, 29. https://doi.org/10.3390/mca27020029

Received: 10 January 2022
Accepted: 15 March 2022
Published: 21 March 2022

1. Introduction

Due to the increase in population growth and groundwater depletion, the demand for fresh and potable water has led to rising growth in desalination plants, especially in arid and semi-arid regions such as the Persian Gulf, Red Sea, and the Gulf of Oman [1]. It has also been estimated that the percentage of water shortage will increase by 60% by the year 2025 [2]. Hence, since about 97.5% of the total volume of the hydrosphere is contained in seas and oceans [3], desalination plants are the most viable solution for today's drinking water problems, however, these plants cause many negative impacts. The effluent from desalination plants, called 'brine', is discharged into the seawater and contains concentrated salt, which is almost double the salinity of the receiving water and ends up adding this salinity to the seawater [1]. Along with this, if a desalination plant is using a multistage flash (MSF) technique, then the brine could also raise turbidity and temperature (Bleninger and Jirka, 2008) [4]. This concentrated brine stream can deteriorate chemical, physical, and biological attributes of the receiving water. Hence, the effect of brine is majorly evident on the environment, especially on flora and fauna. Therefore, many countries like the USA and Europe have made strict regulations for effluent standards [4].

To meet the existing regulations, a diffuser can be placed at the end of the outfall system to dilute the concentrated brine—since in the absence of dilution—brine plume extends its vicinity and will be harmful to the ecosystem [1]. It has also been reported that the discharge of brine using inclined dense jets has been in use since the 1970s, in which dilution and geometry are the major parameters to be considered [5]. Dilution of brine occurs in two steps: (a) Primary dilution, which appears in the near field due to density difference, between the seawater and effluent as well as due to momentum flux and geometry of the outfall; and (b) Natural dilution in the far-field, due to diffusion and mixing [6]. The impact can generally be seen in the range of 300 m from the point of discharge, which is generally the near-field region [2]. Hence, it is important to focus on the near-field region to design the outfall system for greater dilution [7]. Since effluent density varies from the ambient water, which makes the jet rise or fall, when the dense effluent is discharged upwards it is called the negatively buoyant jet. As the jet moves upwards its momentum decreases, which then returns towards the bottom due to its high density after attaining the maximum height. When the effluent's density is lower than the receiving water, and it is discharged downwards, a penetration depth is attained by the jet and the effluent is therefore made to rise, this is known as a positively buoyant jet [8,9].

Extensive studies have been conducted on negatively buoyant jets. Marti et al. (2010) [7] conducted research, in which an angle of 60 degrees was selected with three different Froude number regimes (one-third, two-thirds, and full-flow capacity) and it was found that the Froude numbers below 20 were giving higher dilution than the predicted extrapolation. Zhang et al. (2016) [10] did the numerical investigation for inclined dense jets at a 45° angle and for the study, a large eddy simulation (LES) method was applied along with a Smagorinsky and Dynamic Smagorinsky sub-grid scale (SGS). Later, numerical results including jet trajectory, geometry, and dilution were cross-validated with the experimental results and it was found that LES was able to regenerate the outputs satisfactorily. Shao and Law (2010) [11] studied the behavior of dense jet for angles of 30° and 45° with different densiometric Froude numbers. For the measurement of velocity and concentration, combined Particle Image Velocimetry (PIV) and Planar Laser-Induced Fluorescence (PLIF) were applied. Velocity and concentration profiles were used to find the mixing and diluting parameters as well. It was found that return point dilution, the horizontal distance of return point, terminal height, centerline peak location, and its dilution were correlated to the Froude number. Oliver et al. (2008) [12] investigated the k-ε turbulence model in the standard fluid dynamics package (CFX) and took two approaches, which included, one with the standard form of the model, and another with a calibrated model achieved by adjusting the Schmidt number. After comparing numerical data, experimental data, along with the data obtained from the studies of previous integral models, concluded that the k-ε model was providing better prediction for the trajectory data, except the data for the integrated dilution at the centerline as they were over-predicting the density gradient, which resulted in the under-estimation of the dilution. Palomar et al. (2012) [13] investigated the performances of CORMIX, VISUAL PLUMES, and VISJET models for the inclined dense studies and obtained some significant differences in the dilution prediction. Kikkert et al. (2007) [14] investigated the behavior of negatively buoyant jets with angles ranging from 0° to 75° and Froude numbers ranging from 14 to 99. The results showed good predictions for the outer spread and the maximum height of the outer edge. However, the inner spread was under-estimated and the minimum dilution prediction was conservative. Along with this, previous studies conducted with CorJet and VisJet models were compared with this study, and it was found that these numerical models had under-predicted the horizontal and vertical locations of maximum jet height. Furthermore, CorJet and VisJet were not accurate enough for an integrated dilution prediction compared to analytical solutions and the data obtained in a study by Kikkert et al. (2007) [14]. Jirka (2008) [15] performed a study with smaller angles, such as 30° and 45°, based on laboratory experiments and numerical modeling using the CorJet model. It was found that the lower angle resulted in higher dilution when the bottom slope was taken into consideration, as it provided better offshore transport of the mixed effluent. Kheirkhah Gildeh et al. (2015) [16] performed numerical modeling with 30° and 45° inclined dense jets. Five CFD models including LRR, RNG k-ε, Realizable k-ε, non-linear k-ε, and Launder Gibson were applied, and it was concluded that LRR and realizable k-ε turbulence models resulted in better predictions for mixing and dilution characteristics.

With the development of computing systems in recent years, the application of combined Fuzzy and AI methods has been increasing in engineering problems. Neshat et al. (2012) [17] used ANFIS models for the optimization of concrete mix designs. They found that the ANFIS model can be better than traditional fuzzy systems and non-fuzzy systems. In a study by Nadia et al. (2020) [18], ANFIS was applied for the prediction of the position of the sun in single- and dual-axis solar tracking systems in an attempt to optimize their performance. The results showed a clear advantage of ANFIS over traditional fuzzy methods with high prediction rates and low error values. Heydary et al. (2021) [19] adopted a combined Fuzzy GMDH (i.e., Group Method of Data Handling) Neural Network and Grey Wolf Optimization (GWO) Algorithm to predict the power produced by wind turbines with consideration of supervisory control and data acquisition (SCADA) data. They first applied a combination of K-means and density-based Local Outliers methods (hybrid K-means-LOF) to remove data outliers and the Empirical Mode Decomposition (EMD) method for the decomposition of SCADA data, and then used the GMDH method to predict the future power generation of wind turbines. They found that the performance of a hybrid EMD-FGMDH-GWO can lead to high accuracy, regardless of the time step applied.

Apart from conventional Computational Fluid Dynamics (CFD) and experimental measurements, soft computing methods could be applied to minimize the computational time for the simulation and investment of money on expensive laboratory equipment. Pourtousi et al. (2015) [20] investigated the combination of the CFD and ANFIS methods for the simulation of bubble column hydrodynamics. Previous experimental data were used to validate the CFD model and later these data were used to train the ANFIS model. It was concluded that ANFIS was a promising method for predicting the outputs of bubble column hydrodynamics. Taghavifar et al. (2015) [21] worked on the assessment of heat accumulation in a hydrogen engine, in which the experimental data were compared with the data obtained after CFD modeling to determine the accuracy between the two. Later, the CFD data were fed an ANFIS code to train the model and it was concluded that the ANFIS model with a Triangular membership function had given the highest R-squared (R^2) and lowest root mean squared error (RMSE) value out of other membership functions, and ANFIS was confirmed to be more accurate and simpler than CFD technique in the study. Rezakazemi et al. (2017) [22] evaluated three models, namely ANFIS, ANFIS-PSO, and ANFIS-GA, to determine the performance of hydrogen mixed membranes, in which input parameters such as feed pressure and Nano filter contents were used to evaluate the output parameter (hydrogen gas selectivity). The criteria for investigation of the better model were R^2 and RMSE values and ANFIS-PSO had given better predictability. Amirkhani et al. (2015) [23] studied the performance of ANN and ANFIS models to estimate the inlet air velocity of the chimney. Three days of experimental data were used to train the models and it was found that the ANFIS model's results were closer to the experimental results as its R^2 was higher than ANN. Bonakdari and Zaji (2018) [24] worked on the modified triangular side weir, in which they simulated its discharge coefficient. They studied three different methods of ANFIS, namely ANFIS-GA, ANFIS-PSO, and ANFIS-DE, with combinations of eight different input variables and it was found that ANFIS-DE performed better as it had given lowest the RMSE value compared to ANFIS-GA and ANFIS-PSO. Shabanian et al.

(2017) [25] studied the ANFIS model with eight types of membership functions to predict the hydrogen yield of the jet fuel and efficiency of conversion for a non-catalytic filtration combustion reactor. Later, an imperialist competitive algorithm (ICA) was applied to get the optimized results for the hydrogen yield, which was found to be an efficient algorithm for the combustion process optimization. Apart from a soft computing method, multi-gene genetic programming (MCGP) is also a new approach to predict the output, as shown in the study conducted by Yan and Mohammadian (2019) [26], where MCGP turned out to be a promising method for the prediction of vertical buoyant jets.

Currently, there is a gap in the application of AI methods in the context of inclined dense jets, which are among the most efficient mixing methods. In particular, to the best of the authors' knowledge, no previous study has used an ANFIS model and its variants for negatively buoyant jets with an inclination. Such a method can bridge the gap between AI methods and the simulation of inclined dense jets and can potentially simulate these problems more efficiently than CFD methods. The aim of the current study is to consider the application of new AI methods and the generation of data for the testing and training of these models to find the optimum solutions. The main contribution of this project is to apply new AI methods to simulate and predict the dilution of inclined dense jets in the near-field zone. The proposed approach, as shown in this paper, can accurately and efficiently simulate these jets and contribute towards mitigating the negative environmental impacts of such jets. The discharged effluents can create irreversible damage to the marine environment and aquatic life if the outfall systems are not designed properly. The improper design of the system can also leave toxic contaminants in the coastal area. Hence, it is important to design proper outfall systems with efficient mixing. Furthermore, if the concentration of the effluents is determined before their discharge, then it would be helpful in the implementation of the solutions. The salinity of the discharged effluents can either be predicted by experimental or numerical methods, however, to avoid the cost of experimental equipment and save computational time, artificial intelligence techniques can be implemented in the coastal study. The aim of this research is to investigate the application and performance of a soft computing method with ANFIS, ANFIS-GA, ANFIS-PSO, ANFIS-FFA algorithms for negatively buoyant jets to predict the dilution and mixing characteristics. This is the first study on this topic. Negatively buoyant jets are considered for a wide range of Froude numbers, i.e., 5, 10, 12.5, 15, 17.5, 20, 22.5, 25, 27.5, 30, 32.5, 35, 37.5, 40, 50 and 60 with angles ranging from 20 degrees to 72.5 degrees using realizable k-ε model turbulence model in the OpenFOAM platform [27,28].

2. Materials and Methods

2.1. Dimensional Analysis and Numerical Model

As can be seen in Figure 1, negatively buoyant jets are discharged at an angle Θ and velocity U_o, the density of the ambient water is represented by ρ_a, and density of the jet is represented by ρ_o. It can be observed that $\rho_o > \rho_a$, which makes the jet rise. The diameter of the jet is denoted by D. The terminal height is represented by y_t, which hits the surface at coordinate x_i while the coordinates of peak centerline are represented by x_m and y_m with peak salinity as S_m. Furthermore, jet concentration is represented by C_o. The return point of the jet is represented by x_r and the return salinity value is S_r. For the dimensional analysis a densiometric Froude number is used, which is denoted by the following equation:

$$Fr_d = \frac{U_O}{\sqrt{g_0' D}} \tag{1}$$

$$g_0' = \left(\frac{\Delta\rho_0}{\rho_a}\right) \tag{2}$$

where $\Delta\rho_o = (\rho_o - \rho_a)$ and $g_0{}'$ is the reduced gravitational acceleration.

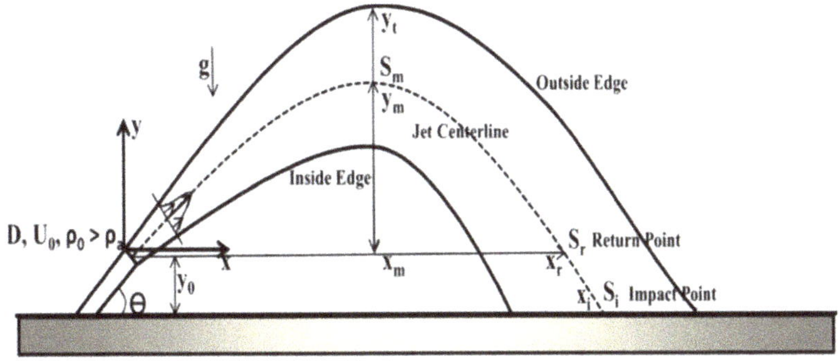

Figure 1. Configuration for Negatively Buoyant Jet (Reprinted with permission from Ref. [16]. Copyright 2022 Springer Nature).

The centerline peak salinity is a function of the Froude number and angle, which can be represented by the following equation:

$$\frac{S_m}{Fr_d} = f(Fr_{d,}\ \theta) \tag{3}$$

For numerical modeling, the following equations were used by Kheirkhah Gildeh et al. (2015) [16]:

- Continuity Equation:

$$\frac{\partial u}{\partial x} + \frac{\partial v}{\partial y} + \frac{\partial w}{\partial z} = 0 \tag{4}$$

- Momentum Equations:

$$\frac{\partial u}{\partial t} + u\frac{\partial u}{\partial x} + v\frac{\partial u}{\partial y} + w\frac{\partial u}{\partial z} = -\frac{1}{\rho}\frac{\partial P}{\partial x} + \frac{\partial}{\partial x}\left(v_{\text{eff}}\left(\frac{\partial u}{\partial x}\right)\right) + \frac{\partial}{\partial y}\left(v_{\text{eff}}\left(\frac{\partial u}{\partial y}\right)\right) + \frac{\partial}{\partial z}\left(v_{\text{eff}}\left(\frac{\partial u}{\partial z}\right)\right) \tag{5}$$

$$\frac{\partial v}{\partial t} + u\frac{\partial v}{\partial x} + v\frac{\partial v}{\partial y} + w\frac{\partial v}{\partial z} = -\frac{1}{\rho}\frac{\partial P}{\partial y} + \frac{\partial}{\partial x}\left(v_{\text{eff}}\left(\frac{\partial v}{\partial x}\right)\right) + \frac{\partial}{\partial y}\left(v_{\text{eff}}\left(\frac{\partial v}{\partial y}\right)\right) + \frac{\partial}{\partial z}\left(v_{\text{eff}}\left(\frac{\partial v}{\partial z}\right)\right) - g\frac{\rho - \rho_o}{\rho} \tag{6}$$

$$\frac{\partial w}{\partial t} + u\frac{\partial w}{\partial x} + v\frac{\partial w}{\partial y} + w\frac{\partial w}{\partial z} = -\frac{1}{\rho}\frac{\partial P}{\partial z} + \frac{\partial}{\partial x}\left(v_{\text{eff}}\left(\frac{\partial w}{\partial x}\right)\right) + \frac{\partial}{\partial y}\left(v_{\text{eff}}\left(\frac{\partial w}{\partial y}\right)\right) + \frac{\partial}{\partial z}\left(v_{\text{eff}}\left(\frac{\partial w}{\partial z}\right)\right) \tag{7}$$

where v_{eff} denotes the effective kinematic viscosity, ρ is the fluid density, ρ_o is the reference fluid density and P represents the fluid pressure. Furthermore, the velocity components in x, y, and z directions are represented by u, v, and w.

- Concentration Equation:

$$\frac{\partial C}{\partial t} + u\frac{\partial C}{\partial x} + v\frac{\partial C}{\partial y} + w\frac{\partial C}{\partial z} = D\left(\frac{\partial^2 C}{\partial x^2} + \frac{\partial^2 C}{\partial y^2} + \frac{\partial^2 C}{\partial z^2}\right) \tag{8}$$

where D is the diffusion coefficient and C denotes the concentration, which in this paper is salinity. For inlet, boundary conditions for velocity in x, y, and z directions are defined as $u = U_0 * \cos(\theta)$, $v = U_0 * \sin(\theta)$ and $w = 0$. While concentration $C = C_0$ and Temperature $T = T_0$ [16].

Initial conditions were assumed as follows: Zero velocity (stagnant) condition was assumed for the flow and the initial concentration of salt was assumed to be zero in the reservoir. As for boundary conditions, a wall function was used for the turbulent quantities at the bottom and a zero-velocity condition (no-slip) was used for the velocity at the bottom wall. Zero shear stress was applied to the top surface and sidewalls, and we developed the condition as assumed for the inlet jet and the values of turbulent quantities such as turbulent kinetic energy and dissipations, which we set according to fully developed pipe flow. The exit of the reservoir was modeled by a zero-gradient condition.

2.2. Data

The data generated from the numerical modeling using the realizable k-ε model in the OpenFOAM platform is used in this part, to train and test the ANFIS and hybrid models. For the soft computing method, two input variables, Froude numbers ranging from 5 to 60 and jet angles ranging from 20 degrees to 72.5, were employed. The aim is to investigate input and output combinations (Table 1) to evaluate the performance of ANFIS, ANFIS-GA, ANFIS-PSO, ANFIS-FFA, and Multivariate regression models. In the present study, the programming language MATLAB is used to design ANFIS [29–31], and the three hybrid models, ANFIS-GA [32,33], ANFIS-PSO [33,34], and, ANFIS-FFA [35–37]. These models are built on the fundamentals of training and testing, which can be seen in Figure 2. The data are divided into two portions of 70% and 30% for training and model validation, respectively, and various error estimates such as RMSE, R2, etc., are measured for evaluation of the model's accuracy.

Table 1. Input-Output combinations.

Combinations	Input 1	Input 2	Output
1	Froude number	Angle	S_m
2	Froude number	Angle	S_r
3	Froude number	Angle	x
4	Froude number	Angle	x_r

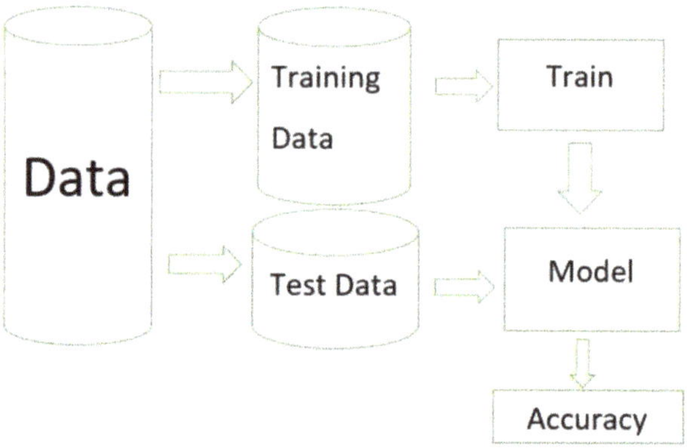

Figure 2. Training data and Test.

2.3. Adaptive Neuro-Fuzzy Inference System (ANFIS)

Adaptive Neuro-Fuzzy Inference System is an artificial intelligence method, applied to solve nonlinear problems. The architecture for ANFIS containing two inputs, one output, *f*, and five layers is illustrated in Figure 3. In the architecture, the Sugeno model with Fuzzy IF-THEN rules is employed. The rules R1 and R2 are shown below:

- R1:

$$\text{If } x_1 = U_1 \text{ and } x_2 = V_1 \tag{9}$$

$$\text{Then } f_1 = s_1 x_1 + t_1 x_2 + r_1 \tag{10}$$

- R2:

$$\text{If } x_1 = U_2 \text{ and } x_2 = V_2 \tag{11}$$

$$\text{Then } f_2 = s_2x_1 + t_2x_2 + r_2 \tag{12}$$

where U_1, U_2 and V_1, V_2 are the membership functions for inputs x_1 and x_2, while $s_1, s_2, t_1, t_2, r_1,$ and r_2 are the adjustable parameters determined during the training process.

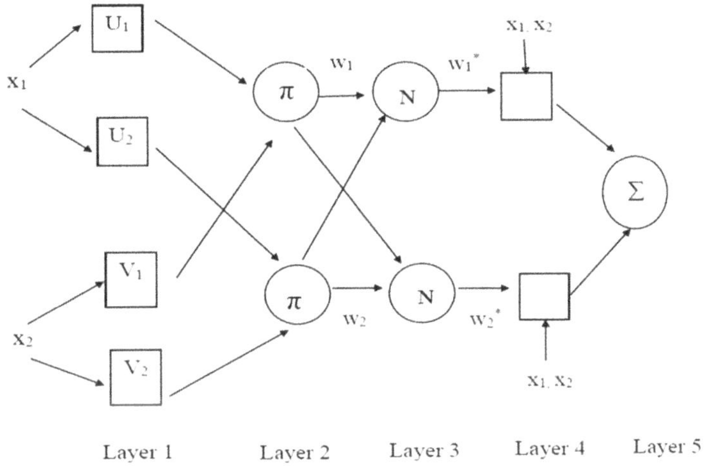

Figure 3. ANFIS structure.

The first layer is the input layer, in which input variables are transferred to the next layer and it is formed by the membership functions of the input variables.

$$O_{1,i} = \mu_{Ui}\,(x_1),\ i = 1,2 \tag{13}$$

$$O_{1,i} = \mu_{Vi}(x_2),\ i = 1,2 \tag{14}$$

The degree of membership functions is represented by μ_{Ui} and μ_{Vi} for the fuzzy sets U_i and V_i, respectively.

In layer two, each node is fixed and non-adaptive, when each node input values are multiplied by each other, weights (w_i) are obtained.

$$O_{2i} = w_i = \mu_{Ui}(x_1) * \mu_{Vi}(x_2),\ i = 1,2 \tag{15}$$

The third layer, which is non-adaptive in nature, is called the rule layer. In this layer, the weight function is normalized as follows:

$$O_{3i} = w_i^* = \frac{w_i}{\sum_i w_i} \tag{16}$$

The fourth layer, which is the layer where defuzzification takes place and the output of the previous layer, is combined with the Sugeno fuzzy rule's function. However, nodes in this layer are adaptive and contain a node function:

$$O_{4i} = w_i^* f_i = w_i^*(s_i x_1 + t_i x_2 + r_i) \tag{17}$$

At layer five, which is the last layer and is called the output layer, the single node will calculate the overall output and will be the summation of all the inputs from the previous layers.

$$O_{5i} = \sum_i w_i^* f_i = \frac{\sum_i w_i f_i}{\sum_i w_i} \quad (18)$$

2.4. Genetic Algorithm (GA)

Genetic Algorithm [32] is the heuristic search algorithm, which can be classed as an evolutionary algorithm (EA) and is based on the concept of natural selection and genetics, where the idea of inheritance, selection, cross-over, and mutation are applied. It is commonly used in various domains such as manufacturing, engineering, science, etc. [38]. Evolutionary algorithms such as GA are applied in conjugation with ANFIS to enhance the accuracy of the method by finding optimal solutions and lowering errors. The genetic algorithm (GA) starts the process of optimization with a random initial population (Figure 4). In GA, a population is a set of individuals, which are present in the workspace. Each individual has a set of parameters (variables), which are called genes, and are joined together to form a chromosome (solution), which could be mutated and altered. These solutions could either be presented in the form of binary coding, i.e., zeros and ones, or in other encoding forms. The criteria for determining the suitability of individuals are set by an evaluation through fitness function as the population is initialized through randomly generated individuals, hence, it is an iterative process [38]. The best suitable individual with a higher fitness value will be chosen from the population to create the new generation and the solutions of this new generation will be used for the next iteration in the same algorithm. The algorithm will be terminated when the produced generations reach the maximum limit, or a satisfactory fitness level is achieved [38]. In this paper, the GA code has been run in MATLAB software and the ANFIS model has been trained to find the mean salinity S_{m}, mean coordinates x_{m} and y_{m}, return salinity S_{r} and return coordinates x_{r} and y_{r}.

2.5. Particle Swarm Optimization (PSO)

PSO begins with random particles in the search space, which looks for optimal solutions, and each particle is associated with a fitness value, which is evaluated by a fitness function. Each particle is influenced by its best achieved individual position and the best position achieved among the group, and for every iteration, the updating of each particle takes place by these two best values. In every iteration particles choose new velocities based on their current velocity and the two mentioned best values. The new velocity and new position can be evaluated by the following equations [24]:

$$v^{\mathrm{i}}[t+1] = wv^{\mathrm{i}}[t] + c_1 r_1 \left(x^{\mathrm{Pbest}}[t] - x^{\mathrm{i}}[t]\right) + c_2 r_2 \left(x^{\mathrm{Gbest}}[t] - x^{\mathrm{i}}[t]\right) \quad (19)$$

$$x^{\mathrm{i}}[t+1] = x^{\mathrm{i}}[t] + v^{\mathrm{i}}[t+1] \quad (20)$$

where, x^{i} and v^{i} are the position and velocity vector for particle i and x^{Pbest} and x^{Gbest} are the best individual position and best position achieved in the group, respectively. c_1 and c_2 are the personal learning and global learning coefficients, respectively, and r_1 and r_2 are the random coefficients. Furthermore, w represents the inertia weight. The PSO algorithm can be seen in Figure 5.

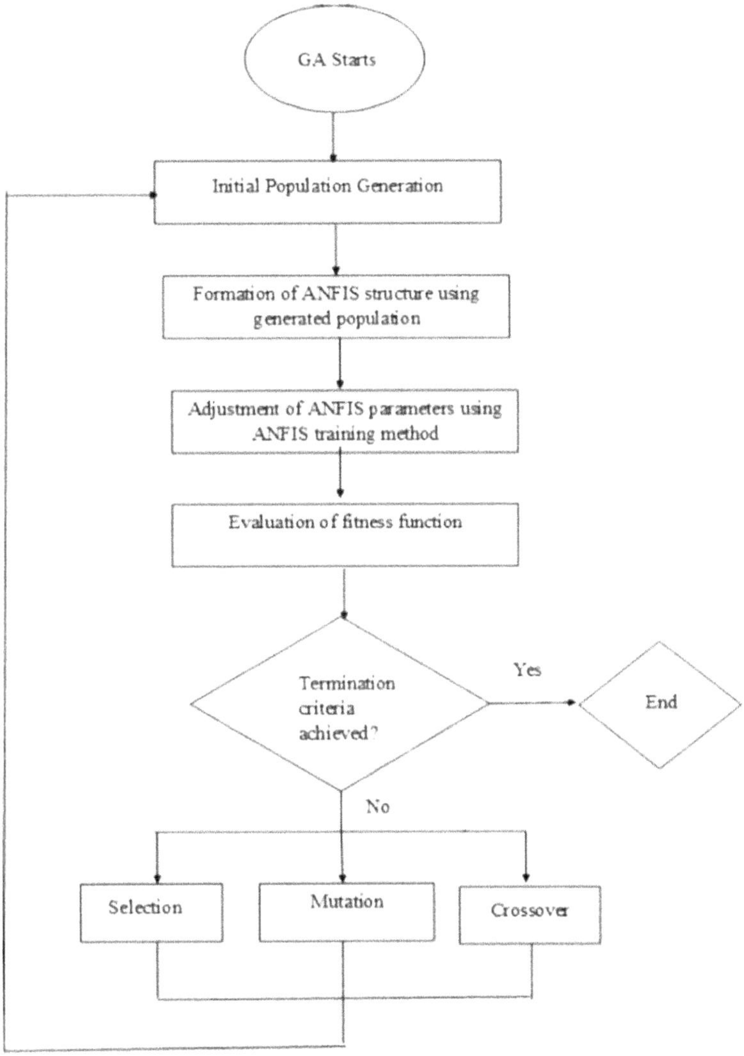

Figure 4. GA Flowchart.

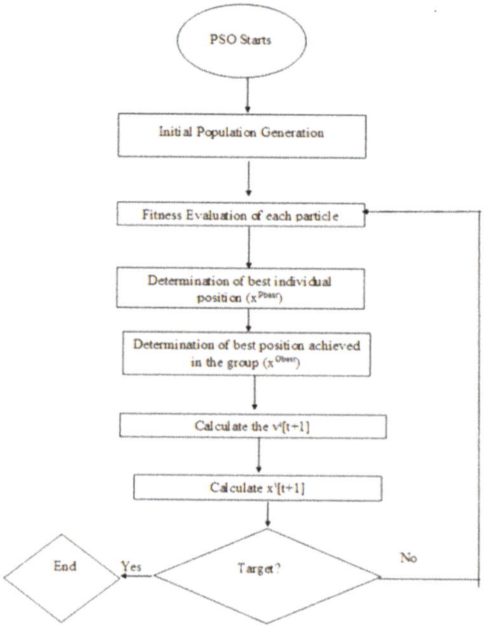

Figure 5. PSO Flowchart.

2.6. Firefly Algorithm (FFA)

The firefly algorithm (FFA) is built on the idea of a relationship between light and fireflies [35]. Based on this relationship, the attractiveness value is directly proportional to the luminosity, hence, it can be calculated by following equations [35,36]:

$$I = I_o e^{-\gamma r^2} \tag{21}$$

$$w(r) = w_0 e^{-\gamma r^2} \tag{22}$$

where $w(r)$ denotes the attractiveness at distance r from the firefly, and I represents the light intensity. I_o and w_0 are the light intensity and attractiveness at distance $r = 0$ and γ is the coefficient of light absorption [35,36].

The distance between the fireflies i and j is represented by r, and can be calculated from the following equation [35,36]:

$$r_{ij} = \| x_i + x_j \| = \sqrt{\sum_{k=1}^{d} (x_{i,k} - x_{j,k})} \tag{23}$$

where x_i and x_j are the locations of fireflies. As fireflies are attracted to one another, the movement for a firefly from one position to another is represented by the following equation [35]:

$$\Delta_{xi} = \beta_0 e^{-\gamma r^2} (x_j - x_i) + \alpha \varepsilon_i \tag{24}$$

where α denotes the randomization coefficient and ε_i represents the random number vector. Furthermore, α varies from 0 and 1. $\beta_0 e^{-\gamma r^2}$ is the attraction term [36]. Figure 6 illustrates the FFA Flowchart.

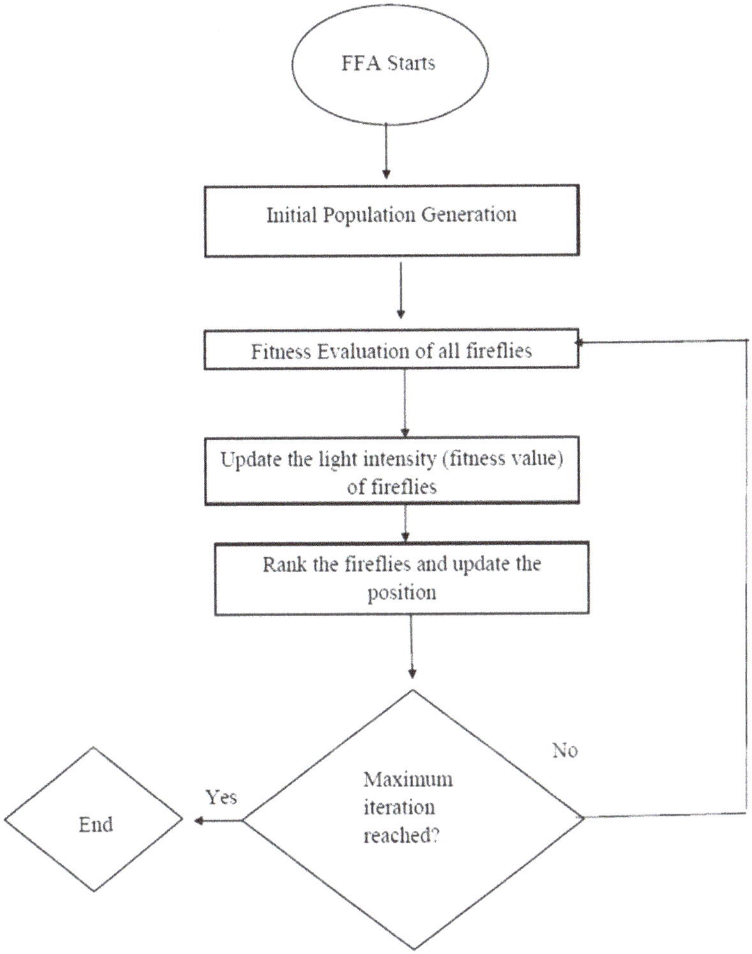

Figure 6. FFA Flowchart.

In this paper, the ANFIS model was integrated with the FFA in order to determine the mean salinity S_m, mean coordinates x_m and y_m, return salinity S_r and return coordinates x_r and y_r. By a trial-and-error method, the values of light absorption coefficient (γ), attraction coefficient base (β_0), and movement coefficient (α) are taken as 0.1, 4, and 0.3, respectively.

2.7. Multivariate Linear Regression Model (MLR)

Multivariate regression analysis is widely used to find a linear relationship between the dependent and multiple independent variables. The data collected from numerical modeling are non-linear, however, to determine the closeness of data with the linearity, multivariate regression analysis has been conducted using Microsoft excel add-ins, which helped to create a model based on least square methods [39]. The generalized equation for MLR can be expressed in the following way [40,41]:

$$Y = \beta_o + \beta_1 X_1 + \beta_2 X_2 \tag{25}$$

where, X_1 and X_2, are the independent variables, which are also called predictor variables. Y is the dependent variable also known as the response variable and n is the number of variables [41].

For multivariate regression, data collected from numerical modeling were divided into training and test data. The equation obtained for the training data set after a regression analysis has been used to generate the predicted test output. The statistical parameters to evaluate the multivariate regression model are the same as the ones used for the ANFIS and hybrid models. The equations mentioned in the statistical parameters section are used for the calculation of the values for regression analysis.

3. Statistical Analysis

To determine the accuracy of ANFIS, ANFIS-GA, ANFIS-PSO, ANFIS-FFA, and multivariate regression models as discussed, statistical parameters of all the models are compared, the statistical parameters taken into consideration are coefficient of determination (R^2), root mean squared error (*RMSE*), mean absolute error (*MAE*), and average absolute deviation ($\delta\%$), i.e., error of the model in percentage [24]. The mentioned parameters can be measured by the following equations:

$$RMSE = \sqrt{\frac{\sum_{i=1}^{N} (O_i - P_i)^2}{N}} \tag{26}$$

$$R^2 = 1 - \frac{\sum_{i=1}^{N} (O_i - P_i)^2}{\sum_{i=1}^{N} (O_i - O_m)^2} \tag{27}$$

$$MAE = \frac{1}{N} \sum_{i=1}^{N} \mid O_i - P_i \mid \tag{28}$$

$$\delta\% = \frac{\sum_{i=1}^{N} \mid O_i - P_i \mid}{\sum_{i=1}^{N} O_i} * 100 \tag{29}$$

where P_i is the predicted value obtained after training the models, O_i is the observed value obtained after numerical modeling on OpenFOAM, O_m is the mean of observed value and N is the number of samples.

4. Results

Overall, 352 data points are obtained from numerical modeling for each of the S_m, S_r, x, x_r, and y outputs. The data are divided into two sections, training data and test data, where training data contain 272 of the total data and test data contain 80 of the total data test data. For the models, two input variables, i.e., the Froude number and the angle are chosen to obtain one output. The targeted outputs are S, S_r, x_m, y_m, x_r. Hence, five sets with different outputs are prepared for all the models, as shown in Table 1.

4.1. *Performance Evaluation for Peak Salinity*

The performance of ANFIS-type models and the Multivariate regression model for peak salinity (S_m) is determined in this section.

4.1.1. ANFIS-Type Models

It can be observed from Table 2 that all the models' RMSE values for training and test data are almost the same, which means none of them are trapped in over-fitting. Furthermore, in Figure 7a–h, targets and outputs coincide reasonably with each other, which confirms the accuracy of the data for the ANFIS and hybrid models. From Table 2, it can be observed that out of all the models, ANFIS-PSO is giving the highest R^2, and the lowest *RMSE*, *MAE*, and $\delta\%$, which are 0.984, 0.589, 0.357, and 5.889% for the test data, making it the most accurate of all.

Table 2. Models' performance evaluation for S_m.

Model	Output	Training Database				Test Database			
		R^2	RMSE	MAE	δ%	R^2	RMSE	MAE	δ%
ANFIS	S_m	0.9505	0.959	0.726	12.160	0.964	0.890	0.709	11.680
ANFIS-GA	S_m	0.947	1.019	0.834	13.974	0.935	1.187	0.866	14.249
ANFIS-PSO	S_m	0.985	0.547	0.336	5.631	0.984	0.589	0.357	5.889
ANFIS-FFA	S_m	0.979	0.643	0.409	6.866	0.975	0.739	0.447	7.367
Multi-variate Regression	S_m	0.594	2.909	2.118	35.482	0.582	3.009	2.245	36.948

4.1.2. Multi-Variate Regression Model

The equation obtained after training the regression model for salinity (S_m) is shown below:

$$S_m = 13.4212 - (0.23703 * Fr) - (0.01983 * \text{Angle}) \quad (30)$$

The S_m equation was used to predict the test outputs and it can be seen from Table 2 that in the regression model there is no over-fitted data as the training and test data sets are showing almost similar statistical parameters, however, overall the regression model had the lowest R^2 value and highest RMSE, MAE and δ %, i.e., in both training and test sets as compared to other models, which made regression model incompatible for predicting peak salinity.

4.2. *Performance Evaluation for Return Salinity*

The performance of ANFIS-type models and the Multivariate regression model for return salinity (S_r) is determined in this section.

4.2.1. ANFIS-Type Models

Table 3 shows the statistical results for S_r and it can be observed that test data for ANFIS-GA, ANFIS-PSO, and ANFIS-FFA are showing almost the same R^2 value, i.e., 0.976, 0.973, and 0.975, yet the lowest RMSE was observed in ANFIS-GA, i.e., 0.471. Hence, the ANFIS-GA model can be considered suitable for predicting the return salinity value. Furthermore, the δ% and MAE are also not high for this model. Notably, the RMSE and R^2 values of the test sets are mainly considered in this thesis work for determining the suitable model.

The graphs in Figure 8a–h shows that targets and outputs are following the same pattern, which again shows that models are properly trained for output S_r.

4.2.2. Multi-Variate Regression Model

The equation obtained after training the regression model for return salinity (S_r) is shown below:

$$S_r = 7.601744 - (0.12495 * Fr) - (0.02693 * \text{Angle}) \quad (31)$$

It can be observed from Table 3 that the parameters for training and test data are almost the same, however, the regression model's performance compared to other models is very poor, as it has the lowest R^2 value, i.e., 0.441 and highest RMSE, MAE values, i.e., 2.265 and 1.429 in the test set, respectively, which shows the model accuracy to predict the data is very low and its test outputs are not near to the outputs obtained from the numerical model.

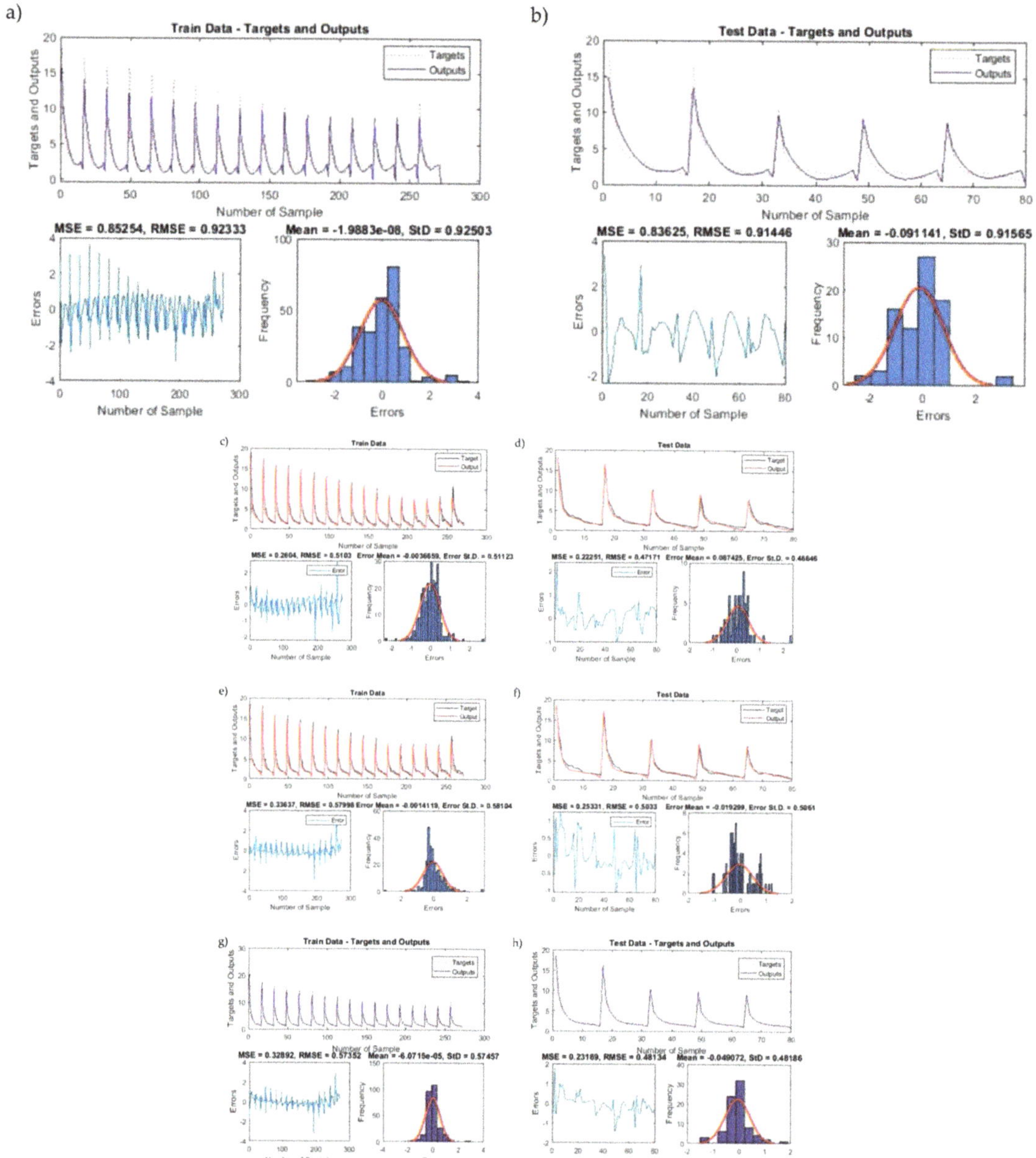

Figure 7. For output S_m, ANFIS Model (**a**) targets and outputs, RMSE, MSE values and frequency vs. errors graphs for train data set (**b**) targets and outputs, RMSE, MSE values and frequency vs. errors graphs for test data set. ANFIS-GA: (**c**) targets and outputs, RMSE, MSE values and frequency vs. errors graphs for train data set (**d**) targets and outputs, RMSE, MSE values, and frequency vs. errors graphs for test data set. ANFIS-PSO: (**e**) targets and outputs, RMSE, MSE values and frequency vs. errors for train data set (**f**) targets and outputs, RMSE, MSE values, and frequency vs. errors for test data set. ANFIS-FFA: (**g**) targets and outputs, RMSE, MSE values and frequency vs. errors for train data set (**h**) targets and outputs, RMSE, MSE values, and frequency vs. errors for test data set.

Table 3. Models' performance evaluation for output S_r.

Model	Output	Training Database				Test Database			
		R^2	RMSE	MAE	δ%	R^2	RMSE	MAE	δ%
ANFIS	S_r	0.893	0.923	0.691	23.870	0.909	0.914	0.684	23.028
ANFIS-GA	S_r	0.967	0.510	0.376	12.981	0.976	0.471	0.350	11.803
ANFIS-PSO	S_r	0.958	0.579	0.404	13.946	0.973	0.503	0.395	13.304
ANFIS-FFA	S_r	0.958	0.573	0.369	12.755	0.975	0.481	0.336	11.331
Multi-variate Regression	S_r	0.455	2.147	1.394	48.104	0.441	2.265	1.429	48.077

4.3. *Performance Evaluation for Peak Coordinate*

The performance of ANFIS-type models and Multivariate regression models for peak coordinate (x_m) is determined in this section.

4.3.1. ANFIS-Type Models

From Table 4, it can be observed that the statistical parameters for output x_m, are showing good accuracy between training and test, which shows none of the models are over-fitted. The models are trained properly, which can be seen from Figure 9a–h. Even though the percentage deviation for ANFIS-FFA's test data is the smallest out of all the models, the statistical parameters of ANFIS-PSO's training data are better when compared to its test data, which makes the model more reliable. Furthermore, the test sets for ANFIS-GA and ANFIS-FFA have the highest R^2, i.e., 0.987, 0.987 with RMSE values, 0.018 and 0.016, respectively. However, considering the RMSE and R^2 values of the test set, it can be seen that ANFIS-FFA is efficient in determining the output x_m as it has the highest R^2 and lowest RMSE, though it also has the lowest MAE and δ%.

Table 4. Models' performance evaluation for output x_m.

Model	Output	Training Database				Test Database			
		R^2	RMSE	MAE	δ%	R^2	RMSE	MAE	δ%
ANFIS	x_m	0.985	0.018	0.013	5.409	0.983	0.018	0.013	5.433
ANFIS-GA	x_m	0.981	0.020	0.015	6.251	0.987	0.018	0.014	5.736
ANFIS-PSO	x_m	0.989	0.015	0.011	4.637	0.976	0.021	0.016	6.250
ANFIS-FFA	x_m	0.987	0.017	0.012	5.011	0.987	0.016	0.011	4.650
Multi-Variate Regression	x_m	0.899	0.058	0.043	17.707	0.892	0.047	0.036	14.348

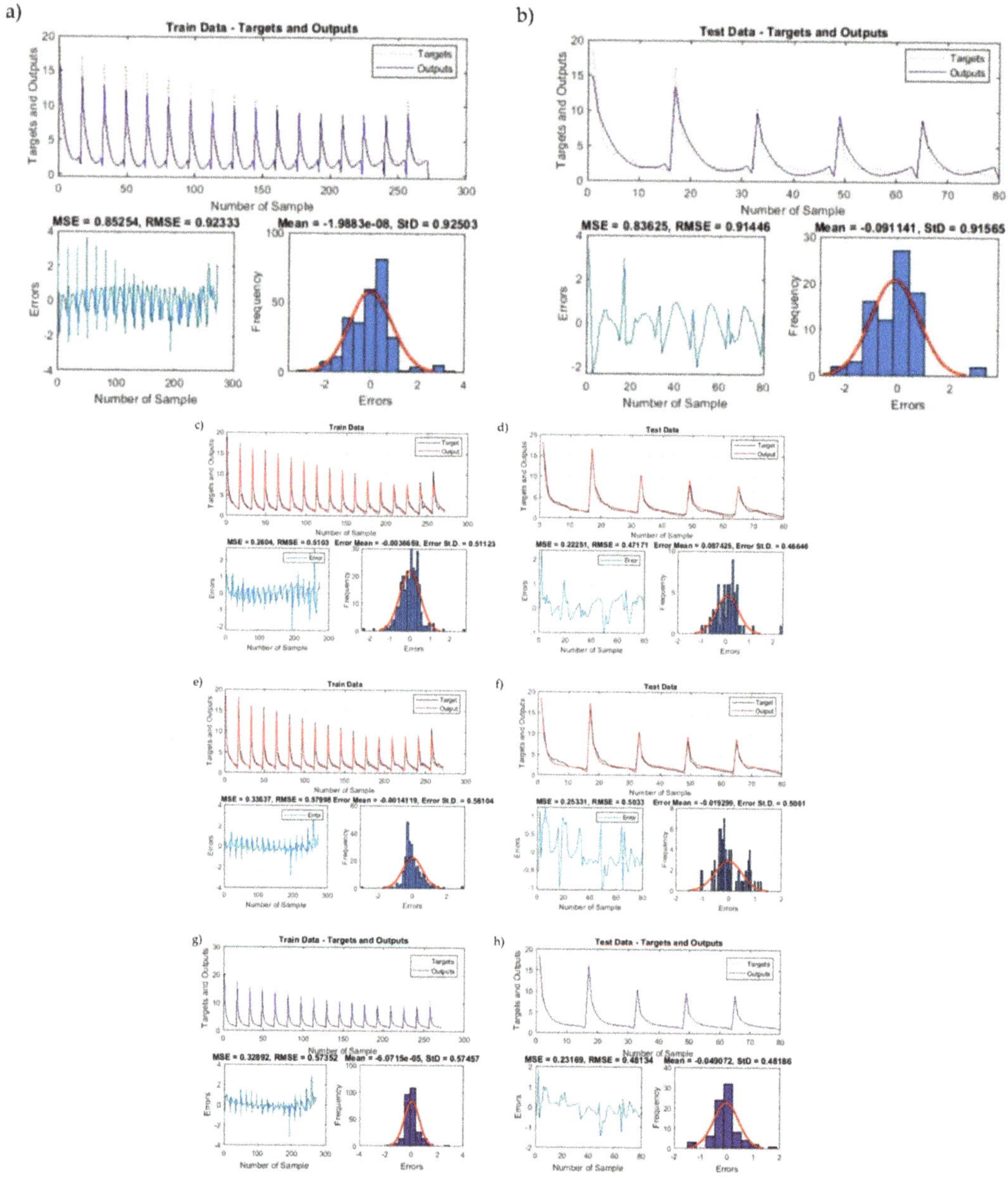

Figure 8. For output S_r, ANFIS Model (**a**) targets and outputs, RMSE, MSE values and frequency vs. errors graphs for train data set (**b**) targets and outputs, RMSE, MSE values and frequency vs. errors graphs for test data set. ANFIS-GA: (**c**) targets and outputs, RMSE, MSE values and frequency vs. errors graphs for train data set (**d**) targets and outputs, RMSE, MSE values, and frequency vs. errors graphs for test data set. ANFIS-PSO: (**e**) targets and outputs, RMSE, MSE values and frequency vs. errors for train data set (**f**) targets and outputs, RMSE, MSE values, and frequency vs. errors for test data set. ANFIS-FFA: (**g**) targets and outputs, RMSE, MSE values and frequency vs. errors for train data set (**h**) targets and outputs, RMSE, MSE values, and frequency vs. errors for test data set.

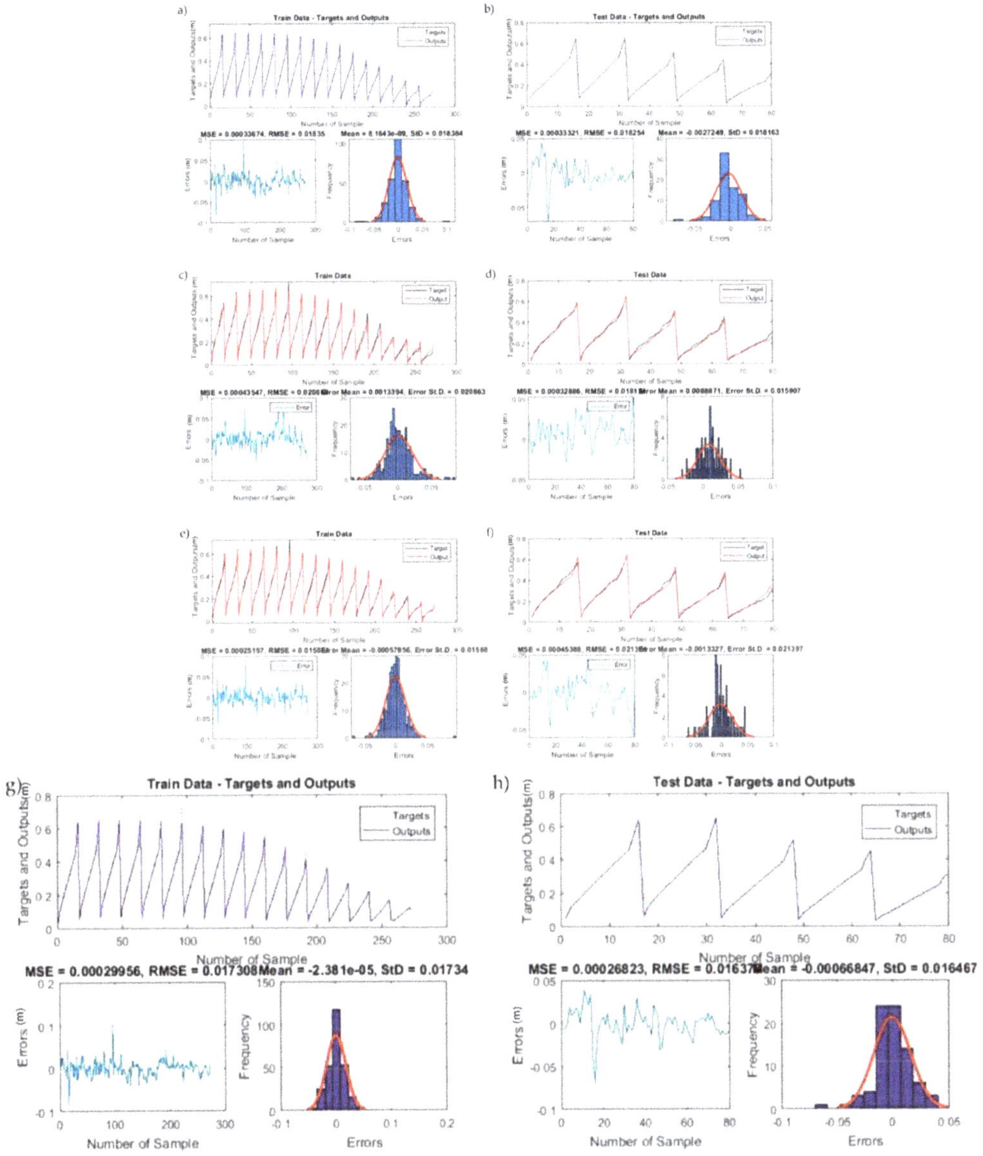

Figure 9. For output x_m, ANFIS Model (**a**) targets and outputs, RMSE, MSE values and frequency vs. errors graphs for train data set (**b**) targets and outputs, RMSE, MSE values and frequency vs. errors graphs for test data set. ANFIS-GA: (**c**) targets and outputs, RMSE, MSE values and frequency vs. errors graphs for train data set (**d**) targets and outputs, RMSE, MSE values, and frequency vs. errors graphs for test data set. ANFIS-PSO: (**e**) targets and outputs, RMSE, MSE values and frequency vs. errors for train data set (**f**) targets and outputs, RMSE, MSE values, and frequency vs. errors for test data set. ANFIS-FFA: (**g**) targets and outputs, RMSE, MSE values and frequency vs. errors for train data set (**h**) targets and outputs, RMSE, MSE values, and frequency vs. errors for test data set.

4.3.2. Multi-Variate Regression Model

To predict, the test output x, the following equation was obtained from the regression analysis of the training set:

$$x_m = 0.261093 + (0.00818 * Fr) - (0.00515*\text{Angle}) \quad (32)$$

Table 4 shows that although the performance of the regression model was satisfactory, it had the lowest R^2 value and highest RMSE, MAE, and δ% in the test set compared to the other models' test sets, which show that the predicted outputs from MLR are not close to the observed numerical outputs. The statistical parameters for the multivariate regression model are calculated by Equations (26)–(29).

4.4. Performance Evaluation for Return Coordinate in x Direction

The performance of ANFIS-type models and the Multivariate regression model for the return coordinate in the x-direction (x_r) is determined in this section.

4.4.1. ANFIS-Type Models

For output x_r, Table 5 shows all the models' test data are close in performance to the training data; in fact, the statistical parameters of the training sets are slightly higher than the test sets, which again shows that models are not over-fitted, and they have good predictability, which can be seen in Figure 10a–h. Out of all the models, ANFIS-PSO had given better results as its test set had a higher R^2 value i.e., 0.985, and lower RMSE, MAE, and δ % i.e., 0.032, 0.021, and 4.639% in the test data, respectively.

Table 5. Performance Evaluation for all the models for output x_r.

Model	Output	Training Database				Test Database			
		R^2	RMSE	MAE	δ%	R^2	RMSE	MAE	δ%
ANFIS	x_r	0.989	0.028	0.021	4.745	0.981	0.035	0.024	5.119
ANFIS-GA	x_r	0.966	0.051	0.039	8.827	0.953	0.056	0.041	8.832
ANFIS-PSO	x_r	0.993	0.022	0.016	3.634	0.985	0.032	0.021	4.639
ANFIS-FFA	x_r	0.986	0.031	0.023	5.258	0.980	0.035	0.026	5.717
Multi-variate Regression	x_r	0.917	0.099	0.076	17.297	0.904	0.079	0.065	13.933

4.4.2. Multi-Variate Regression Model

For output x_r, the multivariate regression model showed quite good results with R^2 as 0.9044, and RMSE, MAE values as 0.079 and 0.065, respectively. The percentage deviation between the observed and predicted test output was 13.93%. The equation used for the predicted output was:

$$x_r = 0.399753 + (0.015318 * Fr) - (0.00804 * \text{Angle}) \quad (33)$$

4.5. Performance Evaluation for Peak Coordinate in y Direction

The performance of ANFIS-type models and the Multivariate regression model for the peak coordinate in the y-direction (y_m) is determined in this section.

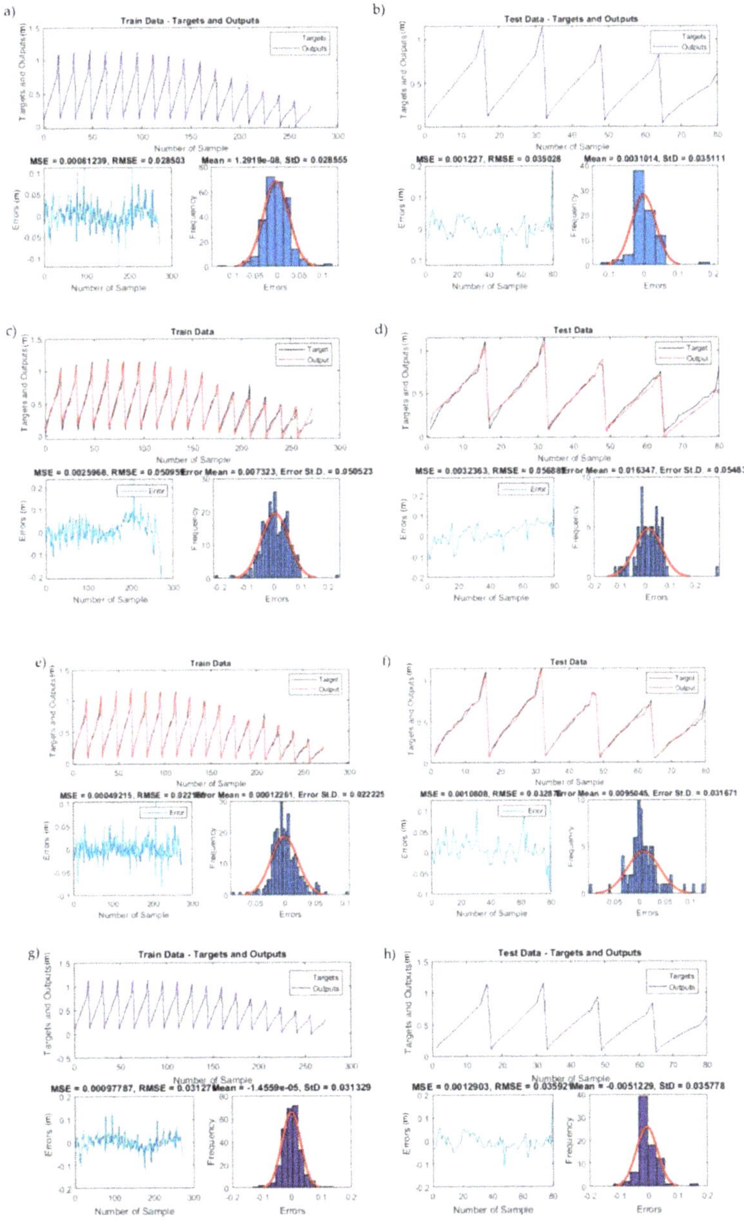

Figure 10. For output x_r, ANFIS Model (**a**) targets and outputs, RMSE, MSE values and frequency vs. errors graphs for train data set (**b**) targets and outputs, RMSE, MSE values and frequency vs. errors graphs for test data set. ANFIS-GA: (**c**) targets and outputs, RMSE, MSE values and frequency vs. errors graphs for train data set (**d**) targets and outputs, RMSE, MSE values, and frequency vs. errors graphs for test data set. ANFIS-PSO: (**e**) targets and outputs, RMSE, MSE values and frequency vs. errors for train data set (**f**) targets and outputs, RMSE, MSE values, and frequency vs. errors for test data set. ANFIS-FFA: (**g**) targets and outputs, RMSE, MSE values and frequency vs. errors for train data set (**h**) targets and outputs, RMSE, MSE values, and frequency vs. errors for test data set.

4.5.1. ANFIS-Type Models

From Table 6 it can be seen that the percentage deviations for both the data sets in ANFIS-GA are 6.093% and 7.472%, which are higher than all the ANFIS-type models. However, ANFIS, ANFIS-PSO, and ANFIS-FFA had given similar R^2 values, i.e., 0.989, 0.986, 0.986, respectively, for their test data sets, which are highest in the test data for all the models. Also, their test sets R^2 value is comparable to their respective training set, which shows the models are trained properly. Along with this, the RMSE values for ANFIS, ANFIS-PSO, ANFIS-FFA are 0.013, 0.014, 0.014, respectively, with the same MAE value of 0.010. Hence, they can be considered for the prediction of y_m. Figure 11a–h shows that models are not over-fitted as targets and outputs coincide with each other.

Table 6. Models' performance evaluation for output y_m.

Model	Output	Training Database				Test Database			
		R^2	RMSE	MAE	δ%	R^2	RMSE	MAE	δ%
ANFIS	y_m	0.985	0.014	0.010	5.014	0.989	0.013	0.010	5.157
ANFIS-GA	y_m	0.979	0.017	0.012	6.093	0.979	0.019	0.014	7.472
ANFIS-PSO	y_m	0.990	0.011	0.008	3.900	0.986	0.014	0.010	5.285
ANFIS-FFA	y_m	0.981	0.016	0.011	5.4363	0.986	0.014	0.010	5.475
Multivariate Regression	y_m	0.855	0.040	0.030	15.016	0.846	0.049	0.038	19.735

4.5.2. Multivariate Regression

The equation obtained after the regression analysis on the training set to obtain the predicted test outputs is:

$$y_m = -0.13746 + (0.007027 * Fr) + (0.003153 * \text{Angle}) \quad (34)$$

It can be deduced from Table 6 that training and test sets show good similarity with each other, which proves that the data is not over-fitted and that statistical parameters' values are lower as compared to ANFIS and hybrid ANFIS models. The gap between the predicted test output and observed test output is visible by the percentage deviation of 19.735%.

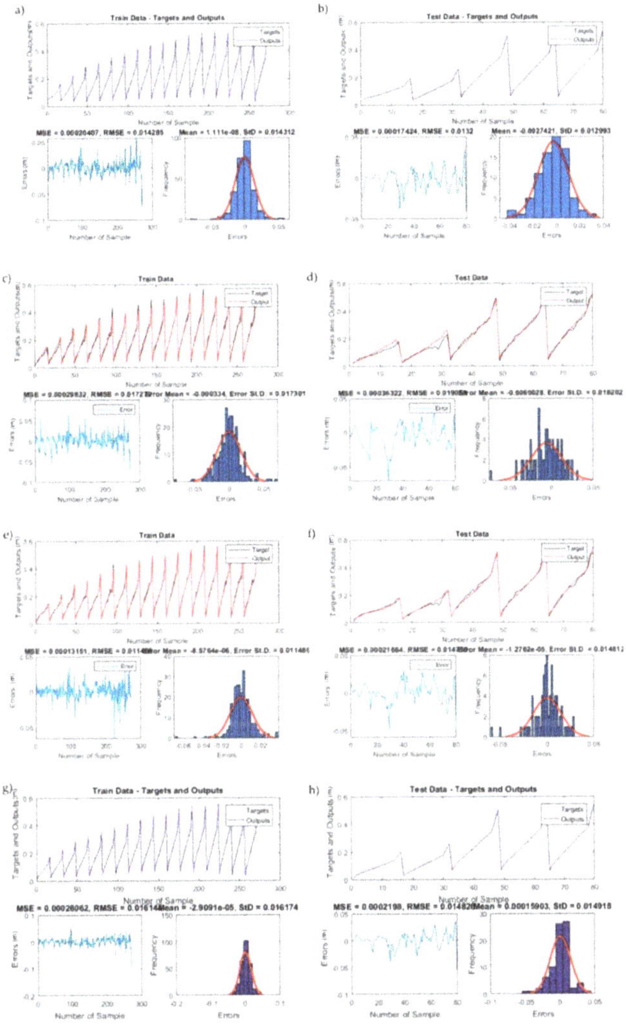

Figure 11. As in Figure 9 for y_m. ANFIS Model (**a**) targets and outputs, RMSE, MSE values and frequency vs. errors graphs for train data set (**b**) targets and outputs, RMSE, MSE values and frequency vs. errors graphs for test data set. ANFIS-GA: (**c**) targets and outputs, RMSE, MSE values and frequency vs. errors graphs for train data set (**d**) targets and outputs, RMSE, MSE values, and frequency vs. errors graphs for test data set. ANFIS-PSO: (**e**) targets and outputs, RMSE, MSE values and frequency vs. errors for train data set (**f**) targets and outputs, RMSE, MSE values, and frequency vs. errors for test data set. ANFIS-FFA: (**g**) targets and outputs, RMSE, MSE values and frequency vs. errors for train data set (**h**) targets and outputs, RMSE, MSE values, and frequency vs. errors for test data set.

5. Discussion

In this paper, the ANFIS model was incorporated with three different algorithms, namely, the Genetic Algorithm, Particle Swarm Optimization, and Firefly Algorithm. Also, to check the linearity of the data, a multivariate linear regression (MLR) was conducted. It was found that the coefficient of determination was too low for MLR, and root mean squared error (RMSE) was too high compared to ANFIS and hybrid ANFIS models. Four statistical indices were measured to determine the efficient model, and they were the coefficient of determination, root mean squared error, mean absolute error, and percentage deviation. Furthermore, to determine the reliable model, the statistical indices of the test set were compared for all the models. It could be seen that, for all the ANFIS-type models, over-fitting was not observed, which showed the models were trained well.

For different outputs, different models showed accuracy. Hence, to evaluate the overall model performance, the average of all the statistical parameters needs to be calculated. It could be seen from Table 7 that ANFIS-PSO and ANFIS-FFA generated the same value for R^2, i.e., 0.980. Also, the RMSE values of these models are lower than the rest of the models, i.e., 0.231 and 0.257, respectively. It could also be seen that the training data of ANFIS-PSO is better compared to ANFIS-FFA as the training set of ANFIS-PSO's R^2 value was 0.983. The model, which showed the poor performance, was MLR as it had the lowest R^2 value, i.e., 0.733, and highest RMSE value, i.e., 1.090. The present study will be helpful in coastal research worldwide it is one of the first studies on artificial intelligence models in negatively buoyant jets for predicting effluent discharges in less computational time as the computational fluid dynamic model had taken approximately three and half days for simulation compared to the ANFIS-type models, which had taken ten to fifteen minutes.

Table 7. Overall performance evaluation for all the test data outputs.

Model	Training Set				Test Set			
	R^2	RMSE	MAE	δ%	R^2	RMSE	MAE	δ%
ANFIS	0.960	0.388	0.292	10.239	0.965	0.374	0.288	10.083
ANFIS-GA	0.968	0.323	0.255	9.625	0.966	0.350	0.257	9.618
ANFIS-PSO	0.983	0.234	0.155	6.349	0.980	0.231	0.159	7.073
ANFIS-FFA	0.978	0.256	0.164	7.065	0.980	0.257	0.166	6.908
Multivariate Regression	0.744	1.051	0.732	26.721	0.733	1.090	0.762	26.608

Additionally, as AI techniques are widely used in the majority of sectors worldwide, it would be interesting to see their implementation in coastal studies. Most of the previous studies have been conducted using computational fluid dynamics models, and this is one of the first studies to have implemented artificial intelligence models in negatively buoyant jets. The study is also an extension of the Kheirkhah Gildeh et al. (2015) [16] study and it can be seen that the ANFIS-type models have performed as well as the CFD model, with less computational time. Although, as mentioned earlier, the present study is one of the first studies using AI models in coastal systems, there are some studies conducted on different coastal topics such as the Bonakdari and Zaji (2018) [24] study on the side weir discharge coefficient, which was modeled using ANFIS-type models. Yan and Mohammadian (2019) [26] studied the Multi-Gene Genetic Programming (MGGP) method for vertical buoyant jets to determine dilution properties. The approach, MGGP in this study, was efficient and accurate, hence, it can also be used to examine the present study on negatively buoyant jets for future work. Furthermore, in future work, several different algorithms such as Differential Evolution (DE), Ant Colony (ACO), Cuckoo Optimization Algorithm (COA) could also be merged with ANFIS to broaden the area of negatively buoyant jet study.

Furthermore, parameters such as temperature, sloping bed, density current, rosette diffuser, effects of stratification on various types of discharges with or without current, effects of secondary flow, various types of jets under wave effect, or current effect can also be considered. Other statistical parameters such as Scatter Index, BIAS, Nash, VAF can also be checked for performance evaluation. Along with this, positive and negative jets can be examined for crossflow using various AI methods such as ANFIS type procedures.

As the errors and performance results in Table 7 show, hybrid models such as ANFIS-PSO and ANFIS-FFA, in comparison with ANFIS, present lower error estimates and overall, more accurate solutions. However, the hybrid models have a higher computational cost. Therefore, the application of these models depends on the desired accuracy and the availability of computing systems. In other words, there is no single answer as to which model is to be selected.

It should be noted that this project did not attempt to train the model and validate it using real-time data, which is the subject of a subsequent study, in order to extend the use of the model to more practical applications.

6. Conclusions

ANFIS model alone and different hybrid models such as ANFIS-GA, ANFIS-PSO, and ANFIS-FFA along with multivariate regression were investigated for the prediction of industrial outfall discharges. In this paper, negatively buoyant jets are focused. Proper outfall design is one of the important factors to mitigate the environmental effects of effluent discharge from desalination plants. With prior knowledge of effluents' discharge characteristics, it is easier to implement proper solutions. Though most of the studies have been conducted using experimental and numerical methods, in this study, the ANFIS type models are examined for lesser computational time and to avoid the cost of an experimental setup. The results showed that ANFIS and hybrid models were trained properly, which could be seen from targets and output graphs in Figures 2–11 as they almost coincided with each other. However, the results showed that the multi-variate regression model was not successful in interpreting the relationship between independent and dependent variables, especially due to the non-linearity of the data. Hence, ANFIS and hybrid models are suitable choices to predict the dilution characteristics of outfall discharges. To determine the efficient model to predict all the outputs mentioned in the previous section, it can be deduced from Table 7, which is generated by averaging all the outputs that ANFIS-PSO has the highest R^2 value, lowest RMSE, and lowest MAE. The other model which showed the same R^2 value to ANFIS-PSO is ANFIS-FFA, however, its RMSE value and MAE value are 0.257 and 0.166, respectively, which are a little higher than ANFIS-PSO. Furthermore, the difference observed between ANFIS-PSO and ANFIS-FFA was in their computational time, which was more for ANFIS-FFA at around 3.5 h compared to ANFIS-PSO, which was 10–15 min. Although, the computational time for all the ANFIS-type models was much less compared to the CFD model, which was approximately around 3.5 days. Overall, these two models, ANFIS-FFA and ANFIS-PAO, can be considered suitable for predicting the effluent discharge, although ANFIS-PSO would be preferable when one considers computational cost as a deciding factor.

The results presented in Table 7 demonstrate that hybrid models (especially ANFIS-PSO and ANFIS-FSA) overall, provide better results with lower errors than ANFIS. However, if one is looking for a quicker answer, ANFIS alone provides results within an acceptable margin of error and can be used.

The use of real-time data for training and validation of the model can be examined in a future study to enhance the practical applicability of the model.

Author Contributions: Conceptualization, A.J., A.M., H.B. and M.S.; Data curation, A.J. and I.B.T.; Formal analysis, A.J.; Funding acquisition, A.M., H.B. and M.S.; Investigation, A.J. and I.B.T.; Methodology, A.J.; Project administration, A.J., A.M. and M.S.; Resources, A.J., A.M., H.B. and M.S.; Software, A.J., A.M. and H.B.; Supervision, A.M., H.B. and M.S.; Validation, A.J. and I.B.T.; Visualization, A.J. and I.B.T.; Writing—original draft, A.J.; Writing—review and editing, A.M., H.B., M.S. and I.B.T. All authors have read and agreed to the published version of the manuscript.

Funding: The research of A.M., H.B., and M.S. was supported by the Natural Sciences and Engineering Research Council of Canada (NSERC).

Data Availability Statement: Additional data will be available upon request.

Conflicts of Interest: The authors declare no conflict of interest.

References

1. Purnama, A.; Shao, D. Modeling brine discharge dispersion from two adjacent desalination outfalls in coastal waters. *Desalination* **2015**, *362*, 68–73. [CrossRef]
2. Alameddine, I.; El-Fadel, M. Brine discharge from desalination plants: A modeling approach to an optimized outfall design. *Desalination* **2007**, *214*, 241–260. [CrossRef]
3. Termeh, S.V.R.; Khosravi, K.; Sartaj, M.; Keesstra, S.D.; Tsai, F.T.C.; Dijksma, R.; Pham, B.T. Optimization of an adaptive neuro-fuzzy inference system for groundwater potential mapping. *Hydrogeol. J.* **2019**, *27*, 2511–2534. [CrossRef]
4. Bleninger, T.; Jirka, G.H. Modelling and environmentally sound management of brine discharges from desalination plants. *Desalination* **2008**, *221*, 585–597. [CrossRef]
5. Christodoulou, G.C.; Papakonstantis, I.G.; Nikiforakis, I.K. Desalination brine disposal by means of negatively buoyant jets. *Desalin. Water Treat.* **2015**, *53*, 3208–3213. [CrossRef]
6. Malcangio, D.; Petrillo, A.F. Modeling of brine outfall at the planning stage of desalination plants. *Desalination* **2010**, *254*, 114–125. [CrossRef]
7. Marti, C.L.; Antenucci, J.P.; Luketina, D.; Okely, P.; Imberger, J. Near-field dilution characteristics of a negatively buoyant hypersaline jet generated by a desalination plant. *J. Hydraul. Eng.* **2010**, *137*, 57–65. [CrossRef]
8. Papakonstantis, I.G.; Christodoulou, G.C.; Papanicolaou, P.N. Inclined negatively buoyant jets 1: Geometrical characteristics. *J. Hydraul. Res.* **2011**, *49*, 3–12. [CrossRef]
9. Papakonstantis, I.G.; Christodoulou, G.C.; Papanicolaou, P.N. Inclined negatively buoyant jets 2: Concentration measurements. *J. Hydraul. Res.* **2011**, *49*, 13–22. [CrossRef]
10. Zhang, S.; Jiang, B.; Law, A.W.K.; Zhao, B. Large eddy simulations of 45 inclined dense jets. *Environ. Fluid Mech.* **2016**, *16*, 101–121. [CrossRef]
11. Shao, D.; Law, A.W.K. Mixing and boundary interactions of 30 and 45 inclined dense jets. *Environ. Fluid Mech.* **2010**, *10*, 521–553. [CrossRef]
12. Oliver, C.J.; Davidson, M.J.; Nokes, R.I. *k-ε* Predictions of the initial mixing of desalination discharges. *Environ. Fluid Mech.* **2008**, *8*, 617. [CrossRef]
13. Palomar, P.; Lara, J.L.; Losada, I.J. Near field brine discharge modeling part 2: Validation of commercial tools. *Desalination* **2012**, *290*, 28–42. [CrossRef]
14. Kikkert, G.A.; Davidson, M.J.; Nokes, R.I. Inclined negatively buoyant discharges. *J. Hydraul. Eng.* **2007**, *133*, 545–554. [CrossRef]
15. Jirka, G.H. Improved discharge configurations for brine effluents from desalination plants. *J. Hydraul. Eng.* **2008**, *134*, 116–120. [CrossRef]
16. Kheirkhah Gildeh, H.; Mohammadian, A.; Nistor, I.; Qiblawey, H. Numerical modeling of 30° and 45° inclined dense turbulent jets in stationary ambient. *Environ. Fluid Mech.* **2015**, *15*, 537–562. [CrossRef]
17. Neshat, M.; Adeli, A.; Sepidnam, G.; Sargolzaei, M. Predication of concrete mix design using adaptive neural fuzzy inference systems and fuzzy inference systems. *Int. J. Adv. Manuf. Technol.* **2012**, *63*, 373–390. [CrossRef]
18. Nadia, A.R.; Isa, N.A.M.; Desa, M.K.M. Efficient single and dual axis solar tracking system controllers based on adaptive neural fuzzy inference system. *J. King Saud Univ. Eng. Sci.* **2020**, *32*, 459–469.
19. Heydari, A.; Majidi Nezhad, M.; Neshat, M.; Garcia, D.A.; Keynia, F.; De Santoli, L.; Tjernberg, L.B. A combined fuzzy GMDH neural network and grey wolf optimization application for wind turbine power production forecasting considering SCADA data. *Energies* **2021**, *14*, 3459. [CrossRef]
20. Pourtousi, M.; Sahu, J.N.; Ganesan, P.; Shamshirband, S.; Redzwan, G. A combination of computational fluid dynamics (CFD) and adaptive neuro-fuzzy system (ANFIS) for prediction of the bubble column hydrodynamics. *Powder Technol.* **2015**, *274*, 466–481. [CrossRef]
21. Taghavifar, H.; Khalilarya, S.; Jafarmadar, S. Adaptive neuro-fuzzy system (ANFIS) based appraisal of accumulated heat from hydrogen-fueled engine. *Int. J. Hydrog. Energy* **2015**, *40*, 8206–8218. [CrossRef]
22. Rezakazemi, M.; Dashti, A.; Asghari, M.; Shirazian, S. H2-selective mixed matrix membranes modeling using ANFIS, PSO-ANFIS, GA-ANFIS. *Int. J. Hydrog. Energy* **2017**, *42*, 15211–15225. [CrossRef]
23. Amirkhani, S.; Nasirivatan, S.; Kasaeian, A.B.; Hajinezhad, A. ANN and ANFIS models to predict the performance of solar chimney power plants. *Renew. Energy* **2015**, *83*, 597–607. [CrossRef]
24. Bonakdari, H.; Zaji, A.H. New type side weir discharge coefficient simulation using three novel hybrid adaptive neuro-fuzzy inference systems. *Appl. Water Sci.* **2018**, *8*, 10. [CrossRef]
25. Shabanian, S.R.; Edrisi, S.; Khoram, F.V. Prediction and optimization of hydrogen yield and energy conversion efficiency in a non-catalytic filtration combustion reactor for jet A and butanol fuels. *Korean J. Chem. Eng.* **2017**, *34*, 2188–2197. [CrossRef]
26. Yan, X.; Mohammadian, A. Multigene Genetic-Programming-Based Models for Initial Dilution of Laterally Confined Vertical Buoyant Jets. *J. Mar. Sci. Eng.* **2019**, *7*, 246. [CrossRef]
27. OpenFOAM Version 2.3.1. Computer Software. Available online: https://openfoam.org/version/2-3-1/ (accessed on 7 January 2022).
28. Greenshields, C.J. *OpenFOAM User Guide*; The OpenFOAM Foundation: London, UK, 2018.
29. Jang, J.S. ANFIS: Adaptive-network-based fuzzy inference system. *IEEE Trans. Syst. Man Cybern.* **1993**, *23*, 665–685. [CrossRef]

30. Ebtehaj, I.; Bonakdari, H. Performance evaluation of adaptive neural fuzzy inference system for sediment transport in sewers. *Water Resour. Manag.* **2014**, *28*, 4765–4779. [CrossRef]
31. Bonakdari, H.; Zaji, A.H.; Shamshirband, S.; Hashim, R.; Petkovic, D. Sensitivity analysis of the discharge coefficient of a modified triangular side weir by adaptive neuro-fuzzy methodology. *Measurement* **2015**, *73*, 74–81. [CrossRef]
32. Holland, J. *Adaptation in Natural and Artificial Systems: An Introductory Analysis with Applications to Biology, Control, and Artificial Intelligence*; MIT Press: Cambridge, MA, USA, 1975.
33. Yarpiz Evolutionary Algorithms. Available online: https://yarpiz.com/477/ypea-yarpiz-evolutionary-algorithms (accessed on 7 January 2022).
34. Kennedy, J.; Eberhart, R. Particle swarm optimization (PSO). In Proceedings of the IEEE International Conference on Neural Networks, Perth, Australia, 27 November–1 December 1995; pp. 1942–1948.
35. Yang, X.S. Firefly algorithm, Levy flights and global optimization. In *Research and Development in Intelligent Systems XXVI*; Springer: London, UK, 2010; pp. 209–218.
36. Yaseen, Z.M.; Ebtehaj, I.; Bonakdari, H.; Deo, R.C.; Mehr, A.D.; Mohtar, W.H.M.W.; Diop, L.; El-Shafie, A.; Singh, V.P. Novel approach for streamflow forecasting using a hybrid ANFIS-FFA model. *J. Hydrol.* **2017**, *554*, 263–276. [CrossRef]
37. Azimi, H.; Bonakdari, H.; Ebtehaj, I.; Michelson, D.G. A combined adaptive neuro-fuzzy inference system–firefly algorithm model for predicting the roller length of a hydraulic jump on a rough channel bed. *Neural Comput. Appl.* **2018**, *29*, 249–258. [CrossRef]
38. Yaghoobi, A.; Bakhshi-Jooybari, M.; Gorji, A.; Baseri, H. Application of adaptive neuro fuzzy inference system and genetic algorithm for pressure path optimization in sheet hydroforming process. *Int. J. Adv. Manuf. Technol.* **2016**, *86*, 2667–2677. [CrossRef]
39. Emadi, D.; Mahfoud, M. Comparison of Artificial Neural Network and Multiple Regression Analysis Techniques in Predicting the Mechanical Properties of A356 Alloy. *Procedia Eng.* **2011**, *10*, 589–594. [CrossRef]
40. Tosun, E.; Aydin, K.; Bilgili, M. Comparison of linear regression and artificial neural network model of a diesel engine fueled with biodiesel-alcohol mixtures. *Alex. Eng. J.* **2016**, *55*, 3081–3089. [CrossRef]
41. Fumo, N.; Biswas, M.R. Regression analysis for prediction of residential energy consumption. *Renew. Sustain. Energy Rev.* **2015**, *47*, 332–343. [CrossRef]

Mathematical and Computational Applications

MDPI

Article

Finding the Conjectured Sequence of Largest Small *n*-Polygons by Numerical Optimization

János D. Pintér [1], Frank J. Kampas [2] and Ignacio Castillo [3,*]

1 Department of Management Science and Information Systems, Rutgers University, 57 US Highway 1, New Brunswick, NJ 08901, USA; jpinter@business.rutgers.edu
2 Physicist at Large Consulting LLC, Bryn Mawr, PA 19010, USA; frank@physicistatlarge.com
3 Lazaridis School of Business and Economics, Wilfrid Laurier University, 75 University Avenue West, Waterloo, ON N2L 3C5, Canada
* Correspondence: icastillo@wlu.ca

Abstract: $LSP(n)$, the largest small polygon with n vertices, is a polygon with a unit diameter that has a maximal of area $A(n)$. It is known that for all odd values $n \geq 3$, $LSP(n)$ is a regular n-polygon; however, this statement is not valid even for values of n. Finding the polygon $LSP(n)$ and $A(n)$ for even values $n \geq 6$ has been a long-standing challenge. In this work, we developed high-precision numerical solution estimates of $A(n)$ for even values $n \geq 4$, using the Mathematica model development environment and the IPOPT local nonlinear optimization solver engine. First, we present a revised (tightened) LSP model that greatly assists in the efficient numerical solution of the model-class considered. This is followed by results for an illustrative sequence of even values of n, up to $n \leq 1000$. Most of the earlier research addressed special cases up to $n \leq 20$, while others obtained numerical optimization results for a range of values from $6 \leq n \leq 100$. The results obtained were used to provide regression model-based estimates of the optimal area sequence $\{A(n)\}$, for even values n of interest, thereby essentially solving the LSP model-class numerically, with demonstrably high precision.

Keywords: nonlinear programming; largest small polygons (LSP); $\{LSP(n)\}$ model-class; optimal area sequence $\{A(n)\}$; revised LSP model; mathematica model development environment; IPOPT solver engine; numerical optimization results and regression model for estimating $\{A(n)\}$

Citation: Pintér, J.D.; Kampas, F.J.; Castillo, I. Finding the Conjectured Sequence of Largest Small *n*-Polygons by Numerical Optimization. *Math. Comput. Appl.* **2022**, *27*, 42. https://doi.org/10.3390/mca27030042

Academic Editor: Oliver Schütze

Received: 16 March 2022
Accepted: 11 May 2022
Published: 16 May 2022

1. Introduction

The diameter of a convex planar polygon is defined as the length of its longest diagonal. The largest small polygon (LSP) with n vertices is defined as a polygon of a unit diameter that has a maximal area. For a given integer $n \geq 3$, we refer to this polygon as $LSP(n)$ with a corresponding area $A(n)$. To illustrate, Figure 1 shows visual representations of our conjectured maximal area $LSP(6)$ and $LSP(18)$.

Nearly a century ago, Reinhardt (1922) [1] proved that for all *odd* values $n \geq 3$, $LSP(n)$ is the regular n-polygon; perhaps surprisingly, the corresponding statement is not valid for *even* values of n. For brevity, here we only refer to the discussion of the low-dimensional cases studied elsewhere. Specifically, Graham (1975) [2] presented an exact solution of the hexagon; Audet et al. 2002 [3] presented the first numerical solutions of the octagon, followed by the exact optimal axially symmetric octagon in Audet et al. (2021) [4]; Henrion and Messine (2013) [5] provided numerical guarantees of global optimality and successfully found the largest small polygons for $n = 10$ and $n = 12$. Mossinghoff (2006) [6] presented some related theoretical background, confirmed earlier best-known results, and gave additional numerical estimates through $n = 20$. In more recent studies, Bingane (2020) [7] gave near-optimal numerical estimates of $A(n)$ with up to $n = 128$, and Pintér (2021) [8] gave globally optimized numerical estimates of $A(n)$ for all even values $6 \leq n \leq 80$, both studies share comparative references to earlier works. These works—and several others cited by both Bingane (2020) [7] and Pintér (2021) [8]—clearly indicate the limitations and varying performance level of the modeling and optimization software packages used earlier.

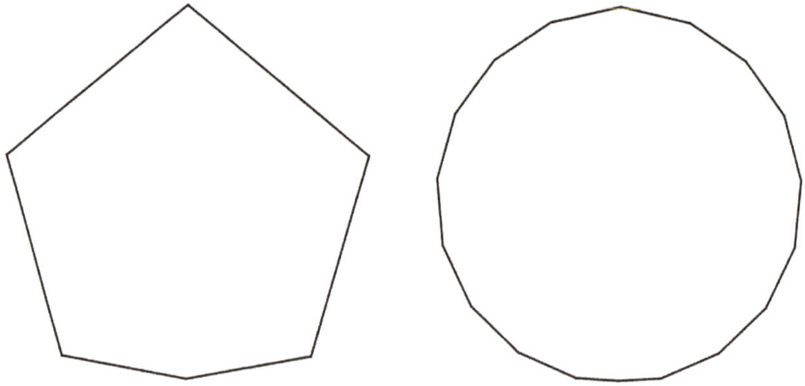

Figure 1. Our conjectured largest small polygons $LSP(6)$ and $LSP(18)$.

Here we present the results of a numerical optimization study aimed at finding conjectured $LSP(n)$ configurations and corresponding values $A(n)$, for a substantial selection of even values covering the range of $4 \leq n \leq 1000$. We propose a revised (tightened) LSP model, then solve model instances using the *Mathematica* model development platform with the callable IPOPT nonlinear optimization solver. Based on the results obtained, we also developed a regression model to estimate $\{A(n)\}$ for all of the even values of n. Since, for large n, the actual calculated optimum estimates $A(n)$ already closely approximate the theoretical limit of $A(\infty) = \pi/4$, our numerical study enables high-precision estimates covering the LSP model-class.

2. Revised LSP Optimization Model

For unambiguity, we consider all $LSP(n)$ instances for even $n \geq 4$ with a fixed position corresponding to the appropriate variants of the instances shown in Figure 1. Following the standard assumptions also postulated by others, each even n-polygon considered here is symmetrical with respect to the diagonal that connects its lowest positioned vertex placed at the origin with its highest vertex. We refer to this configuration as the standard model: it has been used in all topical works referenced by Pintér (2021) [8].

The standard LSP model uses polar coordinates to describe the $LSP(n)$: the vertex i is positioned at the polar radius r_i and at angle θ_i. For unambiguity, we postulate that the vertices $i = 1, \ldots, n-1$ are arranged according to increasing angles θ_i. Placing the last vertex at the origin, we set $r_n = 0$, $\theta_n = \pi$: recall Figure 1 that shows examples of such a configuration. The corresponding standard LSP optimization model is presented next.

Maximize total area of the n-polygon:

$$\max A(n) = \frac{1}{2} \sum_{i=1}^{n-1} r_i r_{i+1} \sin(\theta_{i+1} - \theta_i). \tag{1}$$

Prescribed upper bound for the pairwise distance between vertices i and j:

$$r_i^2 + r_j^2 - 2r_i r_j \cos(\theta_i - \theta_j) \leq 1, \text{ for } 1 \leq i \leq n-2,\ i+1 \leq j \leq n-1. \tag{2}$$

Vertex angle ordering relations:

$$\theta_{i+1} - \theta_i \geq 0, \text{ for } 1 \leq i \leq n-2. \tag{3}$$

Decision variable bounds, and the two fixed variable settings:

$$0 \leq \theta_i \leq \pi \text{ and } 0 \leq r_i \leq 1, \text{ for } 1 \leq i \leq n-1;\ r_n = 0,\ \theta_n = \pi. \quad (4)$$

Based on the structure of the LSP configurations found in all of the earlier studies, next we revise this standard model, by adding the relations shown below.

(i) We postulate bounds on the angle differences, for even n:

$$\theta_{i+1} - \theta_i \geq \frac{\pi}{n}, \text{ for } 1 \leq i \leq n-2, \quad (5)$$

$$\theta_{n/2} = \frac{\pi}{2}. \quad (6)$$

(ii) We postulate the symmetry of the LSP configuration to be found, for even n:

$$r_{n/2+i-1} = r_{n/2-i+1}, \text{ for } 2 \leq i \leq \frac{n}{2}, \quad (7)$$

$$\theta_{n/2+i-1} = \pi - \theta_{n/2-i+1}, \text{ for } 2 \leq i \leq \frac{n}{2}. \quad (8)$$

To illustrate these added constraints, we refer again to Figure 1; further examples will be presented later on. Our preliminary experimentation demonstrated that the symmetry postulates (7)–(8) for even n, despite reducing the number of decision variables, were not useful within our numerical optimization study. However, the bound postulates on the angle differences were useful by effectively tightening the LSP model. As it turns out, the new constraints are essential to guarantee the performance of the *local* solver IPOPT in numerically solving the *global* optimization problem (1)–(4), with the added considerations (5)–(6) for even n.

Observe the potential numerical challenge implied by the nonconvex objective function (1) and constraints (2): the number of these constraints increases *quadratically* as a function of n. While the standard $LSP(n)$ model instances have a unique globally optimal solution, the number of local optima increases with n. Many of the local optima are close in quality to the (numerically estimated, hence only approximately known) global optimum. These features make the $\{LSP(n)\}$ problem-class hard to solve, similarly to many other of the object configuration design and packing problems arising e.g., in computational mathematics, physics, chemistry, biology, as well as across a range of engineering applications.

3. Numerical Results for Even Values $4 \leq n \leq 1000$

The study summarized here was conducted on a laptop PC with the following specifications: Intel Core i7-7700 CPU @ 3.6 GHz (x-64 processor), 16.0 GB RAM, running under the Windows 10 Pro (64-bit) operating system.

To formulate directly scalable LSP model versions, we use the *Mathematica* model development environment (Wolfram Research, 2022a [9]). To handle these models numerically, we use the IPOPT local nonlinear optimization solver engine (Wächter and Laird, 2022 [10]) linked to *Mathematica* (Wolfram Research, 2022b [11]). The result analysis and visualization was also conducted in *Mathematica*.

Since IPOPT is a local scope solver, it requires an initial solution "guess:" hence, it greatly benefits from a good choice of that estimate. Considering the postulated structure of the LSP configurations to be found, for a given n our initial angle settings are chosen as $\theta_i = i(\pi/n)$, for $1 \leq i \leq n$, together with the initial polar radius settings $r_i = 1$ for $1 \leq i \leq n-1$; $r_n = 0$

The numerical results obtained by using *Mathematica* and IPOPT are summarized below. Detailed $LSP(n)$ configurations (listing all decision variable values found and all constraints) can be optionally reported by our *Mathematica* code. All of the optimization results, with related analysis and visualization, are directly cited from the corresponding

Mathematica (notebook) documents: thereby, our study is completely reproducible. The *Mathematica* notebook is available upon request from the authors.

Table 1 summarizes our results for the selected even values of $4 \leq n \leq 1000$. Since the solution times become rather substantial as n increases, we report results only for a representative selection of even values; however, in principle we could handle *all* instances of $LSP(n)$, provably at least up to $n \leq 1000$. For example (see Table 1), the runtime for $n = 10$ is only 0.06 s; for $n = 100$ it is still just 9.44 s; but for $n = 1000$ it becomes 5417.51 s. We did not conduct systematic tests to find the largest possible numerical instance that we could handle using *Mathematica* with IPOPT, noting that all of the modeling systems and optimization engines have their limitations, also depending on the hardware platform and other circumstances.

Table 1. *Mathematica*-IPOPT numerical results for a selection of even values of n.

n	Decision Variables	Constraints	Runtime (Seconds)	Objective Function	Maximum Violation
4	6	5	0.03	0.500000	9.9636×10^{-9}
6	10	14	0.03	0.674981 †	9.9432×10^{-9}
8	14	27	0.05	0.726868 †	9.9236×10^{-9}
10	18	44	0.06	0.749137 †	9.9046×10^{-9}
12	22	65	0.08	0.760730 †	9.8855×10^{-9}
14	26	90	0.11	0.767531 †	9.8663×10^{-9}
16	30	119	0.15	0.771861 †	9.8472×10^{-9}
18	34	152	0.17	0.774788 †	9.8296×10^{-9}
20	38	189	0.23	0.776859 †	9.8101×10^{-9}
24	46	275	0.32	0.779524 †	9.7801×10^{-9}
28	54	377	0.45	0.781111 †	9.7647×10^{-9}
32	62	495	0.68	0.782133 †	9.8456×10^{-9}
36	70	629	0.82	0.782828 †	9.6522×10^{-9}
40	78	779	1.05	0.783323 †	9.6132×10^{-9}
44	86	945	1.34	0.783687 †	9.5741×10^{-9}
48	94	1127	1.53	0.783964 †	9.5352×10^{-9}
52	102	1325	1.85	0.784178 †	9.4967×10^{-9}
56	110	1539	3.04	0.784252 *	9.6754×10^{-9}
60	118	1769	3.34	0.784408 *	9.6548×10^{-9}
70	138	2414	3.92	0.784729 †	9.3199×10^{-9}
80	158	3159	4.98	0.784886 †	9.2227×10^{-9}
90	178	4004	7.59	0.784994 †	9.1242×10^{-9}
100	198	4949	9.44	0.785072 †	9.0264×10^{-9}
110	218	5994	11.18	0.785129 †	8.9291×10^{-9}
120	238	7139	14.21	0.785172 †	8.8306×10^{-9}
130	258	8384	17.52	0.785205	8.7330×10^{-9}
140	278	9729	21.39	0.785232	8.7013×10^{-9}
150	298	11,174	25.11	0.785254	8.8048×10^{-9}
160	318	12,719	29.16	0.785271	8.4389×10^{-9}
180	358	16,109	52.43	0.785298	8.2424×10^{-9}
200	398	19,899	51.31	0.785317	8.0460×10^{-9}
220	438	24,089	64.43	0.785331	7.8491×10^{-9}
240	478	28,679	81.60	0.785342	7.6515×10^{-9}
260	518	33,669	97.73	0.785350	7.6285×10^{-9}
280	558	39,059	118.77	0.785357	7.2925×10^{-9}
300	598	44,849	142.19	0.785362	7.0879×10^{-9}
320	638	51,039	170.00	0.785367	6.8695×10^{-9}
340	678	57,629	202.27	0.785370	6.6526×10^{-9}
360	718	64,619	235.11	0.785373	6.4506×10^{-9}
380	758	72,009	269.55	0.785376	6.2473×10^{-9}
400	798	79,799	316.30	0.785378	6.0432×10^{-9}
420	838	87,989	363.92	0.785380	5.8990×10^{-9}
440	878	96,579	393.75	0.785381	5.6483×10^{-9}

Table 1. *Cont.*

n	Decision Variables	Constraints	Runtime (Seconds)	Objective Function	Maximum Violation
460	918	105,569	464.32	0.785383	5.4201×10^{-9}
480	958	114,959	514.01	0.785384	5.2503×10^{-9}
500	998	124,749	577.96	0.785385	5.0971×10^{-9}
550	1098	150,974	739.51	0.785387	4.4905×10^{-9}
600	1198	179,699	948.26	0.785389	4.0205×10^{-9}
650	1298	210,924	1228.83	0.785391	3.4442×10^{-9}
700	1398	244,649	1460.31	0.785392	2.9080×10^{-9}
750	1498	280,874	1857.58	0.785392	2.3883×10^{-9}
800	1598	319,599	2342.02	0.785393	1.8715×10^{-9}
850	1698	360,824	3177.43	0.785394	1.3518×10^{-9}
900	1798	404,549	3846.06	0.785394	8.2861×10^{-10}
950	1898	450,774	3721.68	0.785395 *	3.0161×10^{-10}
1000	1998	499,499	5417.51	0.785395 *	0.0000×10^{0}

* Instances for which $A(n) \leq \underline{A}_n$*. † Instances $6 \leq n \leq 120$ for which $A(n) \geq A_n$* as reported in Bingane (2020) [7].

The legend used is self-explanatory, "Maximum violation" refers to the maximal constraint violation at the numerical optimal solution. One can verify the *linear* increase in the number of decision variables and the rapid *quadratic* increase in the number of constraints as a function of n (recall the LSP model). The model instance for $n = 1000$ has almost two thousand decision variables and nearly half a million constraints.

For even $n \geq 6$, it was shown in Mossinghoff (2006) [6] and Bingane (2022) [12] that the best known *general* lower bound is given by the diameter graph of an optimal n-polygon that has a cycle of length $n-1$ plus one additional edge from the remaining vertex. As such,

$$A(n) > \underline{A}_n = \frac{n-1}{2}\left(\sin\frac{\pi}{n-1} - \tan\frac{\pi}{2n-2}\right) + \sin\frac{\pi}{2n-2} - \frac{1}{2}\sin\frac{\pi}{n-1}. \tag{9}$$

Our results satisfy this general lower bound except for $A(56)$, $A(60)$, $A(950)$, and $A(1000)$. See Table 1. Indeed, equation (9) provides a very good general lower bound and it could be used as an additional constraint in the LSP optimization model. We note that Mossinghoff (2006) [6] and Bingane (2022) [12] obtained better asymptotic lower bounds by explicit instance constructions, which are not generally applicable as equation (9). In Table 1, we also indicate the instances $6 \leq n \leq 120$ for which $A(n) \geq A_n^*$ (the optimal values of the maximal area problem for even n) as reported in Bingane (2020) [7].

Although Table 1 shows the optimized $A(n)$ values with only six digits after the decimal point, the reported precision in our detailed numerical tests (within *Mathematica*) is set to ten digits after the decimal point. To illustrate, our estimate for $A(1000)$ approximately equals 0.7853949284. Using such higher precision is in line with the required constraint satisfaction level, all in the order of 10^{-9} as shown in the table. The preset 10-digit precision also supports the in-depth comparisons with results obtained earlier. Specifically, our results are in close agreement with or surpass all of the best *numerical* results reported earlier, including Mossinghoff (2006) [6], Bingane (2020) [7], and Pintér (2021) [8], with reported comparisons to all earlier topical works known to us. Our modeling and optimization approach enables solving $LSP(n)$ instances for significantly higher values of n than previously achieved by other researchers using a variety of modeling platforms and solver engines.

Figures 2 and 3 display the solutions found for a selection of even sequences of n, showing all of the pairwise vertex connections. As expected, the configurations found quickly approach the circle, as n increases.

Figure 4 summarizes the difference between the area $A(n)$ of the optimized polygon and $\pi/4$ as a function of n, on a loglog-plot scale. Our calculations reveal a small, but non-negligible difference: the numerically estimated slope of the plot for an even number of sides is approximately -2.04618.

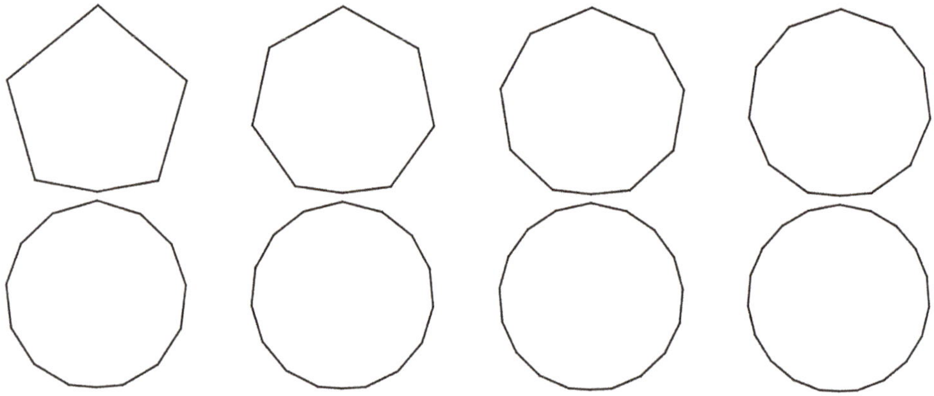

Figure 2. Conjectured largest small polygons $LSP(n)$ for $n = 6, 8, 10, 12, 14, 16, 18, 20$.

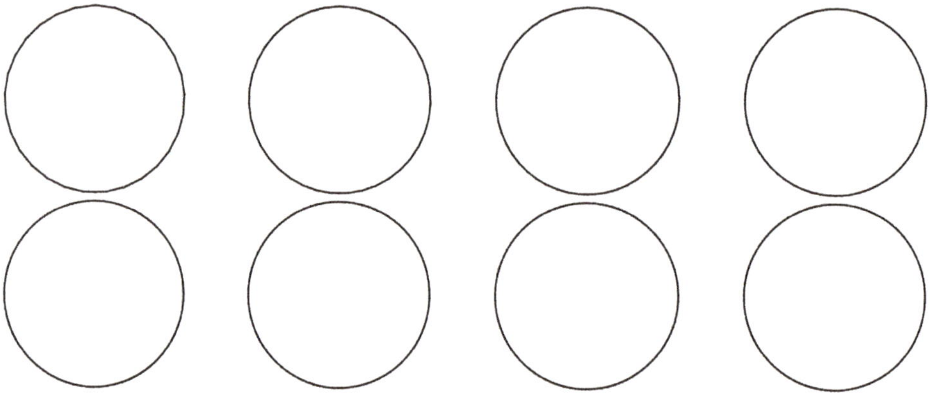

Figure 3. Conjectured largest small polygons $LSP(n)$ for $n = 30, 40, 50, 60, 70, 80, 90, 100$.

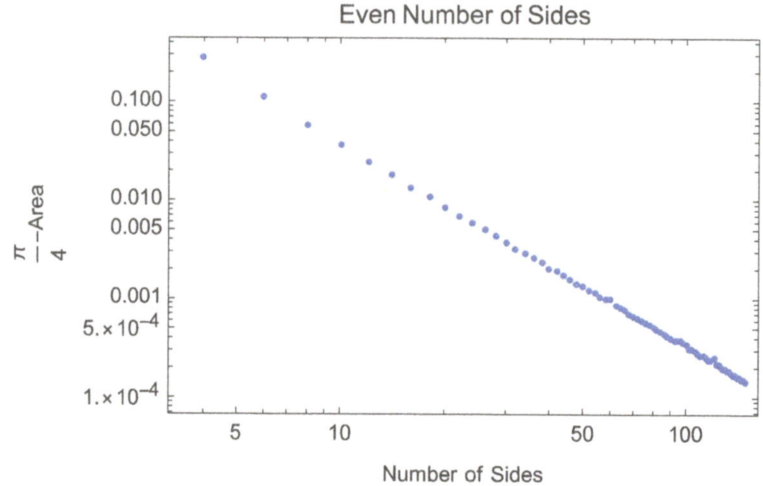

Figure 4. Difference between the area $A(n)$ of the optimized polygon and $\pi/4$, for selected even values $4 \leq n \leq 150$.

We conclude the presentation of numerical results by emphasizing that the tightened LSP model offers advantages over the standard model. Specifically, IPOPT performs well on the tightened model, but it exhibits inferior performance on the standard model for values $6 \leq n \leq 80$, as observed by Pintér (2021) [8]. Figure 5 illustrates the superior IPOPT performance on tightened LSP model-instances up to $n = 300$, when compared to Figure 6 (standard LSP model with the same fixed starting solution as used in the tightened model) and Figure 7 (standard LSP model with n random starting solutions). In the latter case, the solver runtimes also become longer; therefore, we conducted experiments only up to $n = 100$.

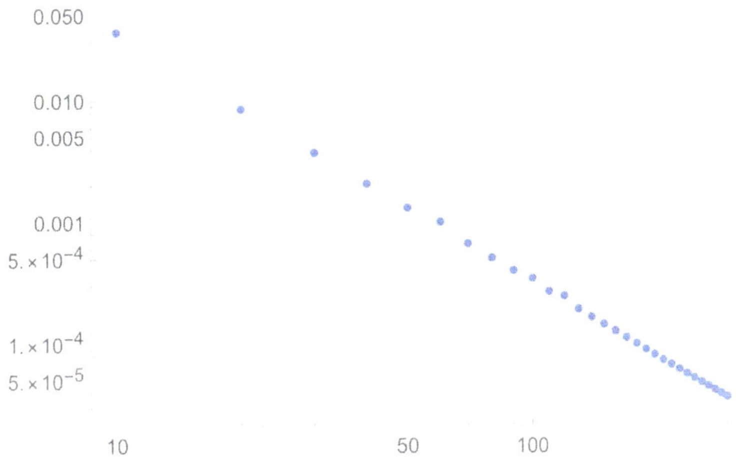

Figure 5. IPOPT performance on tightened LSP model-instances up to $n = 300$. Difference between the area $A(n)$ of the optimized polygon and $\pi/4$.

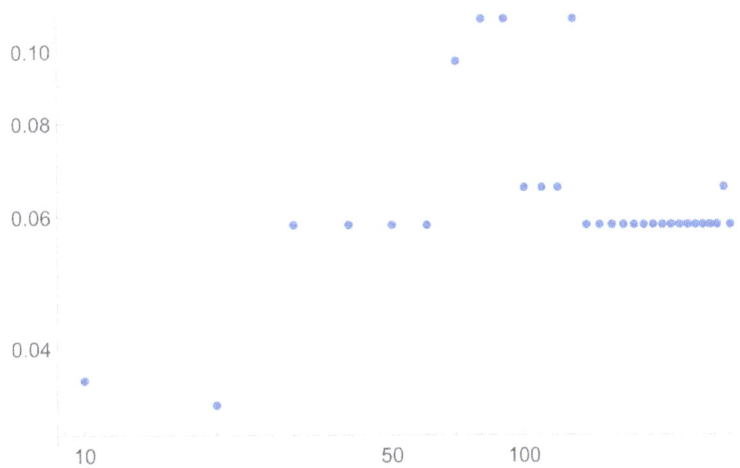

Figure 6. IPOPT performance on standard LSP model-instances, with the proposed starting solution, up to $n = 300$. Difference between the area $A(n)$ of the optimized polygon and $\pi/4$.

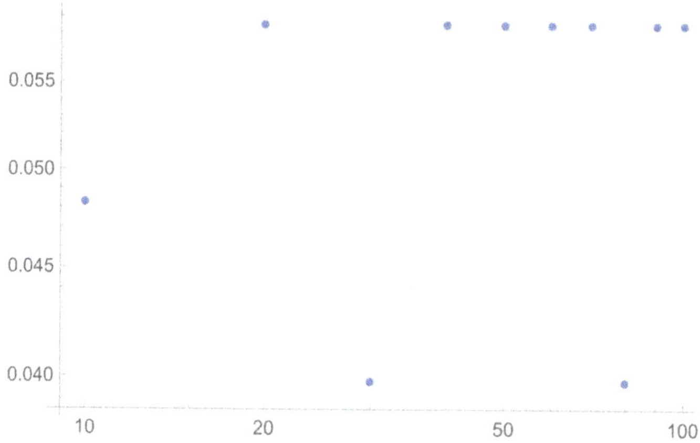

Figure 7. IPOPT performance on standard LSP model-instances, with random starting solutions up to $n = 100$. Difference between the area $A(n)$ of the optimized polygon and $\pi/4$.

4. *LSP(n)* Regression Model for Even n

As it is known, and illustrated by Figures 2 and 3, for large n the optimal $LSP(n)$ configuration approaches the circle with unit diameter; hence the corresponding area limit is $A(\infty) = \pi/4 \sim 0.7853981634$. Comparing this limit value to our optimum estimate obtained for $A(1000) \sim 0.7853949284$, the ratio $A(1000)/(\pi/4)$ approximately equals 0.9999958811. Hence, our $A(1000)$ estimate already leads to a fairly close approximation of the limit value.

Based on this observation and using our numerical results, we developed the following regression model for even values of n.

$$A(n) \sim \frac{\pi}{4} - \frac{5\pi^3}{48n^2} - 3.530190\left(\frac{1}{n^3}\right) - 2.391836\left(\frac{1}{n^4}\right) - 19.489487\left(\frac{1}{n^5}\right). \quad (10)$$

This regression model follows the form outlined in Foster and Szabo (2007) [13] and we received p-values (observed significance levels) well below 0.000001 for all coefficients. This finding indicates that we have very strong statistical evidence suggesting that the regression coefficients are all different from zero. Figure 8 depicts the predicted results using model (10).

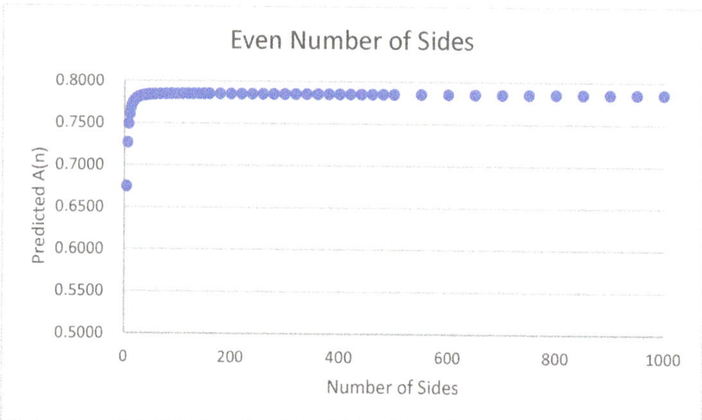

Figure 8. The nonlinear regression model (10) predicting $A(n)$ shown for even $4 \leq n \leq 1000$.

Since the two sequences closely overlap, the preceding figure depicting observed vs. predicted $A(n)$ results become more useful when we zoom in and reduce the ranges considerably. In Figure 9, we display observed vs. predicted $A(n)$ results for $24 \leq n \leq 100$ and $0.779 \leq A(n) \leq 0.785$. The observed vs. predicted sequences appear visually different from each other only with high levels of magnification.

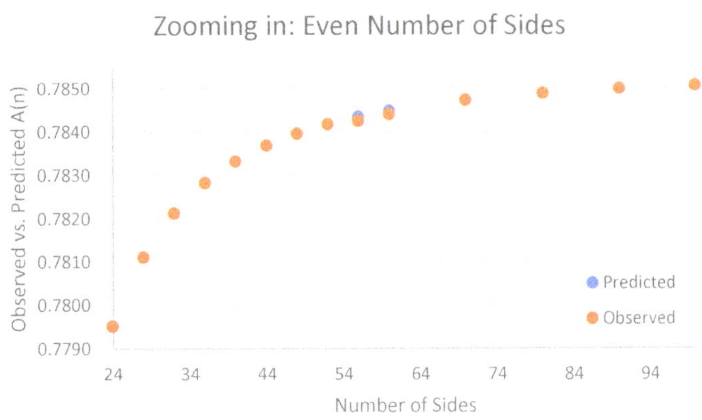

Figure 9. Zooming in: observed vs. predicted $A(n)$ results applying the regression model (10) for even $24 \leq n \leq 100$.

The regression model (10) can be used to directly estimate the selected values from the entire sequence $\{A(n)\}$, including larger values of n which have not been studied earlier and may be out of the range of current optimization solver capabilities. For example,

$$A(2000) \sim 0.7853973555, \quad A(10000) \sim 0.7853981311.$$

It is instructive to compare these estimates to $A(\infty) = \pi/4 \sim 0.7853981634$. Earlier numerical examples, with a different regression model based on results for even $6 \leq n \leq 80$, are presented in Pintér (2021) [8].

5. Conclusions

Our study addresses the problem of numerically finding the sequence of the largest small n-polygons $LSP(n)$ with a unit diameter and maximal area of $A(n)$, in principle aiming for all of the even values of $n \geq 4$. This long-standing mathematical "puzzle" leads to an interesting class of nonlinear (global) optimization problems. We proposed a tightened LSP model and demonstrated its numerical advantages compared to the standard model. Using the *Mathematica* modeling environment with the IPOPT solver option, and our new initial solution estimate, we can find numerical solutions efficiently for a range of even values of n, up to $n \leq 1000$. Our results compare well to all the best results reported earlier for significantly lower values of n. We also propose a regression model that enables the direct estimation of the optimal area sequence $\{A(n)\}$, for even values of n.

The LSP problem-class is one of those entertaining "puzzles" that can be described in a few words yet lead to surprisingly hard theoretical and numerical challenges. Therefore, this model-class—similarly to many other scientifically important configuration design models—can also be used in software benchmarking tests. We think that such problems serve as a significant addendum to the collection of (well-frequented, and often much simpler) unconstrained or box-constrained test problems. For further examples of increasingly hard-to-solve object configuration models, we refer to some of our studies: consult, e.g., Castillo et al. (2008) [14], Pintér et al. (2017) [15], Kampas et al. (2019) [16], Kampas et al. (2020) [17].

Author Contributions: All three authors (J.D.P., F.J.K. and I.C.) contributed to the development of this research study, including conceptualization, methodology, software, numerical tests, and manuscript preparation. All authors have read and agreed to the published version of the manuscript.

Funding: This research received no external funding.

Conflicts of Interest: The authors declare no conflict of interest.

References

1. Reinhardt, K. Extremale polygone gegebenen durchmessers. *Jahresber. Dtsch. Math. Ver.* **1922**, *31*, 251–270.
2. Graham, R.L. The largest small hexagon. *J. Comb. Theory Ser. A* **1975**, *18*, 165–170. [CrossRef]
3. Audet, C.; Hansen, P.; Messine, F.; Xiong, J. The largest small octagon. *J. Comb. Theory Ser. A* **2002**, *98*, 46–59. [CrossRef]
4. Audet, C.; Hansen, P.; Svrtan, D. Using symbolic calculations to determine largest small polygons. *J. Glob. Optim.* **2021**, *81*, 261–268. [CrossRef]
5. Henrion, D.; Messine, F. Finding largest small polygons with GloptiPoly. *J. Glob. Optim.* **2013**, *56*, 1017–1028. [CrossRef]
6. Mossinghoff, M.J. Isodiametric problems for polygons. *Discret. Comput. Geom.* **2006**, *36*, 363–379. [CrossRef]
7. Bingane, C. Largest small polygons: A sequential convex optimization approach. *arXiv* **2020**, arXiv:2009.07893. Available online: https://arxiv.org/abs/2009.07893 (accessed on 14 January 2022).
8. Pintér, J.D. Largest small n-polygons: Numerical optimum estimates for n $\geq$ 6. In *Recent Developments in Numerical Analysis and Optimization (NAO-V, Muscat, Oman, January 2020)*; Al-Baali, M., Grandinetti, L., Purnama, A., Eds.; Springer Nature: New York, NY, USA, 2021; pp. 231–247.
9. Wolfram Research. *Mathematica, Current Release 13.0 (as of January 2022)*; Wolfram Research: Champaign, IL, USA, 2022.
10. Wächter, A.; Laird, C. IPOPT (Interior Point Optimizer), an Open Source Software Package for Large-Scale Nonlinear Optimization. Available online: https://github.com/coin-or/Ipopt (accessed on 14 January 2022).
11. Wolfram Research (2022b), Optimizing with IPOPT. Available online: https://reference.wolfram.com/language/IPOPTLink/tutorial/OptimizingWithIPOPT.html (accessed on 14 January 2022).
12. Bingane, C. Tight bounds on the maximal area of small polygons: Improved Mossinghoff polygons. *Discret. Comput. Geom.* **2022**. [CrossRef]
13. Foster, J.; Szabo, T. Diameter graphs of polygons and the proof of a conjecture of Graham. *J. Comb. Theory Ser. A* **2007**, *114*, 1515–1525. [CrossRef]
14. Castillo, I.; Kampas, F.J.; Pintér, J.D. Solving circle packing problems by global optimization: Numerical results and industrial applications. *Eur. J. Oper. Res.* **2008**, *191*, 786–802. [CrossRef]
15. Pintér, J.D.; Kampas, F.J.; Castillo, I. Globally optimized packings of non-uniform size spheres in R*d*: A computational study. *Optim. Lett.* **2017**, *12*, 585–613. [CrossRef]
16. Kampas, F.J.; Castillo, I.; Pintér, J.D. Optimized ellipse packings in regular polygons. *Optim. Lett.* **2019**, *13*, 1583–1613. [CrossRef]
17. Kampas, F.J.; Pintér, J.D.; Castillo, I. Packing ovals in optimized regular polygons. *J. Glob. Optim.* **2020**, *77*, 175–196. [CrossRef]

MDPI
St. Alban-Anlage 66
4052 Basel
Switzerland
Tel. +41 61 683 77 34
Fax +41 61 302 89 18
www.mdpi.com

Mathematical and Computational Applications Editorial Office
E-mail: mca@mdpi.com
www.mdpi.com/journal/mca

www.ingramcontent.com/pod-product-compliance
Lightning Source LLC
Chambersburg PA
CBHW061451210926
51850CB00008B/1684
9783036579047